T0138400

BIRD WATCH

First edition for North America published in 2011 by
The University of Chicago Press, Chicago 60637
The University of Chicago Press Ltd., London

All inquiries should be addressed to:
The University of Chicago Press
1427 E. 60th Street
Chicago, IL 60637 U.S.A.
www.press.uchicago.edu

20 19 18 17 16 15 14 13 12 11 1 2 3 4 5
ISBN-13: 978-0-226-87226-1 (hardcover)
ISBN-10: 0-226-87226-2 (hardcover)

Library of Congress Cataloging-in-Publication Data

Walters, Martin.
Bird watch : a survey of planet Earth's changing ecosystems / by Martin Walters ; consultant: Jonathan Elphick.
p. cm.
Includes index.
ISBN-13: 978-0-226-87226-1 (hardcover : alk. paper)
ISBN-10: 0-226-87226-2 (hardcover : alk. paper) 1. Rare birds.
2. Birds—Conservation. 3. Birds—Habitat. I. Title.
QL676.7.W35 2011
333.95'822—dc22
2010039328

Conceived, designed, and produced by
Marshall Editions
The Old Brewery
6 Blundell Street
London N7 9BH
www.marshalleditions.com

Publisher: James Ashton-Tyler
Creative Director: Linda Cole
Editorial Team: Graham Bateman (BCS Publishing Ltd), Paul Docherty, Mary Allen, Miranda Smith, Emily Collins
Design Team: Tim Scrivens, Ali Scrivens
Picture Manager: Veneta Bullen
Production: Nikki Ingram

Color separation by Modern Age Repro House Ltd., Hong Kong
Printed in Singapore by Star Standard Industries (PTE) Ltd.
Printed on acid-free paper

Martin Walters (Author) is a writer, editor, and naturalist based in Cambridge, England. He has written and edited many books on natural history, including *Chinese Wildlife: A Visitor's Guide* and *The Illustrated World Encyclopedia of Insects.*

Birdlife International (Consultants) is a global partnership of conservation organizations that strives to conserve birds, their habitats, and global diversity, working with people toward sustainability in the use of natural resources.

Jonathan Elphick (Consultant) is a natural history author and consultant. He has written and edited many books, including *The Birdwatcher's Handbook: A Guide to the Birds of Britain and Ireland*, *Birds: The Art of Ornithology*, and the *Natural History Museum Atlas of Bird Migration*.

▷ *A solitary African Penguin (Spheniscus demersus), now listed as a Vulnerable species, walks on a beach in Cape Town, South Africa.*
◁ *A five-day-old Snowy Plover chick forages on a beach in the U.S.A.*

BIRD WATCH

A SURVEY OF PLANET EARTH'S CHANGING ECOSYSTEMS

MARTIN WALTERS

CONSULTANTS:
BIRDLIFE INTERNATIONAL
JONATHAN ELPHICK

THE UNIVERSITY OF CHICAGO PRESS
CHICAGO AND LONDON

Contents

FOREWORD

It is hard to come to terms with what it really means when a species goes extinct—the loss of a particular life form forever. And nature is facing a wider range of threats than ever before, almost entirely as a result of human activity. Birds are no exception; in fact we know probably more about the threats to birds than those to any other group of plants or animals. Today, over 1,200 species—or 12.5% of the total number of bird species in the world—are threatened with extinction. The natural rate of extinction over time has been less than one species every 100 years, but recently the rate has been at least 50 times that and rising. A total of 56 species were lost during the 20th century, the most costly era for extinctions ever recorded, and the past 30 years alone have seen 19 bird species vanish. In addition, four species are already Extinct in the Wild, with only a few individuals left in zoos and private collections. The latest bird extinction includes the Alaotra Grebe of Madagascar, while the Po'o-uli, a delicate finchlike bird from Hawaii, is almost certainly extinct, both unable to face the invasive competitors and predators introduced by humans.

Not only are rare birds becoming rarer but common birds are also becoming less common. Almost everywhere in the world once familiar birds are now scarce and losing their place in our consciousness. This disappearance of birds across the globe mirrors a deeper loss—that of biodiversity (life in its many forms)—and indicates our failure as humans to live in harmony with the natural environment.

However, the situation is far from lost. Species can be saved and there are many examples of birds that have been brought back against seemingly impossible odds. But it takes timely intervention backed by sound science, adequate resources, political will, and passionate determination. In recent years, at least 16 species would have gone extinct without the conservation programs that were implemented for them. Birds such as Seychelles Magpie-robin and Mauritius Kestrel are still with us because of the work of dedicated conservation groups and individuals. These examples give us the strength to continue the fight.

The BirdLife International Partnership has been at the forefront of bird conservation for many years and we know what is needed to save those in danger. With partner civil society organizations in over 110 countries and territories, we enjoy the strength and cohesion of a global partnership supported by coordinated local presence and action. This, coupled with scientific knowledge, influence with decision-makers, and first-rate conservation experts in the field, means that we understand the issues that face every globally threatened bird, and the main conservation actions that they need. A lack of resources, however, is often the main obstacle to any action being taken.

Our successes have been many but there is still much work to do. This book serves as a timely reminder of the plight of birds across our planet and the pressures that they face. It is easy to despair in the face of such threats, but for us, documenting the decline of species is an important reminder of our responsibility, a tool to sharpen our action planning, and a trigger for even more passionate commitment to making a difference. The BirdLife International Partnership will continue to save species, protect sites, conserve habitats, and empower people in order to ensure that future generations will be able to marvel at the birds showcased on the pages of ***Bird Watch***—hopefully one day, in the not to distant future, out of the danger of extinction.

Marco Lambertini
Chief Executive, BirdLife International

◁ *Black-browed Albatross chicks on nests in the Falkland Islands. Decline in the numbers of this Endangered albatross are attributed to bycatch in longline fishing in open seas.*

THE THREATS TO BIRDS

A HEALTHY POPULATION OF BIRDS, AND A HIGH DIVERSITY OF DIFFERENT BIRD SPECIES, IS A SURE SIGN THAT CONDITIONS ARE GOOD AND THAT A HABITAT IS IN A STABLE AND BALANCED STATE. BIRDS CAN HELP TO WARN US HUMANS OF THE DANGERS OF THE IRRESPONSIBLE EXPLOITATION OF NATURAL HABITATS.

Birds function unwittingly as beacons or barometers—appropriately acting like the proverbial miner's canary. They warn of habitat losses and changes that will ultimately damage us all. Evolved over millions of years and fitted perfectly into their complex habitats, birds are sensitive indicators of problems that affect the natural environment. We disturb this equilibrium at our peril.

In most parts of the world, the natural environment is in grave danger, buckling under the strain of the pressures of modern life, and birds can help us be more aware of this danger, and inspire us to take action to save the situation. By preserving high bird diversity in as many places as possible, which involves maintaining a healthier planet, we are helping not only the living race, but also future generations. Birds are fascinating creatures, as millions of birders and naturalists will attest, but they also have this extra dimension, perhaps more so than any other animal group. They are sending us a message that we must heed.

Following this section, which reviews the threats to birds, this book looks at the major habitats of the world, indicating where they are found, how they are under threat from a wide range of pressures, and highlighting some of the fascinating birds that they support. The third part of the book is a catalog of the world's threatened birds presented in systematic order of families. Each family or group of families is introduced and the threatened species are listed, with an indication of the current category of threat. Selected threatened species are profiled with key information about their features, distribution, and major threats. There follows a section describing how conservation works, with examples of particular aspects, and the role of BirdLife International. Finally, 47 of the world's best birding sites are highlighted, offering a selection representing a wide range of habitats and regions.

◁ *A flock of geese in the mist at sunrise. Most species of geese are migratory, regularly flying long distances between their summer breeding grounds often in the tundra of far northern latitudes to winter feeding grounds farther south on estuaries, mudflats, and coastal grasslands. They are therefore susceptible to changes in two habitats.*

Bird Distribution

Birds play a central role in the habitats in which they live, and the key to their preservation is the protection of these habitats. If we can successfully protect those species that are rare or endangered, the implication is that we are also protecting the whole environment in which they live—from trees and other plants, to mammals, insects, and other groups—to the benefit of all wildlife.

Birds are found throughout the world, and different species have adapted to survive and breed in a wide range of habitats, from the cold polar regions to tropical forests, from the lowlands to the highest mountains, and from the arid deserts to rivers, lakes, and seas. Forests are the most important habitats of all in terms of the number of bird species that they support and more than 75 percent of birds are naturally forest species. Many bird species have managed to adapt to changes to natural habitats brought about by human intervention, from cultural landscapes such as parks and gardens to agricultural habitats; in fact, some 45 percent of birds are now found in such "artificial" habitats. Other important habitats for birds are grassland, savanna, and inland wetlands, each with about 20 percent of species. Conversely, some 40 percent of the world's birds are very specialized in their requirements, and cannot survive if their habitats are degraded or removed.

Some regions are much richer than others, and an analysis of the various biogeographic realms reveals this variation. The richest of these is the Neotropical Realm (the tropics of the New World), which is home to about 36 percent of all known species of land bird—about 3,370 species in all. The next richest is the Afrotropical

▽ *A group of Atlantic Puffins in the Lofoten Islands, Norway. They depend on a good supply of small fish, especially sandeels.*

Distribution of the World's Birds by Geographic Realm and Country

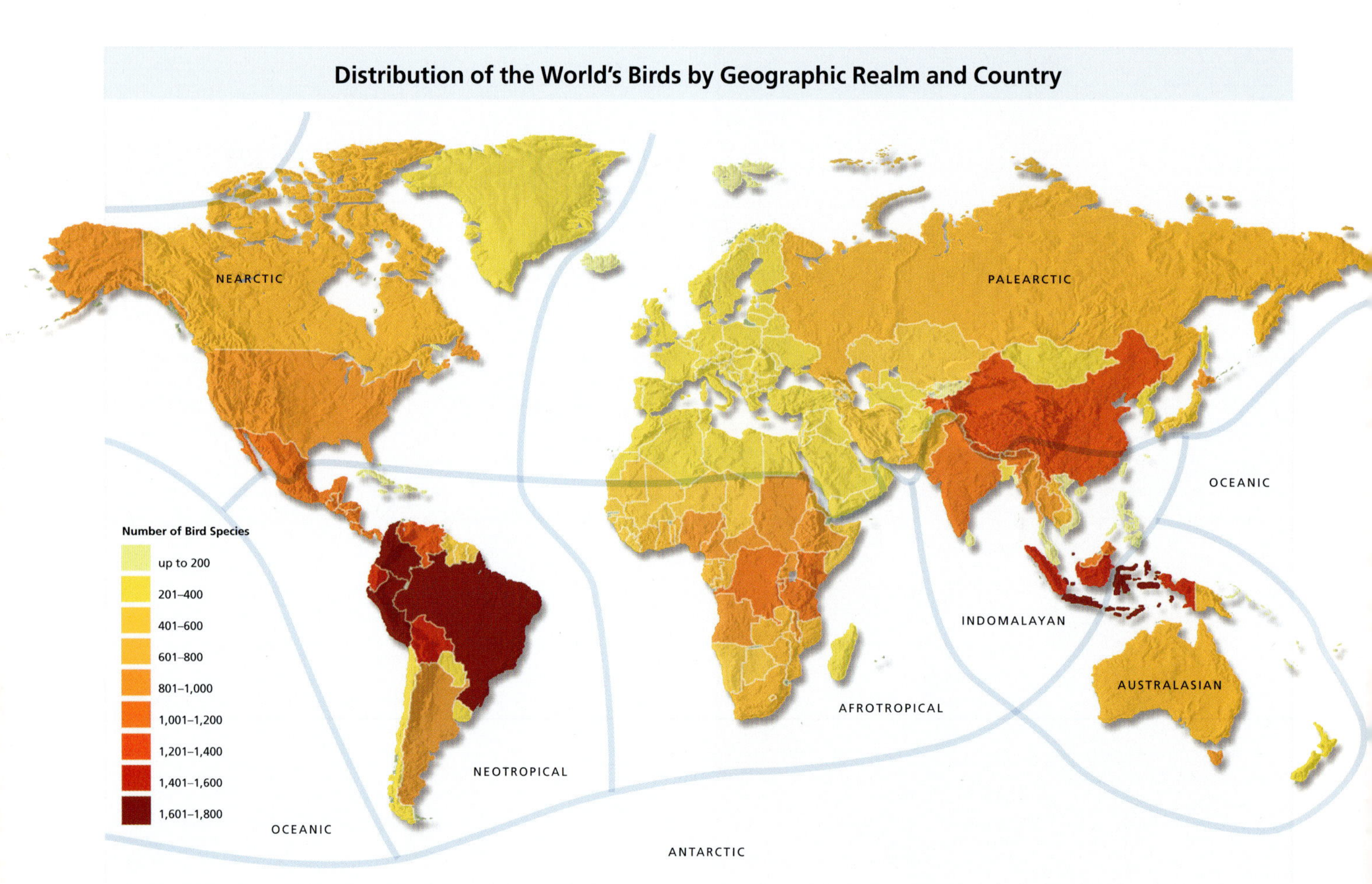

Realm (21 percent; about 1,950 species), followed by the Indomalayan Realm (18 percent; about 1,700 species), Australasian Realm (17 percent; about 1,590 species), Palearctic Realm (10 percent; about 940 species), Nearctic Realm (8 percent; about 730 species), and the Oceanic Realm (2 percent; about 190 species).

Given these distribution statistics, it is perhaps not surprising that the four richest countries for bird diversity are all from tropical South America—Colombia, Peru, Brazil, and Ecuador. These, together with Indonesia, all boast more than 1,500 species of native bird species. Bolivia and Venezuela have more than 1,000 species, and China, India, the Democratic Republic of Congo, Mexico, Tanzania, Kenya, and Argentina also each have about 1,000 species.

◁ A Black-cheeked Woodpecker clings to a tree trunk in the rain forest of Costa Rica, using its stiff tail as a strut.

▷ A Lammergeier coming in to land at a cliff at Giant's Castle in the Drakensberg Mountains, Natal, South Africa. This species has an extremely large range from southern Europe and Africa to India. Although its population is small, at best only 10,000 individuals, it is not believed to approach the thresholds for Vulnerable, so it is listed as Least Concern.

Threats to Wild Birds

For much of their history and evolution, birds have lived in relatively stable conditions, in habitats that for the large part remained intact: these are the conditions to which most have become adapted and upon which they are dependent. For many hundreds of thousands of years nature was untroubled by the activities of people, and this allowed birds and other wildlife to flourish in ecosystems that had attained a natural balance.

Gradually, as human societies spread, bringing agriculture and eventually industry to much of the planet, the pressures on the natural world grew to the point we have reached today at which very few areas remain untouched. In general, such changes have been so rapid that most birds and other wildlife have been unable to adapt fast enough to withstand them. Not all the changes effected by people have been bad for all bird species, and some notably widespread species such as the European Starling, Eurasian Collared-dove, Common Swift, Common Bulbul, and Rose-ringed Parakeet, have taken advantage of human activities, nesting on buildings or feeding in gardens or plantations. However, these opportunistic and adaptable species are the exception. For the majority of birds, the result has been a retreat, with declines in populations and restriction of habitats that have become increasingly fragmented and disturbed.

CRISIS OF NUMBERS

Out of approximately 10,000 species of birds alive today, about one in eight—some 1,240 species—are threatened, and some are facing a very real danger of extinction.

Since 1500, some 134 bird species are known to have gone extinct, and about 190 are so threatened that they are now teetering on the brink of extinction. These latter are classed as Critically Endangered (CR) and require immediate conservation efforts if they are to survive. A further four survive, but only as captive populations, and are classed as Extinct in the Wild (EW). The true rate of extinction may be as high as 150 species lost in the last 500 years, a rate that seems to be accelerating. Of course all the Endangered (E) and Vulnerable (V) species need protection too if they are not themselves to fall into the higher threat category.

Worryingly, there are 838 more species classed as Near Threatened (NT), and these also require careful monitoring and protection to prevent them joining the ranks of the threatened species. Including these species the total number of bird species in the world that require urgent conservation action up to 2,078 (2,082 if those classed as Extinct in the Wild are included).

This century has witnessed the probable extinction in the wild of birds as varied as Spix's Macaw, a large parrot with bright blue-gray plumage, last known from forests in eastern Brazil; the Hawaiian Crow and Po'o-uli, both from Hawaii; the Guam Rail, a flightless victim of predation by introduced snakes on the Pacific island of Guam; and the pretty cinnamon-colored Socorro Dove, last seen in 1972 on Socorro, a Mexican island in the Revillagigedo group.

BirdLife has analyzed the threats faced by Critically Endangered bird species and showed that the most important threats are caused by agriculture and aquaculture (affecting 65 percent of species), followed by the effects of invasive alien species (52 percent), logging (43 percent), and hunting and trapping (41 percent). These factors cause habitat degradation (93 percent of species) and/or direct mortality (64 percent).

The Top Countries Holding the Most Critically Endangered Birds

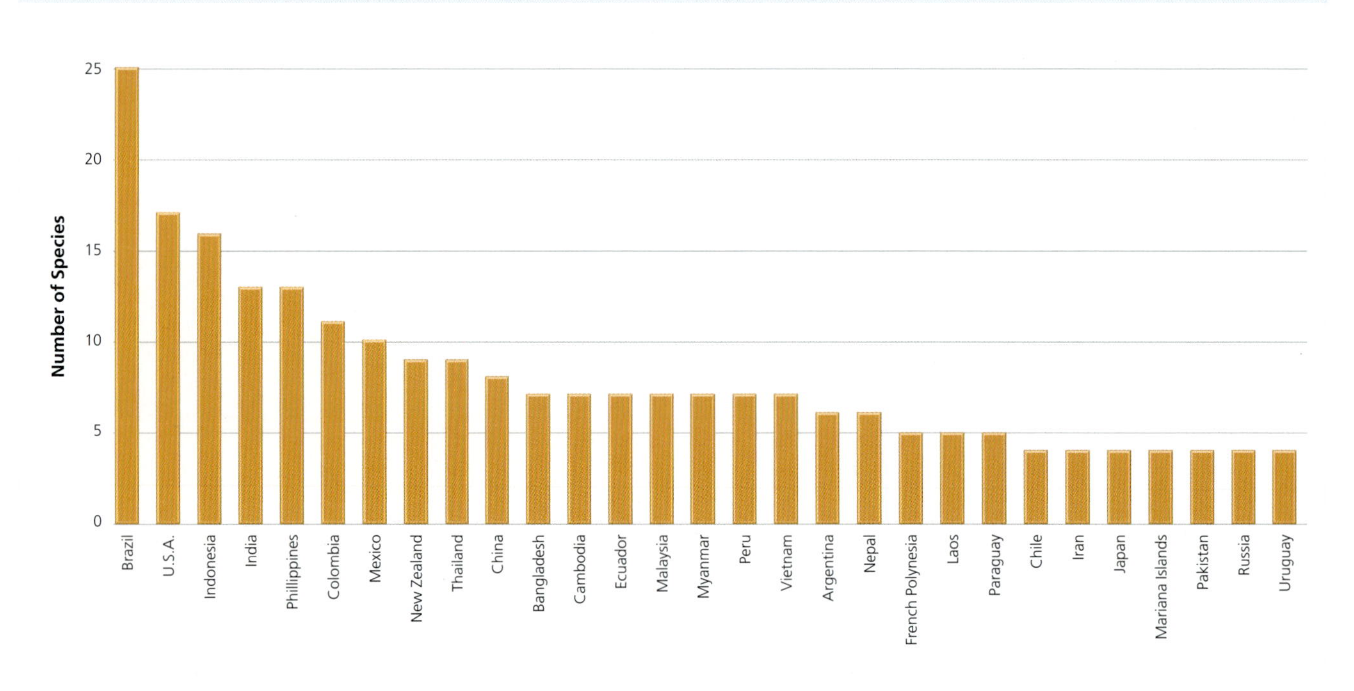

△ *Spix's Macaw is a Critically Endangered parrot with a tiny population in Brazil. Although bred in captivity, it may already be Extinct in the Wild.*

▷ *The Common Bulbul is an example of a bird that has adapted well to a wide range of habitats and human habitations through much of Africa.*

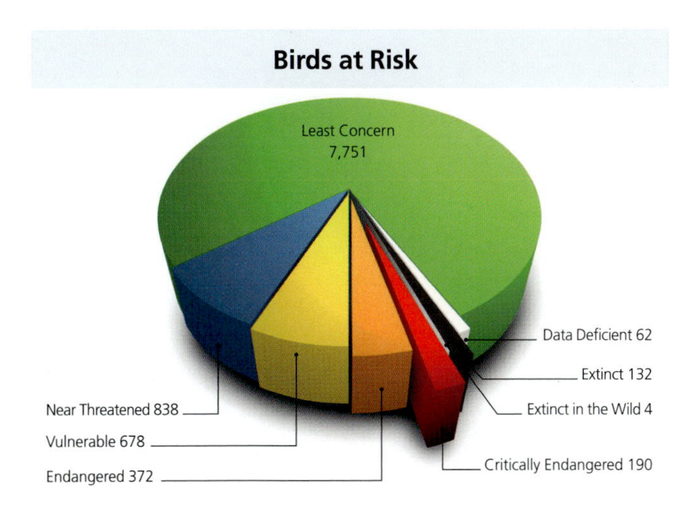

HUNTING AND TRAPPING

Throughout history people have exploited wild birds in a number of different ways—hunting them for food, for their decorative feathers, or for sport, or trapping them to keep as cagebirds, either for their fine songs or attractive plumage, or both. Many species, notably parrots, pheasants, and pigeons, suffer to this day from illegal trapping and trading. While traditional hunting by local people may not have impinged greatly on bird populations in the past, such uses are now mostly no longer sustainable; demand has increased, markets have expanded and become global, roads and tracks have improved access, and techniques for capture have become more sophisticated. The result is that more and more pressure is being put on wild birds and many are unable to withstand it. In the last few decades, over a quarter of the world's bird species have been recorded in international trade, and millions of individual birds are traded every year, many of them tragically dying in transit.

CLIMATE CHANGE

A relatively recent threat to birds comes from climate change, particularly through climate-induced shifts in range, changes in sea level, and increased frequency and severity of storms and floods. Recent years have seen severe fires raging across many regions, notably in southern Australia, California, and Russia, and the increased frequency of such events is possibly due to climate change.

Some species have shifted their ranges to higher altitudes on mountains as the climate has warmed, and there is also evidence, for example in Europe, of range expansion northward. Computer models simulating species richness for Europe suggest that the breeding ranges of European birds will shift northward and eastward by several hundred miles, with uncertain implications. In the Arctic tundra, the favored breeding site for many waders and wildfowl, including the Red-breasted Goose and Spoon-billed Sandpiper, may be reduced as climate change induces a northward expansion of the northern (boreal) forests.

▷ *The Red-breasted Goose, seen here sitting on a nest in the tundra of the Taymyr Peninsula, Siberia, Russia, is listed as Endangered.*

△ *Caged Golden-winged Parakeets illegally captured in the Amazon rain forest await sale in an Iquitos market in Peru.*

▽ *Birds exploit a wide range of habitats, including northern scrub and tundra, as here in the mountains of the northern Yukon Territory, Canada.*

Threats to Habitats

The threats to wildlife and bird species in particular are many and varied. In addition to habitat loss—mainly through the spread of agriculture, pollution, and the effects of climate change (accelerated recently by man-made carbon dioxide and other "greenhouse gas" accumulation)—many rare birds are also affected by introduced predators, and by being trapped or shot. Further threats include housing and other building developments, disturbance, mining, and changes in fire regimes or water management. Destruction and degradation of habitat is the most serious problem of all and affects over 86 percent of all globally threatened bird species.

AGRICULTURE

Expansion of agriculture, and in particular its intensification, is the major cause of habitat loss worldwide. Traditional, less intensive, methods of agriculture are more wildlife friendly, but these are increasingly replaced by high-input cultivation, using irrigation combined with heavy use of artificial fertilizers, weedkillers, and pesticide-treated crops. Where agricultural landscapes featured pockets of woodland, lines of hedgerows, ponds, streams, and marsh, these have often been removed or drained and the fields expanded to maximize the area devoted to crops. Small wonder that the birds and other wildlife have suffered, as seen by declines in European farmland birds. Some wealthy nations have witnessed steep declines in farmland birds because agriculture has not been developed sustainably. About one-third of Europe's Important Bird Areas (IBAs) are threatened by agricultural intensification.

HUMAN POPULATION

Most of the threats to habitats are the result of human activity and as the world's human population continues to grow, such pressures have increased. The U.S. Census Bureau, May 2010 estimate of the global human population stood at over 6.8 billion. Though the rate of growth has slowed in recent decades, the numbers are still rising and are predicted to reach 9 billion between 2040 and 2050. In 1950, the total stood at only 2.5 billion, rising to 6.5 billion by 2005. This rapid rise is very recent in evolutionary terms and the balanced ecosystems of the world reached an equilibrium under conditions of very low human populations; for most of the roughly three million years of human existence, human numbers remained relatively small (about 10 million at the most) with rather little impact on the natural world, except for sustainable hunting and low-impact agriculture. Today, the picture is very different and the effects of human activities (urbanization, intensive agriculture, and much more) on nature can be seen almost everywhere.

TROPICAL FOREST

Deforestation is one of the major problems, especially in the rich habitats of the humid tropics. Tropical forests support around 50 percent of the world's species, including a splendid array of birds, many of which are threatened. Tropical and subtropical forests support 43 percent of all globally threatened bird species.

Precise numbers reflecting deforestation are hard to gather, but some estimates give figures of over 79,000 acres (32,000 hectares) of forest destroyed each day, with a similar area seriously degraded. Burning of timber releases yet more carbon dioxide, and the revealed soils are then exposed to erosion through flash floods and runoff. The FAO (Food and Agriculture Organization of the United Nations) estimates that between 2000 and 2005, more than 26 million acres (10.4 million hectares) of tropical forest were destroyed each year. In some cases, secondary forest or plantations have replaced the original forest, but many birds have been unable to adapt. The main threats to forests are logging, conversion to agriculture, and forest fires, the latter often started deliberately.

In recent years, vast areas of tropical forest, especially in Southeast Asia, have been cleared to make way for extensive plantations of oil palms. What at first may have seemed an environmentally friendly method of producing oil sustainably is now widely seen to have accelerated the destruction of the richest habitats on Earth.

The Amazon rain forest, threatened with deforestation, fire, and climate change could possibly undergo widespread dieback, with a shift to savanna-like scrub in place of the forest, if deforestation reaches 20–30 percent. Worryingly, the figure for the Brazilian Amazon is currently above 17 percent.

TEMPERATE AND NORTHERN FOREST

The forests of temperate and northern regions are not without threats, and most, with the exception of the boreal mainly coniferous forests of the far north, are no longer in a natural state. Indeed, most of the original forest of temperate regions has long been removed to supply timber or to release land for agriculture or building. Yet those that remain, albeit in an altered state, are still home to a range of fascinating bird species.

DESERT

Deserts are threatened particularly from mining, oil exploration and drilling: many deserts have rich deposits of minerals. Access roads impact on the desert vegetation and compact the soil. Many birds of such habitats, such as bustards, sandgrouse, and larks, are wary and easily disturbed.

MOUNTAIN

Compared with lowland regions, many mountain habitats are relatively free from threats. However, pollution and the effects of acid rain have been shown to cause die-back in certain montane forests, notably in coniferous stands in central Europe. In the tropics, even many mountain forests have been affected by timber extraction. Climate change may be forcing some mountain birds to shift their ranges to higher altitudes.

◁ *Encroachment of urban developments into pristine habitats, as here near New Orleans, Louisiana, U.S.A. can have serious impact on local bird populations.*

△ *This rain forest in Sabah, Borneo has been clear-felled in preparation for the planting of oil palms to supply a lucrative crop for the production of biofuel. The impact on local wildlife is enormous.*

GRASSLAND

Natural or semi-natural grassland habitats are all too often replaced by crops, or intensively grazed by livestock, reducing enormously their original biodiversity. The prairies of North America, the steppes of Asia, and many of the varied grass-dominated habitats of South America have been altered in these ways. In Argentina, for example, temperate steppes (*pampas*) and subtropical savannas (*campos*) are some of the country's most threatened terrestrial habitats and most have been converted to grow crops or into rangeland for grazing livestock.

SCRUBLAND AND HEATH

These habitats tend to experience occasional fires, and indeed their constituent plant species have evolved to withstand—and in some cases even to rely on—such events. However, they also sometimes suffer extremely damaging extensive fires that do enormous damage to the local birds and other wildlife.

▷ *A Banded Mongoose (*Mungos mungo*) tackles a bird's egg in Etosha National Park, Namibia.*

WETLAND, COAST, AND OCEANS

Threats to wetland include eutrophication, where freshwater lakes and rivers are affected by runoff of agricultural fertilizers and sewage effluent, resulting in excessive algal growth and loss of more sensitive species. This often leads to declining fish fauna and stocks, as well as health risks from toxic algal blooms. At sea, the effects of increased levels of carbon dioxide increase the acidity of sea water and this reduces the reef-building capacity of corals. Coral reefs are also damaged by pollution and by warmer sea temperatures.

Many seabirds, such as albatrosses and petrels, are accidentally caught by fishing lines and nets in the open sea—so-called "by-catch." The growth of longline fishing is the main problem, and fleets catch 300,000 seabirds every year, along with the fish. Most of these unfortunate birds are dead by the time the catch is hauled in, or so badly injured that they cannot survive. Albatrosses, in particular, are at enormous risk, partly because they share the same Southern Ocean fishing grounds as the fleets and the many illegal fishing boats. Of the 22 species of albatross, eight are classed as Vulnerable, six as Endangered, and three as Critically Endangered, the remaining five being Near Threatened.

△ *This Northern Gannet is a sad casualty of fishing, killed by entanglement in a nylon net, Chesil Beach, Dorset, England.*

ISLANDS

Invasive species (notably introduced predators, but sometimes alien plants) impact on the populations of nearly one-third of threatened

▽ *Dian Chi Lake, Yunnan Province, China's sixth largest freshwater lake, has suffered serious eutrophication due to domestic sewage and agricultural pollution, and requires ecological protection to maintain its biodiversity.*

△ *Thick-billed Guillemots congregating on the sea cliffs of Spitzbergen Island, Svalbard, Norway. This auk has a large circumpolar Arctic range, breeding south to Alaska, Newfoundland, and Labrador. It feeds largely on fish, squid, and crustaceans, and although common has suffered local losses.*

birds, especially those found on isolated oceanic islands. Habitat destruction is also a significant factor on islands, particularly where it involves removal of the native vegetation to which the local birds have become adapted. Rising sea levels as a result of climate change is another concern, inundating coastal habitats. Islands are also at particular risk from coastal pollution—from sewage or oil releases, for example.

THE FUTURE

There is increasingly widespread acceptance that we are living unsustainably (over-using the world's natural resources), and that sincere efforts must be made to reverse this situation. The May 2010 report of the Convention on Biological Diversity (Global Biodiversity Outlook 3) concluded that we are losing biological diversity at a higher rate than ever, with extinction rates possibly up to 1,000 times higher than the historical background rate and that "business as usual is no longer an option if we are to avoid irreversible damage to the life-support systems of our planet." The target set by the world's leaders in 2002 to reduce the rate of biodiversity loss by 2010 has not been achieved, and worse still, some of the pressures have actually increased. On the positive side, more land and sea areas are being protected. Tackling the causes of biodiversity loss must be a higher priority if we are to achieve a healthy planet and a sustainable future. In recognition of this situation, the year 2010 was designated International Year of Biodiversity by the United Nations. The report concluded that action taken over the next decade or two will determine whether a relatively stable environment on which human civilizations have depended for the past 10,000 years will continue beyond the 21st century. A concern is that failure to act may result in ecosystems moving to new, unprecedented and possibly unstable states.

The World Bird Database

BirdLife International has developed sets of vital information known as the World Bird Database (WBDB). The database provides the tools for the BirdLife Partnership to manage, analyze, and report on the state of the world's birds. In addition to detailed information about species, notably the threatened species, the database includes two important categories of defined regions—these are Endemic Bird Areas (EBAs) and Important Bird Areas (IBAs).

ENDEMIC BIRD AREAS (EBAS)

Numbering about 220, Endemic Bird Areas are essentially "hot-spots" of species diversity, sometimes referred to as centers of endemism. An endemic organism (animal or plant) is one that is found only within a particular restricted range, and many of the world's rare bird species meet these requirements.

An EBA is defined on the basis of the areas of overlap between the ranges of the local endemic birds; the number of restricted-range species they contain varies from 2 to 80. Though accounting for only about 5 percent of the land surface of the world, about 2,500 species of bird (or about 26 percent of the total global bird fauna, also called avifauna) are restricted to these special sites. Moreover, since the EBAs also contain many rare or threatened plants and other animal species, and are usually relatively undisturbed, they are vital sites for the conservation of diverse habitats and are therefore considered to be priority sites for conservation. Every bird species with a restricted range of less than 19,000 sq. miles (50,000 sq. km) has been mapped, and the EBAs are defined as areas where the ranges of such endemic species overlap.

Most of the EBAs are found in tropical countries and the majority are dominated by forest, especially lowland tropical forest and moist montane forest. Many are found in mountainous regions or on islands. They vary in size from just 1 sq. mile (2.6 sq. km) to more than 40,000 sq. miles (100,000 sq. km). The following countries have the most EBAs, all of them exceeding ten: Indonesia, Mexico, Brazil, Peru, Colombia, Papua New Guinea, and China.

If protected areas are established and maintained within EBAs, this will help to maximize the protection of rare birds and other organisms, as well as the habitats that support them.

IMPORTANT BIRD AREAS (IBAS)

Important Bird Areas (IBAs) are relatively small areas that have been identified as key sites for conservation. Many are already part of a protected area. Almost 11,000 sites in some 200 countries have now been recognized as IBAs. They are identified, monitored, and protected by national and local organizations, and to qualify each must meet certain internationally agreed criteria based on the occurrence of key bird species that are threatened. Ideally, each IBA should be large enough to support populations of as many key bird species as possible. An IBA should do one or more of the following:

- hold significant numbers of one or more globally threatened species;
- be one of a set of sites that together hold a suite of restricted-range species;
- have exceptionally large numbers of migratory or congregatory species.

Though they are defined by the bird life they support, since birds are effective indicators of general biodiversity, the IBAs also conserve rich habitats with a huge range of other animals and plants.

▽ *The Jamaican Vireo has a restricted range, being endemic to Jamaica, but is not currently threatened.*

▷ *The Cape Vulture is found in South Africa, Botswana, and some other neighboring countries. Listed as Vulnerable, its population is small and declining due to threats that include poisoning and hunting for use in traditional medicines.*

HABITATS

THE INFLUENCE OF HUMANS ON THE REST OF THE NATURAL WORLD HAS BEEN IMMENSE AND GOES BACK SURPRISINGLY FAR INTO THE HISTORY OF OUR SPECIES.

In the past, ecologists used to regard many of the wilderness areas as original and completely untouched by such influences. However, today most ecologists and conservationists accept the view that there are very few, if any, habitats that remain entirely pristine.

Ever since the first people began to migrate from their homelands in Africa, spreading out to colonize the rest of the world, they have had an effect on the habitats that they encountered, creating what many ecologists now refer to as "cultural landscapes." They began as hunter-gatherers, before beginning to develop more complex agricultural techniques such as shifting cultivation, nomadic herding, mixed farming, planting crops, and grazing herds of domestic stock. All of these activities affected, and continue to affect, many of the remaining habitats of the Earth. By about 15,000 years ago, the steadily increasing population of humans had colonized virtually all but the harshest areas of the planet.

While it is easy to see that some habitats, such as the heather moors of northwestern Europe, are essentially created by people and maintained by regular traditional management regimes, it is tempting to regard many other habitats as essentially original. The prairies of North America for example, once thought to be natural, are now regarded as being at least in part the result of thousands of years of regular deliberate burning by Native Americans, partly to assist with the hunting of large mammals such as American Bison (*Bison bison*), and partly to clear land and encourage the fertile growth of grasses. Many other habitats, especially the rich ecosystems of the tropics, give the distinct impression of being in a natural state. However, research has shown that even many tropical forests are younger than 12,000 years old, and that most have a long history of having been altered by people, mainly through cutting and burning. Regeneration can be very rapid in the humid tropics and human influence is quickly obscured by secondary forest growth, to the point where only experts are able to see the signs of historical interference. Recent archaeological research in the tropical forests of the Amazon, Congo, and Southeast Asia suggests that these habitats were all subjected to prehistoric human influences.

There are many reasons why the wildlife habitats that remain are worthy of conservation, for the protection they afford our planet, as well as for the fascinating wildlife that they support. The key to most conservation efforts in the future must also be to take into account the needs of local people, many of whom depend on nature directly for their livelihoods. The challenge is to ensure that such usage is sustainable and not destructive.

△ *Large areas of upland in northern Britain are managed as heather moor, mainly as habitat for the Red Grouse, the subspecies of Willow Ptarmigan that is endemic to Britain and Ireland.*

▽ *While fires are part of the natural cycle of Mediterranean scrub, in modern times degradation of this habitat and fires caused by human activity, as here near Athens in Greece, have caused devastation for people and wildlife.*

WORLDWIDE HABITATS

The major terrestrial habitats of the world, often referred to as biomes, can be conveniently divided into four main types—forests, grasslands, scrub, and deserts. In reality of course the gradations between these are many and they are by no means discrete categories. Each habitat has its own rather distinctive collection of bird species, as well as associated pressures and threats.

The categories mountain, and wetland and coast, as well as all the islands around the world are not strictly comparable with the others, representing as they do locations that each contain a wide range of habitats. However, from the viewpoint of bird distribution and their importance for threatened birds in particular, it makes sense to give them equal prominence.

The map opposite shows the approximate distribution of the major habitats of the world, in terms of what is known as the potential-natural vegetation—that is the plant cover that would exist in each region were it not for the effects of generations of human interference with nature.

Though not the most extensive, the richest habitats of all are those of the tropical forests, found mainly in northern South America, west–central Africa, Southeast Asia and eastern Australia. The bulk of the temperate and northern latitudes are (or would be) dominated by forest, forming a broad girdle around the landmasses, across much of North America, Europe, and Asia. The forests in this girdle are divided into temperate deciduous forests to the south and northern (boreal) forests to the north.

More continental regions carry potential grasslands of various kinds, as the climate here is mainly unsuitable for tree growth, being generally too dry. Regions that are still more arid support desert habitats, notably in North Africa and the Middle East, Central Asia and much of the Australian interior, but also in southwestern North America, southwestern Africa and as a narrow strip along the west coast of South America.

Mediterranean scrub and heathland covers habitats typical of Mediterranean climates as well as heathland and degraded forest, and the frozen wastes of the tundra. Such habitats are often important for birds and other wildlife.

Mountain ranges (see map page 62) alter the climate and have their own complex array of specialized habitats, affected largely by aspect and altitude. Mountain habitats also have their own characteristic bird faunas, including many species that are rare or threatened.

Many birds are adapted to feed in wetland habitats and substantial numbers also breed on or close to the water. Freshwater, saline and coastal habitats are included in this chapter.

Lastly, islands are dealt with as a separate case, as they support many rare bird species, often endemic to a small archipelago, or even to an individual island. Many island species have small populations and are highly sensitive to changes brought about through development, habitat loss or following the introduction, deliberately or accidental, of alien species.

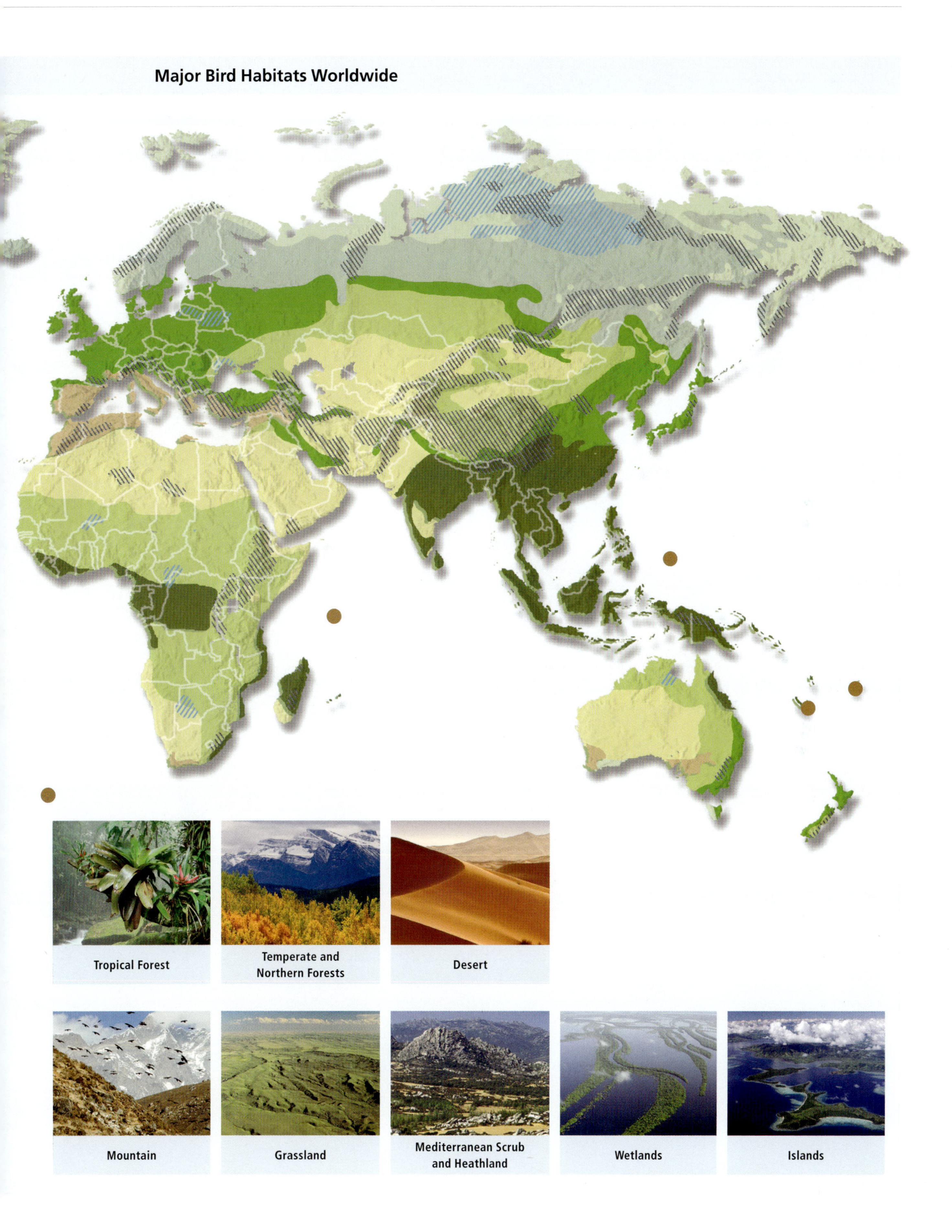
Major Bird Habitats Worldwide
Tropical Forest
Temperate and Northern Forests
Desert
Mountain
Grassland
Mediterranean Scrub and Heathland
Wetlands
Islands

TROPICAL FOREST

TROPICAL RAIN FORESTS ARE FOUND MAINLY IN NORTHERN SOUTH AMERICA (CENTERED ON THE AMAZON BASIN), IN WEST AFRICA, AND IN SOUTHEAST ASIA. CENTRAL AMERICA ALSO HAS IMPORTANT TRACTS OF RAIN FOREST, AS DOES EASTERN AUSTRALIA. THERE ARE A FEW REMAINING POCKETS IN EASTERN INDIA.

Though the tropical rain forests are the most complex and diverse, there are many other types of tropical forest, all home to rare birds and a vast range of animals and plants. Among these are dry forests, seasonal forests, savanna forests, flooded forests, and high-altitude montane cloud forests.

In addition to their value as reservoirs of biodiversity, tropical forests also play a major role in regulating climate and in stabilizing landscapes. They absorb rainfall and channel water into the ground, ultimately feeding streams and rivers, and simultaneously create fertile forest soils and protect these from erosion. Water evaporating from dense tracts of forest forms clouds and replenishes the water cycle, returning as rain both over the forest itself and over nearby agricultural land. The forests also absorb carbon dioxide, a major greenhouse gas, and help to reduce the effects of human-induced global warming. Deforestation is a major driver of climate change for a number of reasons. Huge quantities of greenhouse gases are released during the process of burning the forests, and when reduced in number and density, the forests are less able to remove carbon dioxide from the atmosphere.

Tragically, more than half of all tropical rain forests have been destroyed, reducing their original coverage of some 14 percent of land down to a mere 6 percent. The last 50 years alone have seen a reduction in forest area of more than one-third. At current rates of attrition the tropical rain forests could disappear almost completely within the next hundred years.

◁ *Bromeliads growing alongside a stream in Bocaina National Park, Atlantic Forest, Brazil. Members of the pineapple family, these plants have thick, waxy leaves that form a central bowl adapted to catching rainwater.*

Major Tropical Forests

Tropical forests cover only somewhat less than 6 percent of the surface of the Earth, yet they have been estimated to nurture and house about 50 percent of all species. The variety and numbers of plants and animals are astonishing: 300 species of tree may grow in just 2.5 acres (1 hectare) of tropical rain forest, and as many as 476 have been recorded in the same area of Brazilian Atlantic forest. The numbers of animals, especially insects, are huge, and the forests also support high bird diversity.

CAUSES OF DEFORESTATION

In many countries, tropical forests have been degraded—both by local people for food and firewood and also by international logging and mining operations, and extensive areas have also been cleared for large-scale farming. In recent years, oil-palm plantations have replaced the forest in many regions, as these quickly yield a valuable crop that is partly used as a biofuel petrol substitute. However, tropical forests are not alone in their plight. The native forests of many temperate regions such as North America and Europe have also been cut down and altered over many centuries of development.

Destruction of tropical forests is often a direct result of the construction of roads and wide tracks. A recent study has shown that the best way of preserving diversity in tropical forests is to protect large areas from road building. Roads offer easy access to hunters, loggers, and settlers, and also to invasive species that may affect the natural balance of the habitat. Clearance of forest for agriculture and ranching are the major threats, along with uncontrolled logging for timber and fuel.

Unlike stands of trees in seasonally dry habitats, such as those in regions with a Mediterranean climate, tropical rain forests are essentially wet or damp and normally difficult to burn, and their constituent species have not evolved under conditions of occasional natural fires. Nevertheless, a worry is that as the global climate warms up, some of the tropical forests may become drier and hence be more susceptible to fires.

THE CARBON AND WATER CYCLES

Deforestation is a major driver of climate change, and huge quantities of greenhouse gases are released when forests are burned. As well as absorbing carbon dioxide (a major cause of global warming), healthy forests also return water to the atmosphere, so aiding the natural water cycle. They also stabilize soils and trap water, helping to retain healthy supplies of groundwater, feeding into rivers and streams.

IMPACT ON BIRDS

Though the area occupied by rain forests is huge, these have been destroyed at an alarming rate, mainly to harvest valuable timber, but also to win land for agriculture. Where logging and clearance has ceased, regeneration may be rapid, as long as the soil has not been too badly damaged, and much of the forest that remains is secondary in nature. While such second-growth forest is still valuable for its wildlife and many of the native species may recolonize, some rare species are very sensitive to disturbance and may never return. Continued interference in this richest of bird habitats has resulted in many of the region's bird species being threatened. In Central America and on the islands of the Caribbean, the combined pressures of logging, planting, and tourism have taken their toll on the bird life. In many cases, the rare birds survive mainly in areas that are designated as nature reserves, and ecotourism is playing its part in helping to conserve species.

South America, and parts of Central America, notably Panama, are the richest regions of all in terms of the range and variety of habitats, and also in numbers of bird species—about 3,000 in all. Colombia alone boasts more than 1,870 species of bird—more than any other country in the world. The habitats here are complex and very varied,

Tropical Forest Distribution and Important Bird Areas

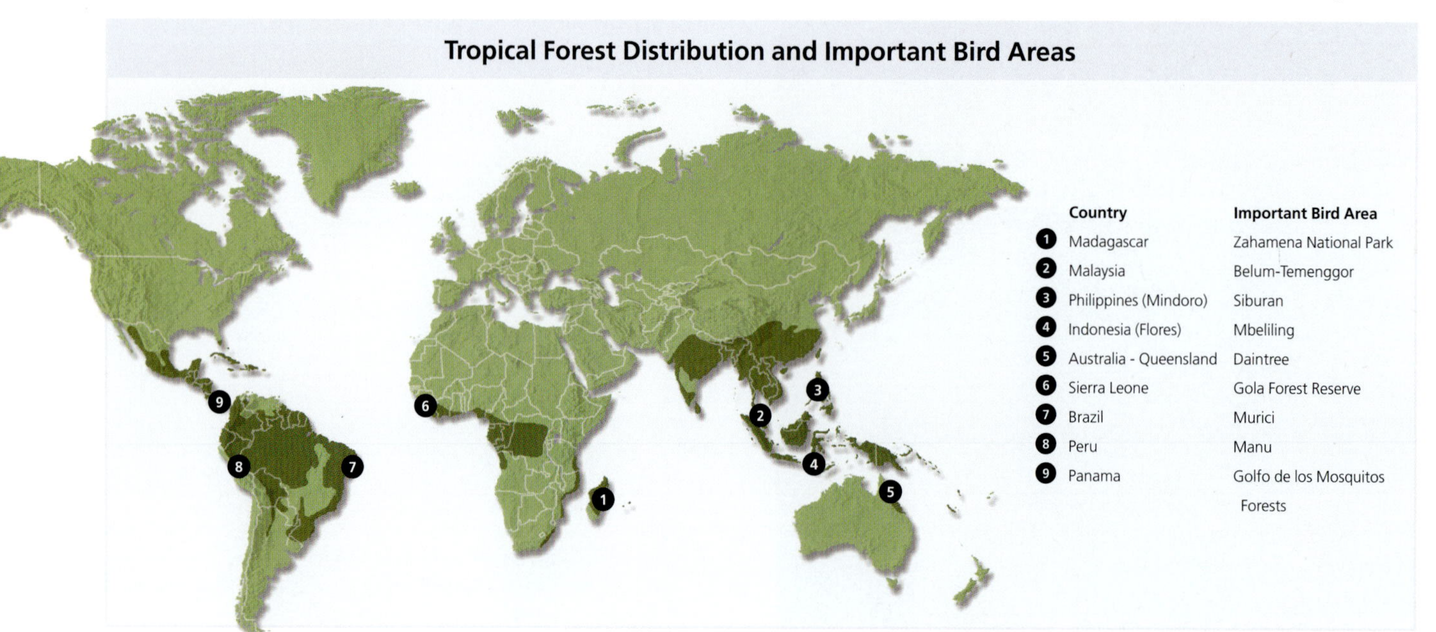

providing a vast range of ecological niches for birds to exploit. In addition, many of these habitats are, or were, very extensive and undisturbed. Sadly, however, human activities such as clearance for building houses and roads, and especially logging of forests for their valuable timber, have fragmented the original vegetation and impacted negatively on many of the region's birds. The rain forests of the Amazon region contain the highest biodiversity of all.

AMAZONIAN FOREST

In South America, the rain forests lie mainly in the north of the continent, centered on the vast area drained by the mighty Amazon River, with its many tributaries, but they are also found scattered north to Mexico. They provide birds with a huge array of habitats and a plethora of feeding and nesting opportunities. The Amazon forest is the largest in the world, based on the Amazon drainage basin and stretching from northern Brazil to Bolivia, eastern Colombia, and eastern Peru. In all, it covers about 40 percent of the continent of South America. The mighty Amazon is fed by more than 1,000 tributaries, many of them, such as the Rio Negro, themselves very long and powerful. These forests are highly diverse, encompassing lowland wet and dry forests, as well as seasonally flooded forest, with fern- and moss-rich cloud forest at high altitudes. Much of it is distinctly layered, with trailing lianas and other vines, such as strangler figs, clambering over the trees, and epiphytes growing on many of the trunks and branches. The warm, moist conditions encourage a vast array of invertebrate life, and luxuriant vegetation with flowers and fruits available all year round—supplies that are exploited by a wide range of birds, many specialized to particular food species.

▽ *Large swaths of rain forest have been cleared, sometimes, as here in Brazil, by deliberate burning. The forest soil is fertile, but is often depleted and washed away by rain storms, leaving wildlife without homes or food.*

△ *A train fully loaded with huge rain forest logs in Cameroon, Central Africa, is traveling the width of the continent to a seaport where the wood will be exported. Demand for furniture and other products made from unsustainably produced wood comes mainly from developed countries.*

DRY TROPICAL FORESTS

In northeastern Brazil, there are forests of a different nature—dry forests that are adapted to a long, dry season of about five months. These dry forests are dominated by trees reaching about 80 ft (25 m), and they often contain many cacti among the ground flora. These are some of the most threatened of the region's forests and only about one-third of the original forest of this type remains. Dry forests are also found in Paraguay and Argentina.

The Atlantic forests of Brazil represent another type of tropical forest, and one that has evolved in isolation from those of the Amazon Basin, resulting in habitats extraordinarily rich in endemic species. These forests once covered about 390,000 sq. miles (1,000,000 sq. km), but now less than 7 percent remain, making these the most threatened of all South America's forests.

AFRICAN RAIN FOREST

The world's second largest area of tropical rain forest is centered on the Congo River of Central Africa and accounts for some 18 percent of the world's rain forest. As with those of the Amazon, these African forests are also under severe threat, not only from logging and clearing for agriculture, but also as a result of human conflict. Hunters target many forest animals, and the trade in so-called "bushmeat" has impacted on many mammals and birds. Deforestation is a very serious issue here and many rare and endangered birds are threatened, mainly through loss of their ideal habitat. Local wars and civil unrest have driven local people further into the forests. As elsewhere in the world, most of the deforestation is by local subsistence farmers who often employ slash-and-burn techniques as a clearing method, collecting wood for fuel and planting crops on the forest soil.

▽ *Large areas of rain forest in this river-valley at Sabah in Borneo, Malaysia have been cleared to make way for oil-palm plantations. Though vast, the extent of the devastation gives little indication of the true cost to wildlife.*

SOUTHEAST ASIAN RAIN FOREST

Indonesia and the Philippines also retain important areas of tropical forest, especially Borneo, which was once virtually covered in forest. These forests of Borneo began to be logged in earnest in the 1980s and 1990s, providing timber for the construction industry and furniture, as well as pulp for paper manufacture—and even chopsticks. Nowadays, huge areas are being planted with oil palms to produce oil that is converted into valuable biofuel. While at first sight this may seem a desirable and sustainable way of replacing petrol derived from crude oil, the forests have suffered enormously

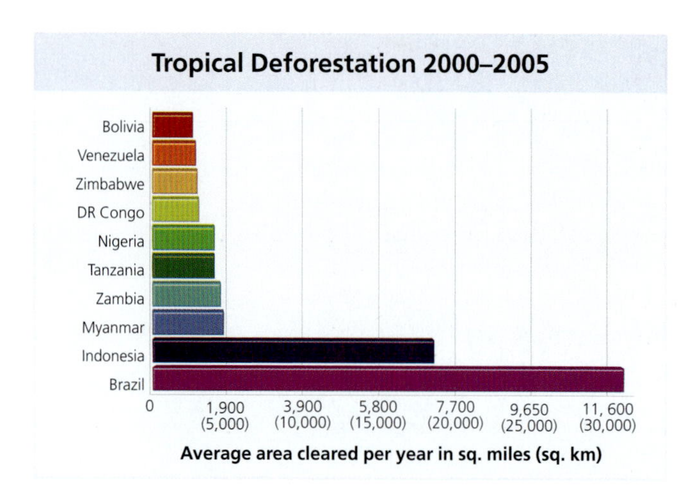

△ *These hillsides north of Quito, Ecuador, South America, have been cleared by logging activity and then replanted for agricultural use by the local farmers. Despite the benefits to the local economy, the negative impact on birds and other animals has been immense.*

as a result of being replaced with this crop. Production of palm oil in Indonesia is expected to reach 25 million acres (10 million hectares) by 2010, up from 1.5 million acres (600,000 hectares) in 1985. Forest has also been cleared for planting soya and for grazing beef cattle.

The remaining forests of Borneo are very diverse. The World Wildlife Fund (WWF) has estimated that they contain about 222 mammal species (44 endemic), 420 birds (37 endemic), and more than 15,000 plants (6,000 endemic). The richest forests of Borneo are those of the lowlands, dominated by trees of the family Dipterocarpaceae, many of which grow up to 165 ft (50 m) tall. Over half of these forests have been cleared and logged, mainly in the last few decades. As well as destroying the habitats, the removal and thinning of these forests has altered the ecology. Fires are now much more frequent and spread quickly, further impinging on the remaining forest areas as well as causing pollution and releasing vast quantities of carbon dioxide into the air, fuelling global warming and climate change. It has been estimated that the forests on damp peaty soils hold over 500 tons of carbon per 2.5 acres (1 hectare), much of which is released when they burn. In their natural state these and other rain forests are relatively immune to destruction by fire, but human interference has rendered them more susceptible.

▷ *Soybean field surrounding only a sliver of Amazon rain forest remains in the middle of this vast swath of soybean plantations. Such cultivated areas support few bird species.*

LOWLAND ATLANTIC FOREST OF BRAZIL

THE ATLANTIC FOREST, OR MATA ATLÂNTICA, STRETCHES ALONG BRAZIL'S ATLANTIC COAST, FROM THE NORTHERN STATE OF RIO GRANDE DO NORTE SOUTH TO RIO GRANDE DO SUL. THE ENTIRE FOREST IS A BIODIVERSITY HOTSPOT, WHERE MANY ENDEMIC SPECIES FACE GREAT THREATS.

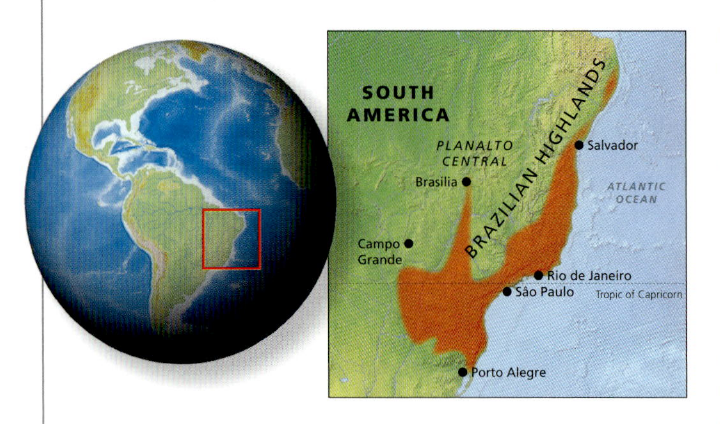

Brazil's Lowland Atlantic Forest is one of the most threatened and least protected forests in the world. Beginning with the creation of sugarcane plantations and later, coffee plantations, this region has been losing habitat for hundreds of years. Now, as Rio de Janeiro and São Paulo expand, it is facing severe pressure from increasing urbanization.

AMAZING WILDLIFE HABITAT

The lowland coastal strip of Brazil's Atlantic Forest is a marvellous haven for wildlife, and the interior forest is also home to many rare animals and plants. Several species of Critically Endangered birds live in the coastal forest, which is about 30–60 miles (50–100 km) deep, and it has been designated one of the most important of the region's Endemic Bird Areas (EBAs). The coastal forest once stretched more than 90,000 sq. miles (230,000 sq. km), but today it covers less than 10 percent of that area. In essence a humid forest, it ranges from coastal scrubby woodland through tall, evergreen rain forest to semi-deciduous forest where there is a distinct dry season. Research has shown that these forests are some of the richest in the world, with an amazing diversity of plants. Over half the tree species are probably endemic, and in some areas there are up to 450 different species in 2.5 acres (1 hectare). Astonishingly, 13 percent of all the threatened birds in the Americas occur in this EBA.

RARE BIRD HAVEN

The forest's bird life is similarly rich and varied and contains no fewer than ten endemic genera that encompass more than 120 species. It is also home to several more threatened birds with a wider range, such as the Purple-winged Ground-dove and Red-browed Amazon. Forest disturbance began with the arrival of European settlers 400 years ago. The forest's soils were fertile and suitable for agriculture, and in many areas crops such as rubber, coffee, and bananas have replaced the original vegetation. The forest has also been logged for timber and removed to allow mining, road building, and housing and hotel developments. The birds that live here have adapted to the particular forest conditions and are specialized to thrive in the native forests. Removal or alteration of these habitats often has dire consequences for them, especially those that are rare and of restricted range.

THE FOREST'S RAREST BIRD

The tiny Kinglet Calyptura is one of the 50 most endangered birds in the world. BirdLife International estimate its population at fewer than 50. It was spotted in 1996 in Serra dos Órgãos, a national park an hour's drive north of Rio de Janeiro, and this was the first

Vital Signs—Lowland Atlantic Forest

Original extent: 476,450 sq. miles (1,234,000 sq. km)
Vegetation remaining: 38,570 sq. miles (99,900 sq. km)
Endemic plant species: 8,000
Endemic threatened birds: 55
Endemic threatened mammals: 21
Endemic threatened amphibians: 14
Extinct species: 1
Human population density: 227 people per sq. mile (87 people per sq. km)
Area protected: 19,460 sq. miles (50,400 sq. km)

sighting for more than 100 years. However, since then there have been no further confirmed sightings. The Kinglet Calyptura has been brought to the brink of extinction as a result of deforestation that has occurred through gold and diamond mining as well as the creation of coffee plantations; these industries are still in operation today. Various plants—including bromeliads, mistletoes, and orchids—are harvested in this region, which may be a factor in reducing the Kinglet's food supply as well as altering the area's habitat structure and its microclimate. The conservation measures that have been utilized up to now involve protecting the remaining small areas of habitat that are suitable for the bird; however, the Kinglet Calyptura's known range is extremely limited.

◁ *This Saffron Toucanet is perched in the Atlantic Forest of southeastern Brazil, although the bird can also be found in nearby areas of Argentina and Paraguay. It feeds mainly on fruits and nests in old woodpecker holes. It is threatened by habitat loss, hunting, and capture for the cagebird trade.*

△ *The Cubatão steel complex juts out from Brazil's endangered Atlantic coastal rain forest. Cubatão, one of the most polluted cities in the world, hosts 24 industries, including steel, oil refining, and fertilizer production that operate outside the sanction of North American and European law.*

The Atlantic Forest's Most Endangered Birds

SCIENTIFIC NAME	COMMON NAME	STATUS	WORLD POPULATION
Calyptura cristata	Kinglet Calyptura	CR	<50
Eleoscytalopus psychopompus	Bahia Tapaculo	CR	50–250
Myrmotherula fluminensis	Rio de Janeiro Antwren	CR	50–250
Merulaxis stresemanni	Stresemann's Bristlefront	CR	50–250
Claravis godefrida	Purple-winged Ground-dove	CR	50–250
Nemosia rourei	Cherry-throated Tanager	CR	200–250
Formicivora littoralis	Restinga Antwren	CR	250–1,000
Crax blumenbachii	Red-billed Curassow	EN	200–250
Cotinga maculata	Banded Cotinga	EN	250–1,000
Touit melanonotus	Brown-backed Parrotlet	EN	250–1,000
Glaucis dohrnii	Hook-billed Hermit	EN	250–1,000
Formicivora erythronotos	Black-hooded Antwren	EN	1,000–2,500
Pyriglena atra	Fringe-backed Fire-eye	EN	1,000–2,500
Myrmeciza ruficauda	Scalloped Antbird	EN	1,000–2,500
Amazona rhodocorytha	Red-browed Amazon	EN	1,000–2,500

RED LIST CR = Critically Endangered EN = Endangered VU = Vulnerable NT = Near Threatened LC = Least Concern

Tropical Forest Endemic Bird Areas

Given the complexity and fertility of tropical forests, it is not surprising that many Endemic Bird Areas (EBAs, see pp. 20–21) have been identified within this habitat.

△ *The Colorful Puffleg hummingbird is Critically Endangered. It lives in an extremely small and decreasing range in the western Andes of southwestern Colombia in South America. Its habitat is being threatened by small-scale logging and cultivation.*

SOUTH AMERICA

In South America one of the most important, but also one of the most threatened, EBAs is the Chocó of western Colombia and Ecuador, along the western slopes of the Andes. This is rain forest par excellence. Some sites here receive up to 630 inches (16,000 mm) of rain annually, making it one of the wettest places on Earth. The vegetation ranges from lowland wet tropical forest to subtropical and montane humid forest at higher altitudes, and wet grassland at higher altitudes still. This EBA is home to the most restricted-range birds of any in the Americas, with more than 50 endemic species. Noteworthy among these are the threatened Plumbeous Forest-falcon, Banded Ground-cuckoo, Turquoise-throated and Colorful Pufflegs, Five-colored Barbet, Long-wattled Umbrellabird, Bicolored Antvireo, Choco Vireo, and many tanagers, including both Gold-ringed and Multicolored types.

INDIA AND INDONESIA

In India, fragments of tropical forest remain, notably in the Western Ghats, hills at the western edge of the Deccan Plateau. This is partly evergreen tropical monsoon forest and partly seasonal moist deciduous forest, with subtropical forest and grassland at higher levels. Threatened birds here include Nilgiri Wood-pigeon and Black-chinned Laughingthrush.

Both Bali and Java were once well covered by tropical forest and these Indonesian islands are also recognized as an EBA. Most of the rare birds are now found in the montane forests, and much of the lowland forest has been cleared. Key species here include Javan Hawk-eagle, Javan Scops-owl, and the Critically Endangered Bali Starling.

AFRICA

In Africa, one of the most important regions is that of the East African coastal forests in Kenya, Somalia, and Tanzania including Zanzibar. Special birds of these semi-evergreen and deciduous forests include the diminutive Sokoke Scops-owl, Sokoke Pipit, Amani Sunbird, and Clarke's Weaver. Madagascar is very rich in wildlife, including a large number of endemic species, and its western dry forests are home to many fascinating birds. This EBA includes patches of seasonally dry deciduous forest. Some of the rare birds here are very local, notably Van Dam's Vanga and White-breasted Mesite.

◁ *The Sokoke Scops-owl is found only in a couple of sites in east Africa: Arabuko-Sokoke Forest, Kenya, and East Usambara Mountains, Tanzania. Removal of timber for firewood and woodcarving has severely fragmented its range, and this delightful owl is currently classified as Endangered.*

△ *Cape Tribulation in northern Queensland lies in the Daintree Rain forest, one of the oldest rain forests on the planet. Despite making up only 0.2 percent of the land mass of Australia, this forest contains 20 percent of all bird species on the continent, as well as 65 percent of Australian butterfly and bat species, and 30 percent of frog, marsupial, and reptile species.*

CHINA

The south of China once held substantial tracts of tropical forest, but centuries of deforestation and rural agriculture have removed much of this original vegetation. The tropical island of Hainan once had extensive lowland evergreen rain forest, now sadly largely deforested. The central mountains of Hainan still retain patches of montane forest, with several rare birds, such as Hainan Partridge, Hainan Leaf-warbler, Pale-capped Pigeon, and the beautiful Blyth's Kingfisher. As elsewhere, forest loss is a major concern. In addition to logging for timber, much original forest here has been replaced by rubber plantations.

AUSTRALIA

In Queensland in eastern Australia, tropical forest once covered much of the lowland and hills, roughly between Townsville and Cooktown. Key species here include the huge flightless Southern Cassowary, the charismatic bowerbirds, and the fascinating honeyeaters. The good news here is that nearly all the remaining forest is protected in the Wet Tropics World Heritage Area, with all logging stopped. However, very little lowland forest remains, and the remnants are mostly very fragmented and disturbed.

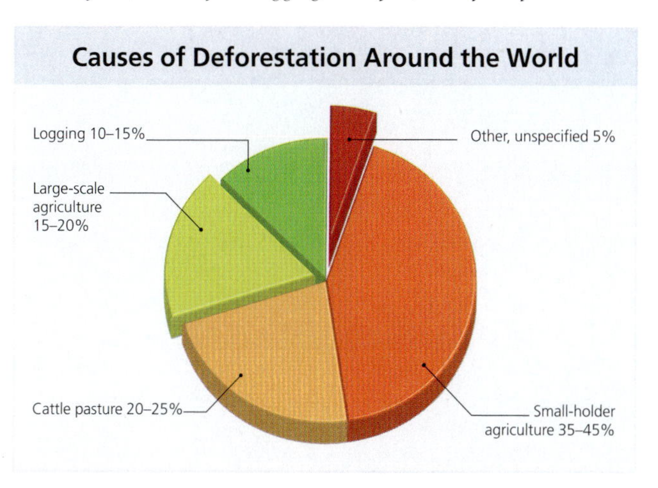

Philippine Eagle

Pithecophaga jefferyi

High in the forest canopy of the island of Mindanao lives one of the world's most impressive and remarkable birds, the magnificent Philippine Eagle, known locally as "Haribon," or "King of Birds."

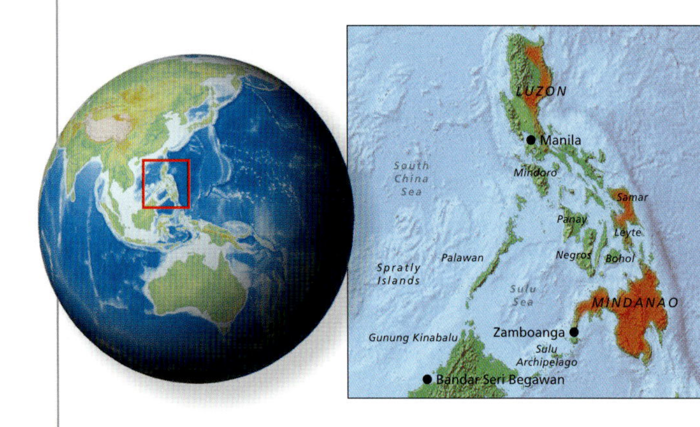

Also known as the Monkey-eating Eagle, the Philippine Eagle feeds on a range of bird and mammal prey, including monkeys, palm civets, and colugos (sometimes confusingly called "flying lemurs"). Sadly, this eagle has become Critically Endangered. It suffers from rapid population crashes combined with a slow reproductive rate, and a resultant very low population—currently estimated at 500 or fewer birds. Endemic to the Philippines, this bird survives only on the islands of Mindanao, Luzon, Samar, and Leyte. Mindanao holds the bulk of the population, with just a handful on the other islands. Its habitat is tropical forest, from lowland up to about 6,000 ft (1,800 m), and it prefers primary rather than regrown secondary forest. Potentially long-lived, like most large raptors, Philippine Eagles breed only slowly. The adults are mature at between five and seven years and pair for life, and the breeding cycle takes two years, with each pair raising only a single offspring. After fledging, which takes about 18 weeks, the young bird remains in the nesting area for about 18 months.

MAIN THREATS

Multiple threats impact on this wonderful bird. As in many tropical habitats, logging and clearance for cultivation has removed large areas of original forest, the bird's favored habitat. It is also subjected to hunting, as well as being accidentally caught in traps that are set on the forest floor in order to catch deer and wild pigs. As a top predator, this eagle may also be suffering from the accumulation of pesticides absorbed from its prey. Studies elsewhere have proved that such concentration of poison in a bird's tissues can reduce reproductive potential, especially in birds of prey, resulting in infertility.

Fortunately, this eagle survives in a number of protected areas, notably in Mt Kitanglad and Mt Apo Natural Parks on Mindanao. There is also a captive breeding center that aims to reintroduce eagles to suitable habitat, so the future for this charismatic bird, though perilous, may not be catastrophic. Mt Apo Natural Park includes the Philippines' highest peak—Mt Apo (9,692 ft; 2,954 m)—and the extinct volcano—Mt Talomo (8,773 ft; 2,674 m)—and covers 178,200 acres (72,113 hectares). The rainfall here is very high, up to 100 in (2,500 mm) average per year, and the original lowland vegetation is rain forest. At higher altitudes this is replaced by lower growth forest rich in epiphytes and mosses, and higher still by scrub and montane grassland. Much of the lowland forest below about 3,300 ft (1,000 m) has been cleared, but remnants survive, mainly on steeper slopes. At Baracatan there is a long-established Philippine Eagle Center from which birdwatchers can catch a glimpse of the eagles.

Although Mt Apo has Natural Heritage Park status, in reality this has afforded its habitat little protection, and management is minimal. Large areas of lowland and montane forest have been destroyed with the result that over half is now deforested and human activities and settlement continue to take their toll.

Mt Apo is included in the Mindanao and Eastern Visayas EBA defined by the presence of a number of endemic forest birds in addition to the Philippine Eagle, including Giant Scops-owl, Silvery Kingfisher, and Celestial Monarch. Many other threatened birds are found here—the Philippine Cockatoo (Critically Endangered); and the Philippine Eagle-owl, Rufous-lored Kingfisher; Philippine Hawk-eagle, Philippine Dwarf-kingfisher, Spotted Imperial-pigeon, and Black-bibbed Cicadabird, all Vulnerable.

▷ *This eagle is famed for its shaggy crest, giving it the appearance of having a mane like a lion.*

▽ *A Philippine Eagle in flight at the Eagle Center at Baracatan on the slopes of Mount Taloma, in the Apo range, Mindanao, Philippines.*

Threatened Birds of South and Central America

Parrots and hummingbirds are among the most beautiful of South America's tropical forest birds, and both these groups contain species that are threatened, partly through habitat loss and hunting for trophies and, with parrots, being trapped for the cagebird trade. Largest of all is the Endangered Hyacinth Macaw, threatened along with several other macaws, notably the vanishingly rare and Critically Endangered Spix's Macaw. Many of the parrots, including the colorful Amazon parrots, are threatened both by habitat loss and by collection as cagebirds. No fewer than 16 species of Amazon parrot are listed as threatened, and of the remaining 14, three are Near Threatened.

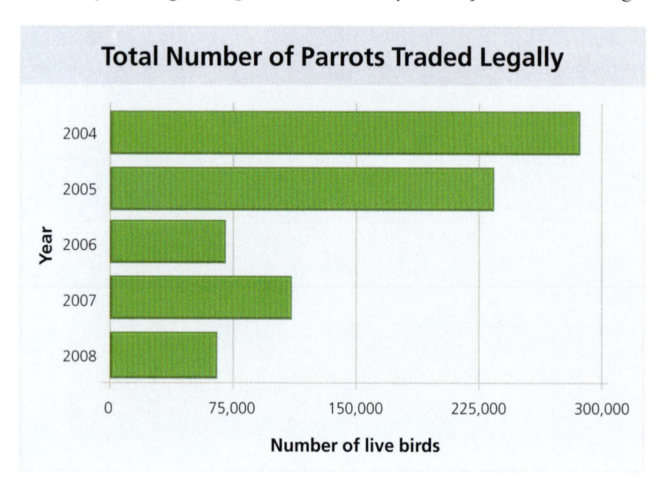

Hummingbirds are among the world's most fascinating birds, with their bright, shiny plumage and astonishing flying agility. Several species have long tail plumes or streamers. The New World tropical forests boast many species of these tiny jewel-like birds, and several are rare. Many have evocative or descriptive names, such as the woodstars, hermits, pufflegs, and sabrewings. Fifteen species of hummingbird are Endangered, nine Critically Endangered, six Vulnerable, and 19 listed as Near Threatened.

Cotingas are brightly colored birds found only in the tropical forests of Central and South America, where they feed mainly on fruit and insects. Of the 96 species in this family, six are Endangered, and one, the Kinglet Calyptura, is Critically Endangered. A further 11 are listed as Vulnerable, including the strange Long-wattled Umbrellabird, with its odd, umbrella-like flat crest. The tanagers are a colorful group particular to this region. Mainly fruit-eating, most of them are found in Central and South America and the West Indies; 22 species of these attractive songbirds are listed as threatened. The tiny Plumbeous Forest-falcon is another rare and Vulnerable bird that inhabits the undisturbed forests of southwestern Colombia and northwestern Ecuador.

The huge Harpy Eagle is a very impressive tropical forest raptor. One of the world's largest and most powerful eagles, this specialist hunter occurs from the Amazon Basin north as far as southern Mexico. The typical habitat is large tracts of lowland tropical forest, normally below about 3,300 ft (1,000 m). It feeds on prey as large as monkeys, sloths, and porcupines as well as other birds and reptiles, often plucking them from the trees using its incredibly powerful, thick legs and feet with sharp talons. The nest is large—up to 5 ft (1.5 m) across—typically built in the crown of a tall forest tree. Though their range is extensive, Harpy Eagles are thinly spread and rare in most regions, with the exception of the Amazon forests of Brazil and Peru where the population is relatively stable. In many parts of its former range it seems to have become extinct, perhaps due to disturbance and deforestation. It is targeted by hunters and, being large and unafraid, it is relatively easy to shoot. The bird is listed as Near Threatened and is undergoing a general decline.

◁ *The Orange-fronted Parakeet is an attractive medium-sized parrot with a wide range from western Mexico south to Costa Rica. Though its population is relatively high and the species remains listed as Least Concern, in certain areas it has suffered local declines through capture for the cagebird trade.*

▽ *Wild birds await their fate in small cages in the Venezuelan forest. Illegal trapping is a major threat to many species, especially in the rich forests of the tropics, and pretty songbirds and parrots are at particular risk of being captured for this cruel trade.*

JEWEL OF THE FOREST

The Resplendent Quetzal is one of Central America's glories—a true jewel of the tropical forest. The female has fairly modest green and gray-brown plumage, with a black-and-white barred tail and red under tail coverts. The male, however, is extraordinary. The upper tail coverts hang down as long streamers to a length of up to 3 ft (1 m), and the bright red belly contrasts with the bright green of most of the rest of its plumage. The Resplendent Quetzal occurs in high-altitude cloud forests in southern Mexico, western Panama, Honduras, Guatemala, El Salvador, Nicaragua, and Costa Rica. It feeds mainly on fruit, supplemented by invertebrates, and small vertebrates such as lizards and frogs. It is not surprising that this species has been persecuted for trade, but reasonable numbers do persist in some areas, especially in reserves and national parks. Deforestation is probably the main threat, and it is listed as Near Threatened.

FUTURE FOR THE TROPICAL FORESTS

Though the threats to the tropical forests are all too real, efforts are being made to protect key sites and to educate people about the importance of maintaining forests. It is largely the wealthier nations that drive the demand for timber. Slowly the message is being heard that timber should be sustainably sourced. If trees are harvested selectively, rather than through clear-felling, the forests have a chance to survive, albeit in an altered state. Though plantations are being established at increased rates—forest cover is actually increasing in some areas—such forests are only attractive to a handful of species and it is the natural or near-natural forests that we must preserve to protect threatened birds and other wildlife.

◁ *A male Resplendent Quetzal emerges from the nest-hole, its long tail streamers folded over its back. Both sexes care for their eggs and young. They typically nest in a hole in a partially rotten tree, as here in a forest in Costa Rica, and require trees in a suitable state of decomposition.*

△ *Canopy walkways such as this in Monteverde Cloud Forest Preserve, Costa Rica, were once used by scientists studying rain forest canopies. Today they are used by ecotourists, while scientists use the more versatile canopy cranes.*

▽ *There are currently five canopy cranes in operation at various rain forest locations around the world, and six in temperate forests.*

TEMPERATE AND NORTHERN FOREST

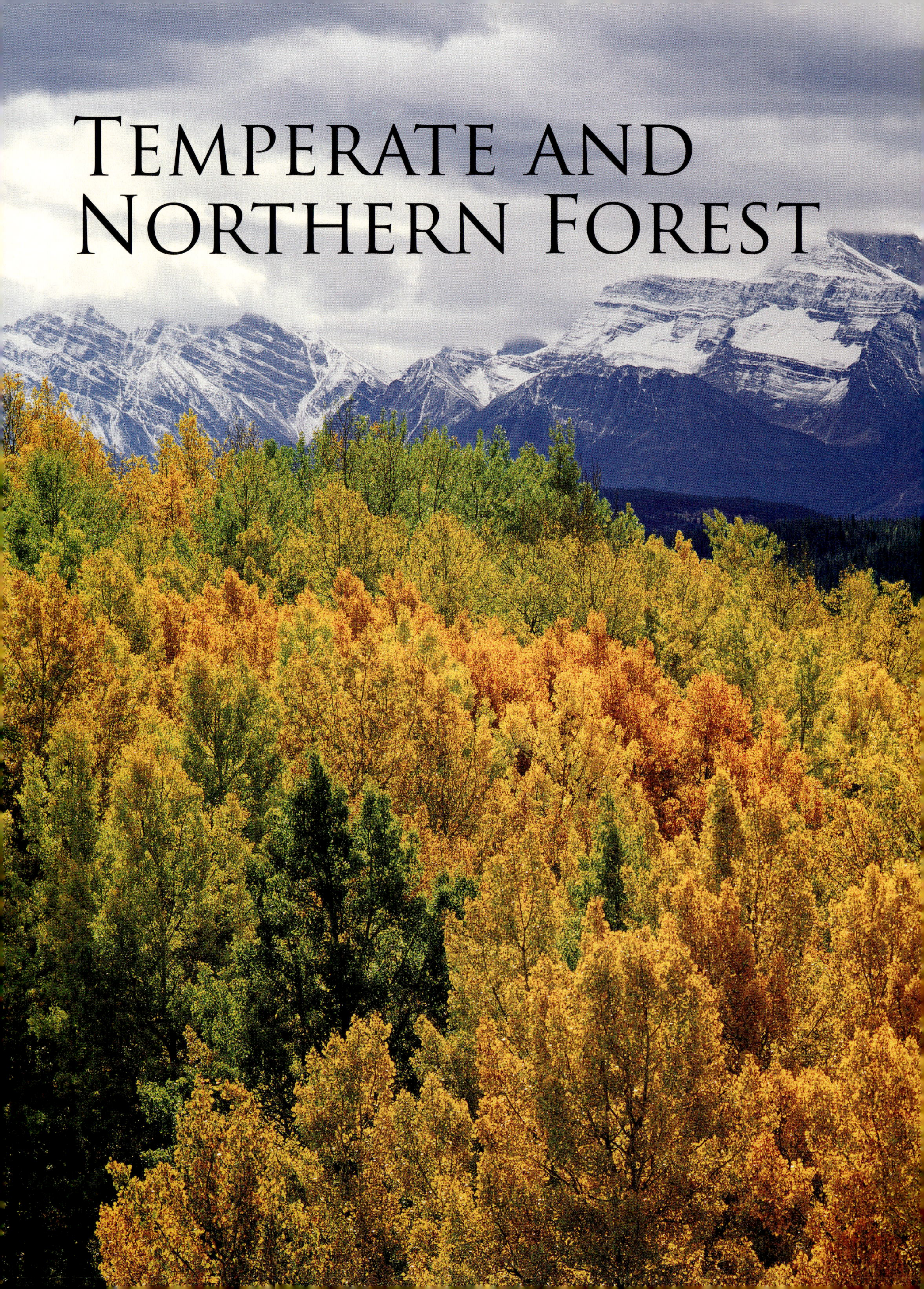

TEMPERATE AND NORTHERN FORESTS ARE MOSTLY FOUND IN NORTH AMERICA, EUROPE AND IN NORTHEASTERN ASIA. THESE FORESTS ARE MUCH LESS SPECIES-RICH THAN THE TROPICAL FORESTS, BUT THEY ARE HOME TO MANY BIRDS, SUCH AS WOODPECKERS, OWLS, WARBLERS, AND PHEASANTS, AS WELL AS WOODLAND RAPTORS SUCH AS GOSHAWKS AND SPARROWHAWKS.

There are also areas of temperate forest in eastern Australia and Tasmania that have their own special bird life. Even in tropical and subtropical regions, however, temperate forests can be found on the flanks of mountains where the altitude reduces the ambient temperature. Such habitats are mentioned on pp. 59–71. Here the discussion mainly concerns temperate and northern forests in the lowlands or hilly country.

While the deforestation of the tropics is a major ecological concern, the area once occupied by temperate forest has also been hugely reduced over many centuries because these forests tend to coincide with heavily populated regions, especially in North America, Europe, and China. Clearance for timber and agricultural and housing development are major threats, as well as human disturbance, and hardly any truly original temperate forest now remains, most having been cleared, interfered with, or managed for timber extraction down the ages. In addition to these direct effects, many of these forests have been affected by pollution via acid rain resulting from the burning of fossil fuels. There is also increasing evidence that global warming is slowly changing the composition of certain forests, such as along the southern border of the northern (boreal) forests, with deciduous trees able to grow and compete further north as the climate becomes more suitable for them.

◁ *This aspen* (Populus tremuloides) *forest, with Mount Christie and Mount Brussels as its backdrop, is in Jasper National Park, Alberta, Canada.*

MAJOR TEMPERATE AND NORTHERN FORESTS

NORTHERN OR BOREAL FORESTS

These forests lie roughly between latitudes 50°N and 60°N, occupying a belt between the temperate forests and the treeless tundra. They are composed mainly of coniferous trees and are more uniform and less structured than the temperate forests. These dark boreal forests are also known as taiga, a word derived from Turkic or Mongolian.

Spruce, fir, pine, and larch are the main components of the temperate and northern forests that cover large swaths of northern North America and northern Europe and Asia. Though these forests yield valuable timber and are subject to extensive logging, they still cover the largest area of any major terrestrial habitat. The belt of the taiga stretches from northeastern Europe all the way across Russia to the Pacific coast, and across North America from Alaska in the west to Newfoundland in the east. Large parts of this forest lie within the Arctic Circle, and across much of it the average annual temperature is below zero. The average January temperature is often as low as –58°F (–50°C), though the summers can be surprisingly warm, averaging about 68°F (20°C).

TREE TYPES

In northern Europe the boreal forest is usually dominated by Norway Spruce (*Picea abies*) and Scots Pine (*Pinus sylvestris*). Siberian Larch (*Larix sibirica*), Siberian Fir (*Abies sibirica*), Siberian Pine (*Pinus sibirica*), and Siberian Spruce (*Picea obovata*) are important towards the east. Other species of spruce, pine, and fir dominate in North America. Where wind, fire, or felling open up the dark forests, birch and poplar quickly germinate and form more open successional stands, gradually developing into the typical climax forest of pine and spruce. Boreal forests are home to many birds, including woodpeckers, owls (including the very large Great Gray Owl and the Northern Hawk Owl), and hawks. Many birds such as crossbills, grosbeaks, and nutcrackers specialize on feeding on the seeds of cones, and warblers take advantage of the abundant insect life in spring and summer. Crossbills use their uniquely crossed beaks to tweak the seeds from cones. Many birds of the northern forests are migrants, feeding and breeding here in the spring and summer, then moving south to overwinter in warmer regions. There are conservation issues connected with such migratory movements as deteriorating conditions in their wintering grounds affect their survival, threatening a number of species.

TEMPERATE FORESTS

Though these are not watertight categories, it is useful to distinguish three main types of temperate forest—cool temperate forests; warm temperate forests; and oceanic temperate forests.

COOL TEMPERATE FORESTS

Temperate forests once covered much of Europe, eastern North America, and eastern Asia, but only scattered fragments remain today. Mainly dominated by deciduous broadleaved trees, they respond to the seasons with a burst of growth in spring and summer, followed by leaf loss in the fall, and a period of dormancy over the winter. The summers are relatively cool and the winters not extreme, though frost and snow are regular. Annual rainfall is moderate, between 20 in (500 mm) and 60 in (1,500 mm), evenly distributed, often with a slight peak in the summer.

Leaf fall produces abundant leaf litter and accumulated humus on the forest floor, supporting invertebrates such as worms, ants, beetles, and spiders that provide the diet for insectivorous birds and mammals. Woodpeckers extract insects and their larvae from tree bark, while many woodland birds, such as warblers and tits, feed mainly on caterpillars and other insects.

Broadleaved deciduous forest is the natural vegetation of much of the lowland and hill country in Europe. Characteristic dominant

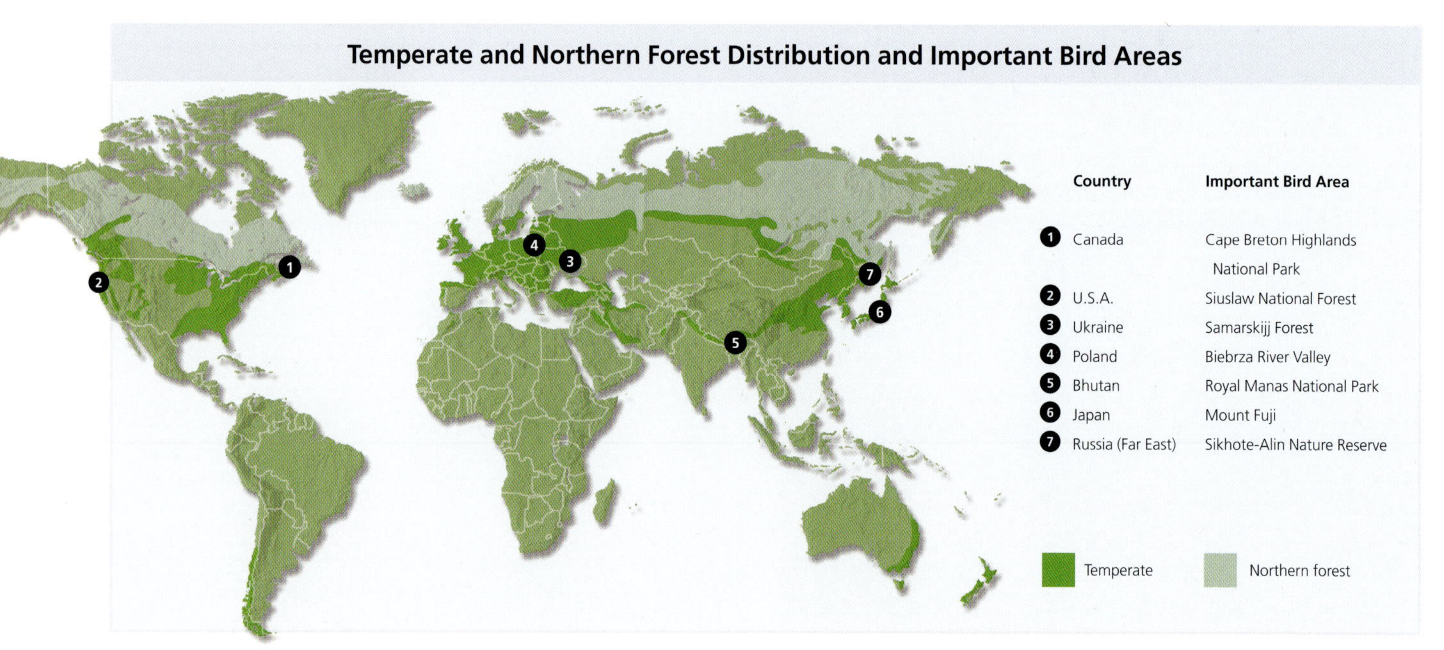

trees of these habitats are species of oak (*Quercus*), elm (*Ulmus*), lime (*Tilia*), maple (*Acer*), and beech (*Fagus*), along with hornbeam (*Carpinus*), ash (*Fraxinus*), birch (*Betula),* and alder (*Alnus*), with different species or combinations dominating the woods depending on the soil type and microclimate. On dry soils, pines often thrive, either in relatively pure stands or mixed with the broadleaved trees. Beech woods are particularly well developed in central Europe, especially on the slopes of the major mountain ranges, such as the Alps, Pyrenees, Apennines, and Carpathians.

The North American and Asian cool temperate forests are richer in species than those of Europe, which is related both to the topography and to the glacial history. In Europe, the Pyrenees, the Alps, and the Mediterranean Sea acted as barriers to species retreating from the advances of glaciers during the the last Ice Age, whereas in North America and Asia, species were able to retreat south and then re-invade as the ice moved north again. In the northern U.S.A. and Canada in particular, these forests are famous for their amazing displays of fall colors that are contributed to most spectacularly by species such as Sugar Maple (*Acer saccharum*). In Korea, and around the border between northeastern China and Russia, the winters are rather colder and the deciduous trees tend to be mixed with conifers such as Korean Pine (*Pinus koraiensis*).

TEMPERATE LIFE PATTERNS

Life in the cool temperate forests is highly seasonal, and by shedding their leaves the deciduous trees reduce water loss and metabolic activity through the cold winters. Spring brings a burst of activity with green shoots and surges in insect life—many forest birds time their breeding to take advantage of these food sources. Many temperate forest birds migrate to warmer latitudes during the winter thereby avoiding dangerous food shortages, while others—including owls and woodpeckers—are mainly resident throughout the entire year.

Typically these forests show distinct layers. The tallest trees reach 66 ft (20 m) to 165 ft (50 m), and below these grow shrubs of up to about 13 ft (4 m), with a herb layer at ground level. Many plants of the herb layer grow and flower in spring, taking advantage of the light before the main trees are fully in leaf.

VIRGIN FORESTS

In parts of central and eastern Europe there are surviving areas of near-natural "virgin" mixed forests, such as the Bavarian Forest straddling the German-Czech border, and the Bialowieza Forest in east Poland and adjacent Belarus. Sometimes referred to as "original forests," these are much richer than managed woodland, though it is doubtful that even these relict forests have remained unaltered throughout their long history. A trudge through some of these forests gives one the impression of the complexity and diversity of the original, more extensive, forest. Instead of selective logging and coppicing as in managed forest, in these "natural" habitats old trees die, fall, and provide a multitude of microhabitats and a more diverse flora and fauna.

▽ *Baxter State Park lies in north–central Maine within the northern deciduous forest region of the North American continent and experiences the cool, moist climate typical of this area. Common birds in the park include warblers, thrushes, and flycatchers as well as owls, hawks, and ducks and other wetland birds.*

△ *The breeding habitat of Kirtland's Warbler in Michigan has declined by 33 percent since the 1960s, yet its population of singing males has increased from about 200 to 1,500 since the 1980s. It remains Near Threatened.*

COOL FOREST BIRDS

Warblers, woodpeckers, and flycatchers are some of the more typical birds of the cool temperate forests. They take full advantage of the vast numbers of invertebrates, especially caterpillars and other insect larvae. Interestingly, the warblers of Europe and Asia (the Old World Warblers) belong to a different family (Sylviidae) from those of North America (Parulidae, the New World Warblers). However, species in both these families show very similar behavior and morphology, adaptations to allow them to exploit fully the similar conditions. Both warbler families are well represented in this habitat, and many have attractive songs that carry clearly in the forests, and they typically migrate to warmer climates for the winter. The Near Threatened Kirtland's Warbler is a rare North American bird that breeds only in very specific Jack Pine (*Pinus banksiana*) habitats, mainly in north and central Michigan. It requires young pines on well-drained sandy soil. Happily, the population of this woodland warbler is increasing.

WARM TEMPERATE FORESTS

In central China, deciduous broadleaved forests dominated by oaks, maples, limes, and chestnuts once covered much of the lowland and hills, but most have long since been cleared, especially those in lowland sites. Further south, evergreen oaks, chestnuts, and other broadleaved trees appear in some of the richest of all temperate forests. Some of these forest remnants are home to the rare and Vulnerable Brown Eared-pheasant and Gray-sided Thrush.

In southeastern North America, the remnants of the natural forests are also very rich in species. The southern Appalachians in particular retain patches of splendid temperate forests with a mixed tree layer of species such as Tulip Tree (*Liriodendron tulipifera*), in addition to limes, beeches, and chestnuts. The trees include species of oak and hickories, alongside beech, chestnut, buckeye, and basswood, with attractive shrubs such as magnolias, dogwoods, and rhododendrons. Toward the west, the woods become gradually drier, often dominated by oaks and hickories, before merging into the grassland habitats of the prairie.

CORK WOODS

Much of the Mediterrranean region once supported evergreen oak forests, but these have been so degraded that few remain intact, being mainly replaced by shrub habitats such as *maquis*, maintained largely through grazing and fire. In Portugal and Spain, there are special types of open warm temperate woods that have been maintained and sustainably managed for centuries. These are known locally as *montado* in Portugal and *dehesa* in Spain and the main trees are evergreen cork-oaks and holm-oaks. These forests are grazed, mainly by goats, sheep, and pigs, and sometimes cultivated, and the cork-oaks stripped on a rotational basis for cork. The *dehesas* are rich in wildlife and are home to some fascinating birds such as the Azure-winged Magpie and European Roller, as well as raptors including the Black-shouldered Kite. In some areas, rarer raptors such as Spanish Imperial Eagle and Bonelli's Eagle still hunt over these unusual woodland habitats.

LAUREL AND EUCALYPTUS FORESTS

Pockets of warm temperate forest also persist in the Caucasus and around the eastern shore of the Black Sea. A highly unusual type of temperate forest is the laurel forest of the Canary Islands. On La Palma and Tenerife especially there are some fine remnants of this habitat, dominated by the laurel, *Laurus azorica*. Both Dark-tailed Laurel Pigeon and the Endangered White-tailed Laurel Pigeon are resident, as is the Island Canary, ancestor of the Domestic Canary.

Warm temperate forests dominated by eucalyptus species are found scattered in southeastern Australian coastal regions. In the Blue Mountains, not far from Sydney, these forests are very varied and rich, with many species of eucalyptus trees. This is also the habitat of the rare Wollemi Pine (*Wollemia nobilis*), a relict species discovered only in 1994, growing in a deep sandstone gorge. Previously known only from fossils, this unusual conifer is related to monkey-puzzles (*Araucaria*).

▷ *Valley hillsides with temperate forest in China.*

▽ *The Brown Eared-pheasant of northern China has stable populations in protected areas, but outside these areas is threatened by deforestation for agriculture and urban development, and habitat degradation due to logging and livestock-grazing. It is ranked as Vulnerable.*

△ *Pumalín Park, in the north of southern Chile's fjordland, is the largest privately owned national park in the world, covering more than 1,250 sq. miles (3,200 sq. km). It consists largely of evergreen broadleaved forest, also known as temperate rain forest. These remarkably wet, original forests reach to the ocean, which is rare for such wet forest. The park protects some of the last remaining stands of Alerce (*Fitzroya cupressoides*), an unusual and long-lived conifer.*

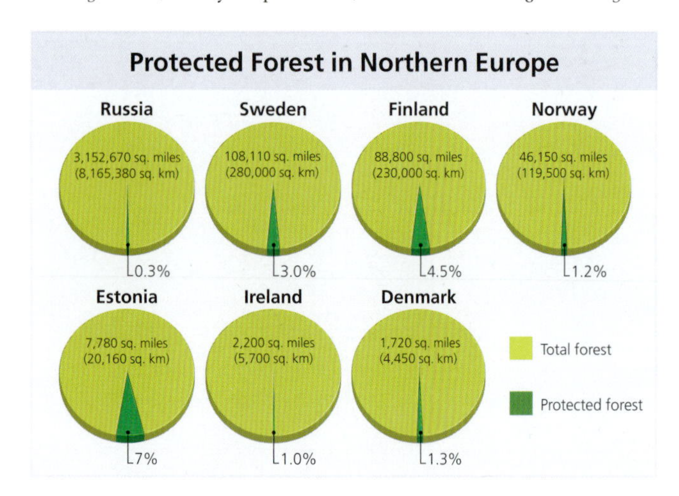

OCEANIC TEMPERATE FORESTS

In certain temperate areas exposed to high rainfall and humidity, notably on coastal mountain slopes that experience moist onshore winds, a special kind of temperate rain forest has developed. The main areas for this habitat are northwestern North America, western Chile and the South Island of New Zealand. Rainfall in such sites tends to be fairly evenly distributed through the year and is often above 160 in (4,000 mm) annually.

Such habitats are found along the Pacific coast of North America, from northern California to the Gulf of Alaska. Forests dominated by the giant Coast Redwoods (*Sequoia sempervirens*) occur in southern Oregon and northern California. The rainfall is high, up to 120 in (3,000 mm) per year, and there are also regular coastal fogs. Douglas Fir (*Pseudotsuga menziesii*) tends to dominate in the Coast Ranges, Olympic Mountains, and northern Cascade Range, up to about 3,300 ft (1,000 m). Both these species grow to over 330 ft (100 m) tall.

In the southern Andes of Chile and Argentina, the forests are rather special. Here there are temperate rain forests dominated by deciduous trees, such as species of *Nothofagus*, with unusual conifers such as monkey-puzzle (*Araucaria*) in some places.

Western Himalayas Endemic Bird Area

THE WESTERN HIMALAYAS HAVE BEEN RECOGNIZED AS AN ENDEMIC BIRD AREA, OR CENTER OF ENDEMISM. THIS EBA FOLLOWS THE HIMALAYAN MOUNTAIN FLANKS FROM WESTERN NEPAL THROUGH NORTHWESTERN INDIA AND INTO NORTHERN AND WESTERN PAKISTAN.

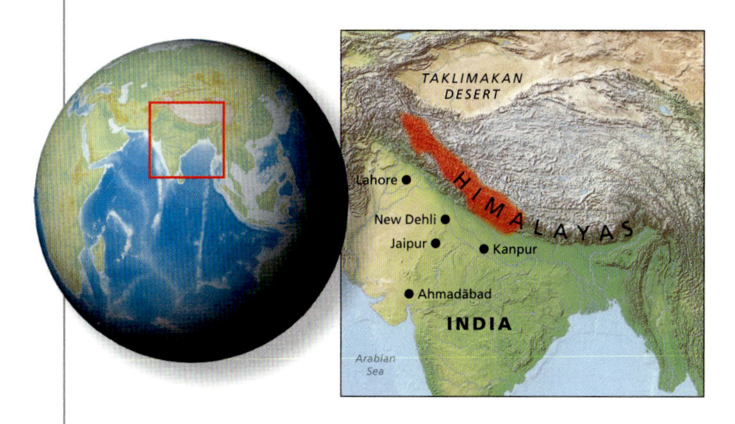

The remaining temperate forests found here consist of a mixture of coniferous and broadleaved trees. Much of this forest has been lost through logging for fuel and timber and there is also considerable overgrazing in some areas. Nevertheless, there are regions that still retain good forest cover, notably in northwestern India and Nepal, and these habitats support no fewer than 11 bird species that are found nowhere else, including several threatened species.

In Pakistan, much of the forest has been lost, but one area—the Palas Valley—still boasts relatively undisturbed forest, the largest single tract of temperate forest in Pakistan. This scenic valley is a stronghold of the Western Tragopan, listed as Vulnerable, and this is its largest known population. With its gray-black, white-spotted plumage and bright red upper breast and collar, the male is most striking. This bird is featured as a "flagship species" for the Himalayan Jungle Project, an initiative promoting its conservation and the development of sustainable forest use by local people. The total population of this rare pheasant is estimated at about 5,000, with perhaps 325 pairs in the Palas Valley. Also found in this region is the Cheer Pheasant, another Vulnerable species. The rare, dainty Kashmir Flycatcher, has a very small range, centered in Kashmir; it is also listed as Vulnerable. This beautiful red-throated species breeds in mixed deciduous forests with a dense shrub layer, between about 6,000 ft (1,800 m) and 8,900 ft (2,700 m). The birds migrate to southern India and Sri Lanka for the winter. Tytler's Leaf-warbler inhabits coniferous forest where it is under threat from logging. Classed as Near Threatened, this species winters in the Western Ghats where it is similarly affected by incursions into the forests. Other interesting birds found here are the Kashmir Nuthatch, Spectacled Finch, and Orange Bullfinch.

△ *The Satyr Tragopan has a Himalayan range stretching from India to Nepal, Bhutan, and China. Threats that are affecting this species include hunting and habitat degradation. It is Near Threatened.*

Western Himalayas' Threatened and Local Birds

	SCIENTIFIC NAME	COMMON NAME	STATUS	WORLD POPULATION
■	*Ophrysia superciliosa*	Himalayan Quail	CR	<50 (last recorded c.1889)
■	*Tragopan melanocephalus*	Western Tragopan	VU	5,000
■	*Catreus wallichi*	Cheer Pheasant	VU	4,000–6,000
■	*Ficedula subrubra*	Kashmir Flycatcher	VU	2,500–10,000
■	*Phylloscopus tytleri*	Tytler's Leaf-warbler	NT	Unknown
■	*Aegithalos leucogenys*	White-cheeked Tit	LC	Unknown
■	*Aegithalos niveogularis*	White-throated Tit	LC	Unknown

RED LIST ■ CR = Critically Endangered ■ EN = Endangered ■ VU = Vulnerable ■ NT = Near Threatened ■ LC = Least Concern

SPOTTED OWL

STRIX OCCIDENTALIS

THE SPOTTED OWL HAS TAKEN ON ICONIC STATUS BEING AT THE CENTER OF EFFORTS TO PRESERVE OLD-GROWTH TEMPERATE CONIFEROUS FORESTS IN THE PACIFIC NORTHWEST OF THE U.S.A. AND CANADA.

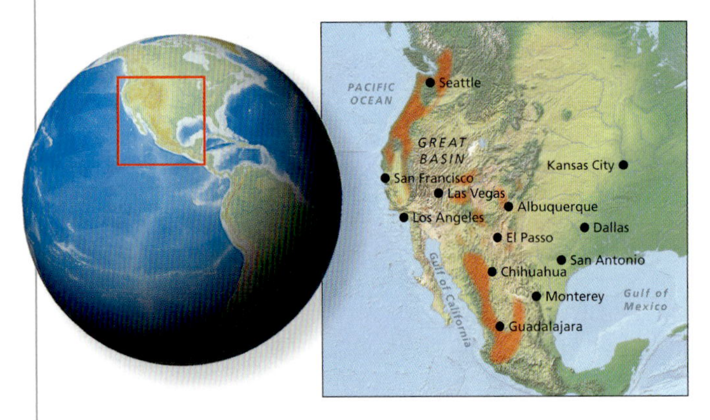

Superbly adapted to undisturbed forest, this owl feeds on mice, voles, flying squirrels, other small mammals, birds, and insects. Its forest home is dominated mainly by tall coniferous trees such as the Douglas Fir (*Pseudotsuga menziesii*) and the Western Hemlock (*Tsuga heterophylla*), many of which are 350 to 750 years old.

A single pair of Spotted Owls requires over 2,500 acres (1,000 hectares) of unspoilt old-growth forest for successful breeding. As logging has decimated its habitat, this owl has declined rapidly and is now listed as Near Threatened, with a population of about 15,000. The northern subspecies (*Strix occidentalis caurina*), which is the owl in this particular habitat, is down to only about 3,750 pairs, and is declining still further, numbers having fallen by more than 50 percent in a decade.

This once very extensive forest has suffered greatly as a result of timber extraction. In Washington State, for example, more than 90 percent has been logged or cleared for development. Clear-felling obviously ruins the owl's habitat, but even selective felling disturbs it and renders the forest more vulnerable to storm damage and fires. In recent years this owl has also begun to face competition from the Barred Owl (*Strix varia*), a more common species. The latter is expanding its range from the east, into western Canada and the northwestern U.S.A.

The conflict between logging companies keen to extract valuable timber and conservationists protecting the owl and its diminishing habitat have at times exploded into almost warlike proportions, with bumper stickers appearing that bear slogans such as "Kill a Spotted Owl—Save a Logger." Though the tension seems to have eased, the Northern Spotted Owl continues to decline.

▷ *The Northern Spotted Owl faces threats from habitat loss caused by logging, and competition from the Barred Owl.*

△ *The Swift Parrot of southeastern Australia is classified as Endangered. Its Tasmanian eucalypt forest habitat is subject to clearance for agriculture, residential development, plantation timber, sawlog production, and clear-felling for woodchips. Its estimated population in 2001 of 2,500 mature individuals was half the estimated number in 1988.*

In New Zealand there are large areas with oceanic temperate forests, especially in the Fjordland region in the South Island, and substantial parts of this fascinating habitat are protected in reserves and National Parks. Species of *Nothofagus* such as Southern and Silver Beech are characteristic of these moss- and fern-rich forests, as are the conifers Kahikatea (*Dacrycarpus dacrydioides*) and Totara (*Podocarpus totara*). Two striking rare birds of this region are the Endangered Takahe (the largest member of the rail family) and the Kakapo, a Critically Endangered parrot, the heaviest member of the parrot family that is unusual in being both nocturnal and flightless.

In southeastern Australia, there are also considerable areas of temperate rain forest, especially on mountain slopes not far from the coast, in a region stretching from southeastern Queensland south to central coastal New South Wales, with emergent trees such as booyong (*Argyrodendron*) and the Red Cedar (*Toona ciliata*). These habitats are important refuges for a number of rare birds including the Red Goshawk (Vulnerable), Regent Honeyeater and Eastern Bristlebird (both Endangered), as well as the Near Threatened Albert's Lyrebird. The eucalypt forests of Tasmania are the haunt of the Swift Parrot and Forty-spotted Pardalote (both Endangered), as well as the Critically Endangered Orange-bellied Parrot.

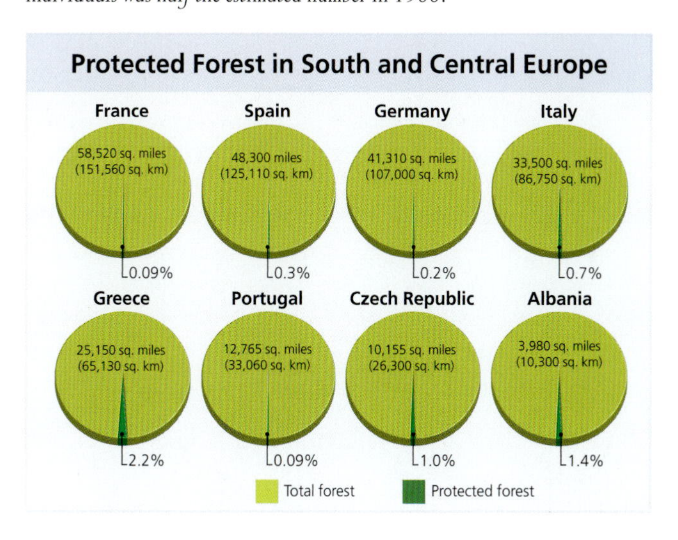

Desert

DESERTS ARE AMONG THE LEAST HOSPITABLE, AND MOST CHALLENGING, OF ALL TERRESTRIAL HABITATS, AND RELATIVELY FEW SPECIES HAVE EVOLVED TO EXPLOIT THEM. THE EXTREMES OF TEMPERATURE AND GENERAL SCARCITY OF WATER ARE NOT FAVORABLE FOR MOST PLANT LIFE. THOSE ORGANISMS THAT CAN SURVIVE HAVE TO BE HIGHLY SPECIALIZED.

In total, deserts account for about one-third of the land surface of the Earth. Most deserts lie between latitudes 15° and 40° north and south of the Equator and have climates that are dominated by high-pressure systems that block moist depressions from entering the region. While high daytime temperatures are a feature of most deserts, they can drop dramatically at night as the clear skies allow rapid nocturnal heat loss. The temperature range may be as much as 54°F (30°C) in 24 hours. Some deserts, such as those of Central Asia, also experience very cold winters. Certain deserts are incredibly arid. Parts of the Sahara Desert in North Africa and the Atacama Desert of coastal Chile, for example, receive hardly any rain at all, and most deserts receive only 10 in (250 mm) per year or less.

Desert plants have various adaptations to prevent excessive water loss. Some have succulent tissues that store moisture, while others have reduced, waxy leaves or spines that reduce evaporation and also deter grazing animals. Desert annuals survive the driest conditions as seeds, germinating and growing rapidly to flower when the rains finally arrive. In true deserts, perennial plants tend to grow only along watercourses and the landscape is extremely barren, usually dominated by sand, rocks, or gravel. At their margins, deserts often grade into savanna or scrub, via a belt of intermediate habitat, usually referred to as semi-desert. In semi-deserts, specialized perennials are able to survive and are not restricted to watercourses. Birds adapted to desert or semi-desert conditions include raptors such as certain vultures, wheatears, sandgrouse, doves, larks, and finches. Hummingbirds and sunbirds take advantage of nectar-bearing flowers, especially in shrubby semi-desert sites.

◁ *The Sahara Desert is a harsh environment, but it is home to more than 300 species of bird, including raptors such as eagles and vultures, as well as bustards, sandgrouse, and specialist passerines such as larks and wheatears. Many migrant birds also pass over the desert, stopping off to refuel at oases.*

Major Deserts of the World

Deserts tend to be subject to less threat than other habitats, and many are in fact increasing in area as the adjacent grass and shrubland is degraded by overgrazing. Extraction of underground water for irrigation allows crops to be grown on former desert soil, reducing suitable habitat for the native wildlife. Many deserts hold important deposits of minerals, and mining operations impact on desert habitats in some areas. Large quantities of borax were extracted from the Great Basin Desert (U.S.A.) and the Atacama (Chile) yields sodium nitrate, used as fertilizer and to make explosives.

Excluding the icy "deserts" of the Arctic and Antarctic, deserts and semi-deserts are found mostly in the interiors of continents, in areas where the climate is very dry. There are major deserts in North, Central, and South America, in Africa, Asia, and in Australia.

△ *The Atacama Desert is one of the driest places in the world. Its rare and endemic birds include the Chilean Woodstar, Coastal and Thick-billed Miners, White-throated Earthcreeper, Slender-billed Finch, and Drab Seedeater.*

NORTH AND CENTRAL AMERICA

In the southern U.S.A., the major deserts are in southern California, Nevada, Arizona, and New Mexico, and in Mexico mainly in the states of Sonora and Chihuahua. The Sonoran Desert of southern California, southwestern Arizona, and northwestern Mexico is the hottest of the North American deserts, and many cacti grow here, including the tall, branching Saguaro (*Carnegiea gigantea*), along with drought-tolerant woody shrubs such as Mesquite (*Prosopis glandulosa*) and Desert Ironwood (*Olneya tesota*). Occasional winter storms bring a flush of flowering annuals. Birds found in this desert include the strange roadrunners, Costa's Hummingbird, the Cactus Wren, the Gila Woodpecker, and the endemic Rufous-winged and Five-striped Sparrows. As in many deserts, threats come from a variety of sources. Trampling and overgrazing by cattle are a problem, as are pumping of groundwater and the seeding of fodder grasses that have replaced the original desert scrub in many sites. The Great Basin Desert (mainly in Nevada and Utah) is a cold desert, with cold, snowy winters. One of the most common birds of this desert is the Western Meadowlark, while another is the American White Pelican—about 15 percent of the population breed at the desert's Great Salt Lake. Golden Eagles and Swainson's Hawk hunt in the region and Sharp-tailed Grouse and Greater Roadrunner are also frequently seen. Drainage and the use of pesticides have affected pelican populations, though numbers have recovered with stricter environmental control.

SOUTH AMERICA

Lying as a narrow strip along the arid coast of southern Peru and northern Chile, the Atacama Desert is the driest of all deserts, and also the driest place on Earth, with rainfall generally averaging less than ⅔ in (15 mm) a year, and with some spots never having recorded rain. Even here, however, especially in the large salt basins,

Desert and Dry Scrubland Distribution and Important Bird Areas

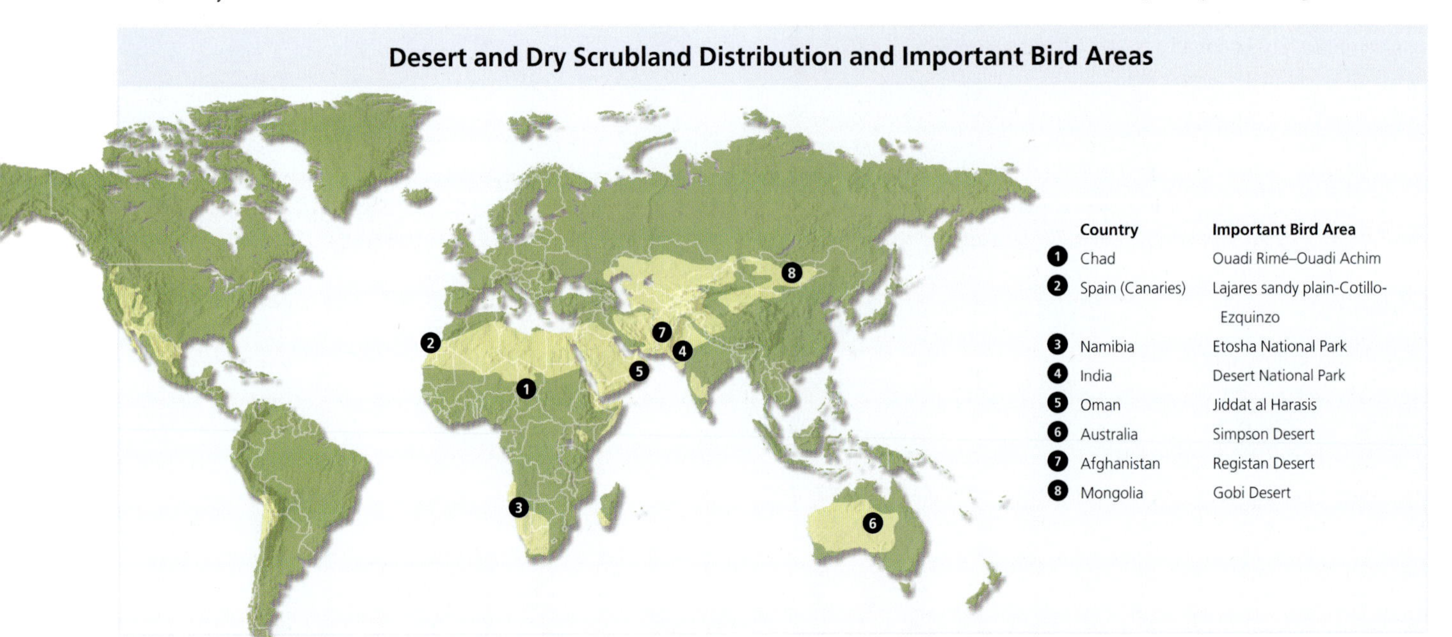

	Country	Important Bird Area
1	Chad	Ouadi Rimé–Ouadi Achim
2	Spain (Canaries)	Lajares sandy plain-Cotillo-Ezquinzo
3	Namibia	Etosha National Park
4	India	Desert National Park
5	Oman	Jiddat al Harasis
6	Australia	Simpson Desert
7	Afghanistan	Registan Desert
8	Mongolia	Gobi Desert

△ *The Sossusvlei region of the Namib Desert has the highest sand dunes in the world. It also has areas of saltpans that collect water in pools in exceptionally rainy seasons. Some 180 species of bird occur in the Namib Desert, including Ostrich, Secretary Bird, Karoo Long-billed Lark, Dune and Gray's Larks, Bank Cormorant, and Rüppell's Bustard. Birds of prey include Lappet-faced Vulture, classified as Vulnerable, and Black-chested Snake-eagle.*

occasional storms create saline lakes. At higher altitude, three rare species of flamingo take advantage of such seasonal habitats: the Andean Flamingo (Vulnerable), and the Chilean and Puna Flamingos (both Near Threatened). Endemic birds here include the Great Inca-finch and Pied-crested Tit-tyrant. The Endangered Chilean Woodstar is a pretty hummingbird with a very small population in only a couple of desert river valleys in northern Chile. Threats to the Atacama Desert come mainly from urbanization, mining, and grazing. Many new roads have been built in recent years to provide access to mining centers, causing local pollution and disturbance. The Patagonian Desert, developed in the rain-shadow of the Andes in southern Argentina, is a cold semi-desert.

AFRICA

The Sahara in North Africa is the largest of all the hot deserts, covering an area of more than 3,500,000 sq. miles (9 million sq. km). It connects to the east with the Arabian and Iranian deserts and the Thar Desert of Pakistan and India. The Arabian Desert covers much of the Arabian Peninsula—an area of over 900,000 sq. miles (2.3 million sq. km). In southern Africa the main deserts are the Kalahari and Namib.

Stretching some 1,200 miles (2,000 km) along the coast of southwestern Africa, the Namib Desert features mainly gravel plains, with large sand dunes rising as high as 1,000 ft (300 m). Rainfall is between ⅔ in (15 mm) and 4 in (100 mm) per year, but much of the moisture comes as sea mists, fog, and dew. Characteristic birds here are larks, including the Karoo Long-billed Lark, Dune Lark, and Gray's Lark. Mining activities are a serious threat in parts of the Namib. In addition to the effects of the excavations, the regular traffic of heavy vehicles compacts the soil, affecting the vegetation and other wildlife. Extraction of ground water for domestic use and for the mining industry has also caused a drop in the water table in some places and this has often been accompanied by the construction of pipelines and electricity pylons.

ASIA

In Central Asia, in an area north of the Himalayas, lie the cold Taklimakan and Gobi Deserts. The mainly sandy Taklimakan Desert stretches a huge distance between the Tibetan Plateau and the Tien Shan Mountains of China's Xinjiang Province.

The Taklimakan is China's largest and driest desert, covering over 285,000 sq. miles (740,000 sq. km), and has very sparse vegetation. One of the rarities of this habitat is the Near Threatened Xinjiang Ground-jay. Intensive grazing by camels and goats as well as irrigation of the land have caused its decline in some areas, though it is still common in the central desert. The Vulnerable Houbara Bustard can be seen, along with more common species such as the Black-bellied Sandgrouse and the Desert and Isabelline Wheatears.

To the east of the Caspian Sea, the Kara Kum Desert is dominated by dry clay soils and dune systems. The vegetation is sparse, yet many birds live here. In addition to the Houbara Bustard, notable species include the Desert Lark, Desert Wheatear, Desert Warbler, Asian Desert Sparrow, and Turkestan Ground-jay. Raptors—the Golden and Steppe eagles and the Short-toed Snake-eagle—hunt over the desert, as does the rare Saker Falcon. Both the Black-bellied and Pin-tailed Sandgrouse are found here, as is the Cream-colored Courser. Sandgrouse show remarkable adaptations to the arid conditions, making long flights to water sources both to drink and to transport water in their absorbent belly feathers back to their chicks. These deserts suffer overgrazing by livestock and also from irrigation-based cultivation, notably of cotton.

The Thar Desert covers about 170,000 sq. miles (450,000 sq. km), centered on the Indian state of Rajasthan and extending into southeastern Pakistan. This is classic sandgrouse country and is also home to the Endangered Great Indian Bustard, now reduced to fewer than 1,000 birds. This splendid bird stalks the arid grassland and scrub but, though legally protected, it is severely affected by hunting for both sport and food, as well as by disturbance and loss of suitable habitat. Irrigation has resulted in the conversion of much of the semi-desert to agriculture. This desert also contains India's largest salt lake, Sambhar Lake, winter home for thousands of waterfowl, Dalmatian Pelicans (Vulnerable), and large flocks of Greater Flamingo and Common and Demoiselle Cranes. Threatened by regular salt extraction, the lake is being actively promoted for ecotourism.

AUSTRALIA

Much of Australia's interior consists of desert or semi-desert and the Great Sandy Desert of western Australia is huge—covering some 320,000 sq. miles (820,000 sq. km). Tough grasses and drought-resistant shrubs dominate, with occasional hardy trees. Red sandy soils are typical, with some dune formation. This desert is best known for the amazing red sandstone mound of Uluru, sacred to native Australians. Rare birds of this desert include the Princess Parrot. Classified as Near Threatened, this pretty species has a small and declining population, mainly in central and western desert areas where it feeds on seeds and fruits. Introduced domestic cats and Red Foxes (*Vulpes vulpes*) may be one reason for its decline.

SAHEL MIGRANTS

Changes to the environment in desert and semi-desert regions can have severe impacts on the local birds, including those species that

▽ The Demoiselle Crane arrives in the Thar Desert during winter, stopping over during its tough migration, after having flown over the Himalayas at altitudes of up to 26,000 ft (8,000 m) to reach its wintering grounds. This bird is classified as Least Concern.

MOUNTAIN

IN A WORLD WHERE NATURAL HABITATS ARE INCREASINGLY THREATENED BY A RANGE OF ISSUES FROM DIRECT DAMAGE TO POLLUTION AND CLIMATE CHANGE, WILDLIFE REFUGES ARE BECOMING FEW AND FAR BETWEEN. MOUNTAINS CANNOT BE EXPLOITED TO THE SAME EXTENT AS LOW-LYING LAND AND THEY OFFER REFUGE FOR ALL TYPES OF WILDLIFE.

In lowland sites agriculture, housing, and industry continue to take precedence over nature. Mountains offer vitally important refuges to birds in particular. Many of the most important wildlife reserves are to be found in upland and mountain sites, and many rare bird species find refuge in mountain habitats.

The particular habitats that mountains contain are many and varied. On a typical mountain, the natural habitats change gradually from the lower slopes with increasing altitude. This is mainly because temperature falls at a rate of roughly 0.9–1.8°F (0.5–1°C) with every 330 ft (100 m) increase in altitude. Other climatic and geological conditions such as rainfall and aspect (such as north- or south-facing), acidity and structure of the rocks and soils, and topography all have their influence on the habitats that develop. In a tropical mountain range, there may be tropical broadleaved forest on the lowest foothills, with bands of subtropical and temperate broadleaved forest above, coniferous forests at higher altitude, followed by shrubby communities, and finally high-altitude (often tussocky) grassland and alpine rocky habitats.

◁ *Relatively few birds are adapted to live at high altitudes. Acrobatic and gregarious, choughs are very much at home in such habitats, and are here seen against the backdrop of Mount Everest in the Himalayas.*

Mountain Ranges

Mountains are vitally important refuges for birds in tropical regions. The warm climates here mean that complex habitats including forests exist at high altitudes on uneven or steeply sloping ground that is less accessible to people. While much of the natural vegetation of the lowland and hills has been removed or altered by human activities, many mountain ranges still retain considerable patches of natural or near-natural habitats, from coniferous and broadleaved forest, to bamboo-rich scrub and rocky grassland.

MOUNTAINOUS VARIETY

The Andes of South America and the Himalayas in Asia consist of vast tracts of largely interconnected high ground with generally few extensive lowland "barriers" between the peaks and ranges. In other regions, such as parts of Africa and to some extent China, the mountain ranges are much more isolated from each other, appearing as isolated peaks, rather like islands set in a "sea" of lowland, much of it cultivated. The flora and fauna of these mountains have evolved in relative separation, similar to those of oceanic islands. This has resulted in many mountains or mountain chains being rich in endemic species—often referred to as "centers of endemism." It is not, therefore, surprising that many of the Endemic Bird Areas (EBAs) recognized by BirdLife International are centered on tropical mountain ranges.

On some tropical mountains, notably in Africa, the uppermost "elfin" woodland consists of low-growing, often twisted trees rich in mosses and lichens. Another unusual habitat of these mountains is dominated by rosette-forming species that grow very tall when in flower. In Africa, these are often from the daisy or lobelia families and are quite unlike their lowland relatives in growth form. Some South American mountains have similar habitats featuring plants from the pineapple and pea families. The rosette growth habit seems to protect the plants from night frosts, though the days can be quite hot.

ADAPTING TO HABITAT DIVERSITY

Being highly mobile creatures, birds can easily take advantage of the variety of habitats offered by mountain sites, sometimes moving between different types to follow seasonal changes such as flowering and fruiting, often migrating to lower or higher levels to avoid dry or cooler seasons. Birds of prey such as condors, Old World vultures, and many eagles hunt in mountainous country where the long vistas and (at higher altitudes) sparse vegetation facilitate location of their quarry. Some eagles and many hawks are agile hunters in forest habitats using their relatively shorter wings and long tails to give them great maneuverability. Hummingbirds and sunbirds are often found in mountain sites, and occupy similar ecological niches in the Americas and Old World respectively. Some hummingbirds become torpid during the cold nights—exhibiting a kind of nightly hibernation—and become active again during the relatively hot daytime when they resume feeding from nectar-rich mountain flowers.

DISTANT PEAKS

In Africa especially, many of the mountain ranges are widely separated, such as those close to the Albertine Rift Valley in eastern Zaire, Rwanda, Burundi, southwestern Uganda, and western Tanzania. The altitude ranges from about 6,600 ft (2,000 m) to 16,400 ft (5,000 m) and the habitats are very diverse, ranging from lowland and montane forest to elfin and bamboo forests above 8,000 ft (2,400 m), and topped by moorland scrub and grassland above 11,500 ft (3,500 m). Rare birds found here include the Congo Bay-owl, Itombwe Nightjar, and Prigogine's Greenbul, all of which are Endangered, and the Albertine Owlet, African Green Broadbill, Yellow-crested Helmet-shrike, Rockefeller's Sunbird, and Chapin's Flycatcher, all of which are Vulnerable. Three of the most famous reserves here are the Virunga National Park in Zaire,

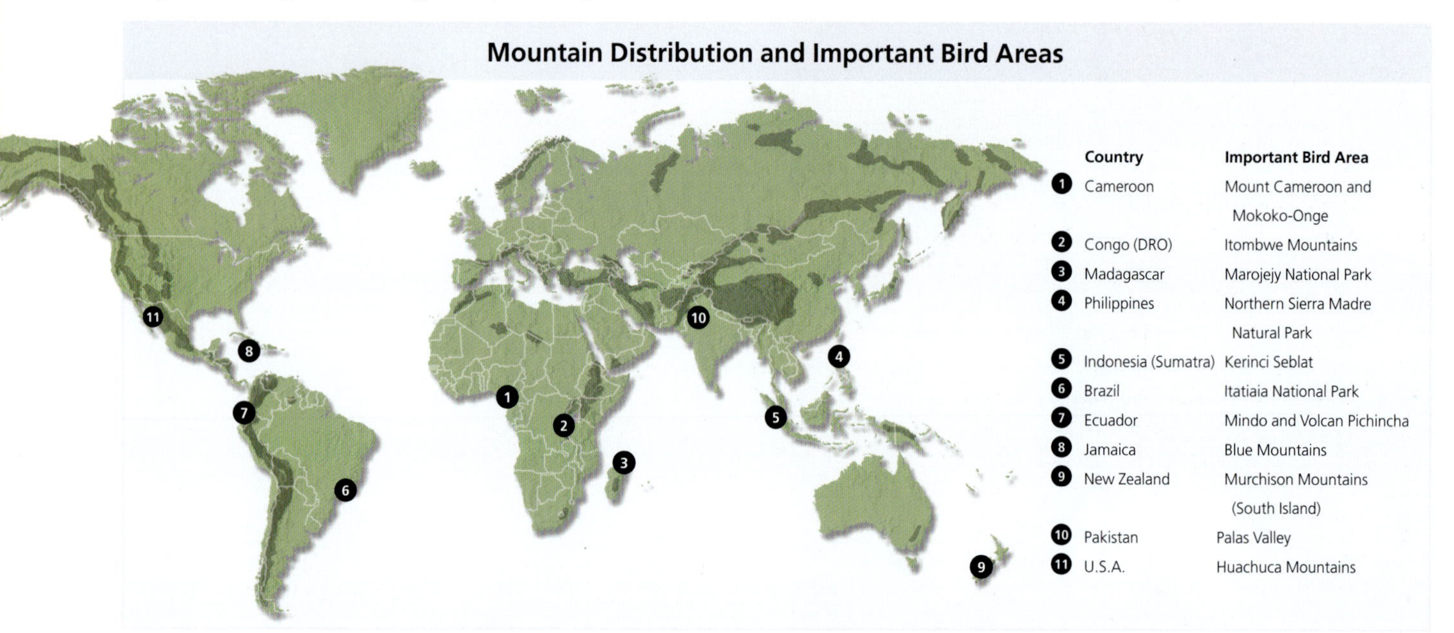

△ *The stunningly colorful Himalayan Monal, a member of the pheasant family, ventures above the Himalayan tree-line to wander on the grassy slopes during the summer months. During winter, this high-altitude species shelters from the weather lower down in coniferous and mixed forests.*

and Uganda's Bwindi Impenetrable National Park and Mgahinga Gorilla National Park. Together these parks hold most of the world's Mountain Gorillas, as well as many varied and fascinating birds, several of them rare. The Udzungwa Forest-partridge, for example, is an Endangered species found only at three sites in the Udzungwa Highlands and Rubeho Mountains in central Tanzania. It forages for invertebrates in montane evergreen forest between 4,250 ft (1,300 m) and 7,900 ft (2,400 m). With a population of only about 3,500, it is at risk mainly from hunting and human disturbance.

VOLCANIC PEAKS

In western Africa, some of the most fascinating rare birds can be found in the volcanic mountains of northwestern Cameroon. Here, the lowland rain forest grades into montane forest with bamboo in some sites above about 8,200 ft (2,500 m). The Vulnerable Mount Cameroon Speirops and Endangered Mount Cameroon Francolin are both endemic to Mount Cameroon, which, at 13,435 ft (4,095 m), is the highest mountain in West Africa and also an active volcano. The beautiful Bannerman's Turaco and Banded Wattle-eye (a dainty black-and-white flycatcher) are found only in certain montane forests in this region, and both birds are listed as Endangered—they are threatened, as are many others, by the fragmentation of their habitat.

The East Usambara Mountains of Tanzania

The East Usambara Mountains in northeastern Tanzania are a tremendously important ecological region. Their relative isolation has allowed the evolution of a number of fascinating endemic species and they are recognized as one of the world's biodiversity hotspots.

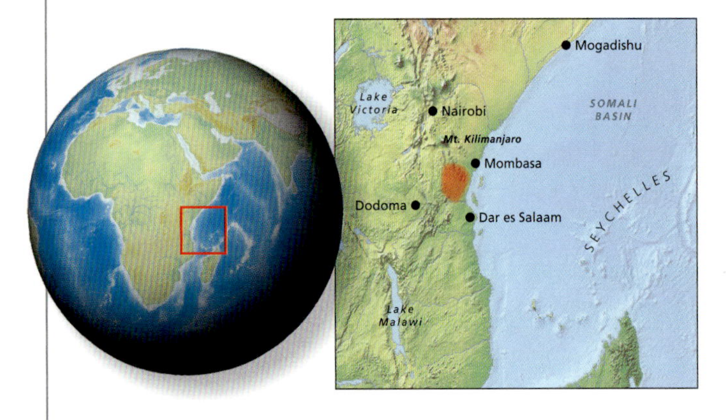

These mountains, part of an EBA, are one of Africa's prime sites for globally threatened birds. Though relatively small, they are rather isolated from Tanzania's other mountain ranges and it is partly this separation that has resulted in the evolution of so many endemic species of both plants and animals. This small range reaches only to just over 4,900 ft (1,500 m) and is about 22 miles (35 km) from the coast, in eastern Tanzania. As they are so close to the coast, these mountains enjoy high rainfall. The main rainy season is March to May, with another peak in October to December, but the drier seasons also receive occasional rain. On the main plateau the annual rainfall can be as great as 80 in (2,000 mm).

The East Usambaras have retained large areas of lowland and submontane forest, though plantations of coffee, tea, and teak have replaced much of the original forest. The lowland forest is semideciduous, with tall emergent trees reaching 115 ft (35 m). The submontane forests are evergreen and in some the tallest trees reach 210 ft (65 m), with a second tree layer to about 100 ft (30 m). These latter forests are very luxuriant and species rich. They contain many endemic species and also plants of medicinal and horticultural significance. For example, several species have edible fruits, and these forests contain no fewer than eight species of African violet (*Saintpaulia*). Many endemic birds also live in the forests and notable threatened species are two rare owls—the large Usambara Eagle-owl, classified as Vulnerable, and the tiny Sokoke Scops-owl, classified as Endangered.

Other Endangered species found here are the Usambara Akalat, Usambara Weaver, and the pretty Amani Sunbird. Dapple-throat and Swynnerton's Robin—both classified as Vulnerable—and the beautiful Near Threatened Fischer's Turaco (see opposite), also live in these lush forests.

AMANI FOREST RESERVE

The Amani Forest Reserve is an important protected area within the East Usambaras. It incorporates the Amani Botanic Garden, established by Germany in 1902. Until 191,4 many exotic species were introduced to form what is now essentially an arboretum with one of the largest collections in tropical Africa. Unfortunately, some of the exotic species have become invasive in the natural forest, reducing their diversity. Nevertheless, the reserve offers visitors wonderful opportunities for experiencing some of East Africa's finest forests and seeing rare birds, butterflies, and plants.

Key Features of East Usambara Mountains

Location:	Tanzania, Tanga Region
Area:	104,804 acres (42,413 hectares)
Altitude:	500–5,000 ft (150 –1,525 m)
Habitat:	Tropical forest
Threats:	Logging (now ceased), cultivation of cardamom, tea and other crops, reduced rainfall and invasion of the non-native tree *Maesopsis eminii*

◁ *Early morning light reveals the cloud forests, also known as fog forests, of the Usambara Mountains, northeastern Tanzania. These forests are home to many endemic species, including several endemic bird species.*

△ *Fischer's Turaco has a population of about 1,500 birds in the Usambara Mountains. During the 1980s and 1990s, hundreds of these birds were trapped for the cagebird market, and this trade, though reduced, is still a real threat.*

Threatened Birds of the Usambara Mountains

	SCIENTIFIC NAME	COMMON NAME	STATUS	WORLD POPULATION
■	*Artisornis moreaui*	Long-billed Tailorbird	CR	50–250
■	*Otus ireneae*	Sokoke Scops-owl	EN	2,500
■	*Ploceus nicolli*	Usambara Weaver	EN	1,000–2,500
■	*Sheppardia montana*	Usambara Akalat	EN	28,000
■	*Anthreptes pallidigaster*	Amani Sunbird	EN	2,500–10,000
■	*Bubo vosseleri*	Usambara Eagle-owl	VU	2,500–10,000
■	*Modulatrix orostruthus*	Dapple-throat	VU	10,000–20,000
■	*Swynnertonia swynnertoni*	Swynnerton's Robin	VU	2,500–10,000
■	*Anthreptes rubritorques*	Banded Sunbird	VU	2,500–10,000
■	*Tauraco fischeri*	Fischer's Turaco	NT	2,500–10,000
■	*Sheppardia gunningi*	East Coast Akalat	NT	10,000–20,000

RED LIST ■ CR = Critically Endangered ■ EN = Endangered ■ VU = Vulnerable ■ NT = Near Threatened ■ LC = Least Concern

SACRED MOUNTAINS

The mountains of China have some remarkable endemic and rare birds. In western Sichuan, there are a number of important mountain ranges, including the Min Shan and Qionglai Shan, famous as the habitat of the Giant Panda. The reserves set up to conserve this charismatic mammal also protect a number of rare birds, such as the Vulnerable Chinese Monal and Sichuan Jay, and the Near Threatened White Eared-pheasant. The holy mountain of Emeishan is also in Sichuan Province. This World Heritage Site, sacred to Buddhism, rises to 10,170 ft (3,100 m) above sea level and has splendid remnants of subtropical forest, with fascinating local birds, most notably the pretty Omei Shan Liocichla, a Vulnerable babbler restricted to this and nearby ranges. To the south, the province of Yunnan once had extensive mountain forests, now largely removed by decades of logging for timber or fuel. Some of the pine forest remnants are home to the Giant and Yunnan Nuthatch, while the Vulnerable White-speckled Laughingthrush inhabits bamboo-dominated forest above about 9,800 ft (3,000 m). The Lu Liang Mountains of central Shanxi Province support populations of the Vulnerable Brown Eared-pheasant, a local endemic. Further south still, in Guangxi Province, Damingshan has montane forest and rocky gorges. It is famous as

△ *The Caucasian Snowcock is endemic to the Caucasus Mountains, particularly the west of the range, where it breeds at altitudes from 6,500 to 13,000 ft (2,000 to 4,000 m) on bare, rocky sites.*

a site for the secretive Endangered White-eared Night-heron. Also found here are the Vulnerable Silver Oriole, along with more common species such as the Red-tailed Laughingthrush, Green-billed Malkoha, Small Niltava, and the attractive Gould's Sunbird.

SOUTH AMERICA'S BACKBONE

The Andes Mountains stretch as a backbonelike chain along the entire western coast of South America, from Colombia in the north, down to the southern tip of the continent at Cape Horn. Though the prevailing climate of much of this region is tropical, the hills and mountains offer local temperate and even cold high alpine conditions, with a band of rather cool, highland steppes along the eastern edges of the Andes. Many birds have adapted to exploit the varied habitats found in these mountain ranges.

In the northern central Andes of Colombia and Ecuador there are montane evergreen forests and also elfin cloud forests above about 8,200 ft (2,500 m) and these support a good number of endemic and threatened bird species. Antpittas are well represented here with the Endangered Brown-banded Antpitta, as well as the Vulnerable Giant and Bicolored Antpittas. Notable, too, are a number of rare parrots. The Rusty-faced Parrot (Vulnerable) has a restricted range in the cloud forest and scrub, where it feeds on seeds, flowers, and fruit. Three threatened parrots inhabit the cloud forests: the Indigo-winged and Yellow-eared Parrots (both Critically Endangered), and the Vulnerable Golden-plumed Parakeet. Both parrots have tiny ranges in Colombia. The Indigo-winged Parrot suffers mainly from selective logging of mature trees, on which the parrots rely for nesting in natural cavities. Yellow-eared Parrots depend on Wax Palms (*Ceroxylon quindiuense*) for food, roosting, and nesting. Cattle browse on young palms and local people also use fronds for Palm Sunday celebrations. These pressures reduce the regeneration of these slow-growing palms.

The high Andes from Peru, through Bolivia and into northern Argentina experience altitudes from about 3,300 ft (1,000 m) in the valleys up to 15,100 ft (4,600 m) on the altiplano. The climate is rather dry, with temperate forests on the slopes, cactus-rich scrub in the valleys, and puna and open rocky grassland above.

▽ *Torres del Paine National Park, in southern Chilean Patagonia, is home to about 100 species, ranging from the Black-chested Buzzard-eagle and the Andean Condor that hunt on the lower mountain slopes, to the Black-necked Swan, Chilean Flamingo, and Spectacled Duck that live on the lakes below.*

California Condor

Gymnogyps californianus

This magnificent bird of prey has become a symbol of conservation efforts in North America. At the same time it has highlighted the expense and difficulties involved in returning an almost extinct species to its natural habitat.

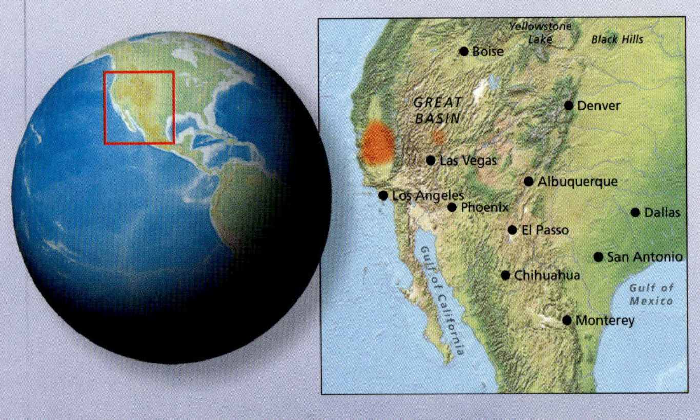

Like its cousin the Andean Condor, the California Condor is huge and impressive. Now listed as Critically Endangered, this massive raptor once haunted the canyons and valleys of western North America, from Baja California, Mexico in the south to British Columbia in the north. It declined steadily, reaching a low point of just 22 birds in 1981. By 1987, the last few wild birds had been captured to join a captive breeding program in a last-ditch attempt to rescue this magnificent bird from extinction.

By 2003, 85 had been reintroduced to the wild in California and Arizona, increasing to 130 by 2006. Condors live for 50 years or more (in captivity) and do not breed until at least six years old, so building up a population is a slow process. A single egg is laid and if it hatches successfully the parent birds attend the young bird even after it has fledged. The preoccupied parents are unable to breed again for the rest of that year and the following year.

Captive rearing involves feeding the growing birds without letting them see their human foster parents. The use of a glove in the rough form of an adult condor's head is one trick used in this process. A better arrangement is to let adult captive condors raise the chicks, thus avoiding any damaging imprinting to humans. When the birds are ready to be released, this must be done with great care, using specially constructed release pens in which the birds can be kept and provided with food for months, with previously released individuals able to make occasional visits before the captives are finally given their freedom. Major food sources for California Condors are dead whales and sea-lions washed up on the coast though, like vultures, they are attracted to any carrion, including dead livestock or dead mammals such as deer. They usually detect their quarry by following other scavengers such as Turkey Vultures.

▽ *This group of adults and juvenile California Condors in Zion National Park, Utah, has been bred in captivity and released wearing radio transmitters, so that scientists can track and protect them.*

Fall and Rise of the California Condor Population

Number of birds: 0, 50, 100, 150, 200

Year: 1930, 1981, 1987, 1992, 1996, 1999, 2003, 2006, 2009, 2010

In captivity — In the wild

FREEDOM OF INFORMATION

Released individuals carry global positioning system (GPS) devices that transmit data to research computers 12 times each day. This has yielded much valuable information about their movements and also allows rapid recovery of dead or sickly individuals. Poisoning by lead ingested with meat from the carcasses of shot deer and other mammals seems to be a continuing problem for the condors. In October 2007, California passed a bill prohibiting the use of lead bullets so it is hoped that this threat will gradually diminish.

The number of California Condors living wild had reached about 187 in 2010, with 161 in captivity. They have been reintroduced to the Big Sur coastal region of central California and also to the coastal mountains of southern California and to the Grand Canyon of Arizona, as well as northern Baja California.

▽ *This California Condor chick in San Diego Wild Animal Park is being persuaded to feed using a glove puppet in the guise of one of its parents so that vital food can be delivered directly into its immature mouth.*

BIRDS IN THE HIGH ANDES

Some of the valleys and streams of the Andes hold patches of forest dominated by *Polylepis*, a tree in the rose family that grows at higher altitudes than any other flowering plant tree. These special mountain habitats are home to a number of local birds including the Endangered Red-fronted Macaw, Wedge-tailed Hillstar (a Near Threatened hummingbird), Bolivian Blackbird, and Gray-crested Finch. Along with many parrots, the Red-fronted Macaw has suffered through being trapped for the cagebird trade as well as from habitat loss and being shot as a pest in maize and peanut fields. The Endangered Cochabamba Mountain-finch inhabits mixed forest near the Bolivian city of Cochabamba. One of the rarest Andean mountain birds is the Critically Endangered Royal Cinclodes, restricted to patches of humid woodland (typically *Polylepis*) and scrub between 11,500 ft (3,500 m) and 15,800 ft (4,800 m), where it forages in the mossy leaf litter. Fire and grazing have combined to alter the ground flora and prevent tree regeneration. Conservation measures have focused on planting other native trees as sources of firewood and also introducing fuel-efficient stoves to reduce demand for fuelwood.

In the Andes of Peru, above 6,600 ft (2,000 m), dry, rocky areas with low rainfall are dominated by montane scrub and cacti, with pockets of woodland in damper sites. Several of the bird species here are threatened, including the Vulnerable Taczanowski's Tinamou, Apurimac Spinetail, and White-cheeked Cotinga, the Endangered White-browed Tit-spinetail, Gray-bellied Comet, and Ash-breasted Tit-tyrant, and the Critically Endangered Royal Cinclodes. Three more Vulnerable birds found here are the Golden-plumed Parakeet, Creamy-breasted Canastero, and White-tailed Shrike-tyrant.

THE ANDEAN CONDOR

One of the most magnificent birds of the Andes is undoubtedly the huge, majestic Andean Condor. Though it has a wide range, it is classed as Near Threatened and is declining in most regions, mainly due to direct persecution—it is unfairly accused of attacking livestock. It feeds mainly on carrion, yet this has also resulted in deaths as the birds sometimes feed on carcasses that have been deliberately poisoned, usually to kill mammalian predators such as foxes. The Andean Condor, like its much rarer relative the California Condor, is the subject of a number of attempts at re-introduction from captive-bred stock, with some success. Such projects have

△ *The magnificent Andean Condor is the world's largest flying bird, with a wingspan of 9–10½ ft (2.8–3.2 m). Once aloft, it rarely flaps its wings, relying instead on thermals to power its flight.*

been carried out in Colombia, Venezuela, and Argentina. Ecotourism is also helping by encouraging local people to protect this "must-see" New World vulture, the symbol of the Andes.

CLIMATE CHANGE AND MOUNTAIN HABITATS

As the average global temperatures increase, changes are being seen in many habitats, including those of mountain sites. This effect is perhaps most clearly seen by the retreat of most of the world's glaciers. Such changes affect not only the wildlife, but may also have consequences for river flow and land use in the lowlands. Mountain glaciers store huge quantities of water, then release a proportion in the warmer, drier seasons, feeding river systems into the valleys and plains. Climate change may upset the stability of this system and cause floods and unreliable water levels in the rivers. Such effects are already being seen in some areas, such as in the Himalayas and Andes.

Glaciers the world over have been in retreat for decades, not only in northern ranges such as the Alps and Rockies, but also on tropical mountains such as the Chacaltaya Glacier in Bolivia and most famously on Mount Kilimanjaro in northern Tanzania. Since about 1850, most glaciers have retreated, and the speed of their retreat has increased markedly since about 1980, due largely to human-induced global warming.

Some high-altitude montane forests, such as the spruce and fir forests of the southern Appalachians, are particularly vulnerable to climate change. As the temperature rises, species adapted to living in these habitats may have nowhere to retreat and be driven to extinction. At the same time, habitats at lower altitudes move upward, narrowing the areas occupied by higher altitude habitats such as these high-altitude forests and also alpine meadows and scrub.

▷ *The Pedersen Glacier in Alaska has retreated dramatically. The top image was taken in 1920, the lower one in 2005. Note that grasses and even some pioneer trees now grow on land once dominated by ice. Glaciers are sensitive to climate change and most have retreated in recent decades.*

GRASSLAND

Grasslands are areas where the land is dominated by grasses and non-woody herbaceous plants. Large swaths are found on all continents except for Antarctica. In temperate regions the grasses are mainly perennial, while in warmer parts annuals predominate. All grasslands support a huge range of mammals, birds, and insects.

The question of whether grasslands are natural habitats is even more vexed than the position of forests. Even in areas of extensive grassland habitats, such as the prairie of North America or the steppe of central Asia, the degree to which natural habitats have been affected by the activities of people is much debated. There is increasing evidence, for example, that early nomadic people altered such habitats deliberately, especially through removal of woody plants and promotion of grasslands by regular burning—a process that almost certainly expanded and promoted grassland in the central plains of North America, and elsewhere. Despite the influence of people and domesticated livestock, there are definite regions where the climate and soils favor grass-dominated habitats over woodland or desert. Such regions include the centers of the larger continents at temperate latitudes, where the conditions are too extreme to support closed woodland or forest. One region of steppe grassland extends from eastern Europe through Russia to Mongolia, while the prairie is a comparable habitat in the central states of North America; parts of continental Argentina, Uruguay, and southern Brazil support another kind of temperate grassland known as pampas. There are also some areas of arguably natural grassland in New Zealand, especially in the southeast of the South Island. Subtropical and tropical regions also feature some habitats in which grassland dominates, most notably in the open savannas of eastern Africa.

It is convenient to divide grassland habitats into those occurring in temperate regions and those found in the tropics and subtropics.

◁ *The Grasslands National Park in southern Saskatchewan, Canada protects one of the few remaining areas of undisturbed short-grass prairie in the country. American Bison have been re-introduced here and the birds include the Sage Grouse, Sharp-tailed Grouse, and Burrowing Owl.*

GRASSLANDS AROUND THE WORLD

TEMPERATE GRASSLANDS

In temperate grasslands, the summers are usually quite warm, with mean temperatures of 68–75°F (20–24°C), though the winters can be cold with some regions having two or three months with mean temperatures below freezing. Rainfall is usually between 8 and 30 inches a year (200 and 750 mm a year), generally in spring and summer, and with regular snow in winter. Trees find life hard under such conditions and perennial grasses tend to dominate, along with perennial herbs that usually flower in the spring or fall. Many of these herbs have showy flowers that attract pollinating insects.

The soils of these grasslands are usually rather deep and fertile, a fact that has not gone unnoticed by farmers. Such soils are ideal for growing domesticated grass species—grain crops such as wheat, corn (maize), and barley, and it is no surprise that much of the wild grassland now lies under intensive cultivation, with large areas devoted to huge-scale, highly mechanized cereal production, involving the use of chemical fertilizers and pesticides. Such monocultures are damaging to the native wildlife, including birds, and the chemicals often affect the birds either directly or by reducing prey numbers and diversity.

△ *Tufted grasses dominate in some types of grassland as here in southern Chile. Similar habitats can be found in Patagonia. Special birds include the Patagonian Yellow-finch, Patagonian Mockingbird, and White-throated Caracara.*

The prairie of North America once covered vast areas of the central plains, from Saskatchewan and Manitoba in southern Canada through the central U.S. states south to Oklahoma. Once covering about 1.35 million sq. miles (3.5 million sq. km), these rolling grasslands were grazed by herds of American Bison, Pronghorn Antelopes, and deer, the former both eventually hunted to the very brink of extinction. Many of the prairie grasses are deep-rooted and have evolved to regenerate following occasional fires, whether natural or deliberate. The type of natural grassland varies, mainly with the rainfall, and ranges from short-grass prairie that occurs mainly in the west, with rainfall about 10 in (250 mm), to tall-grass prairie found mainly in the east, with rainfall in the range 20–40 in (500–1,000 mm). Much of the prairie is now grazed by cattle or planted with crops. Typical prairie birds are meadowlarks, grouse such as sage-grouse and prairie-chickens, and raptors including Turkey Vultures, while Burrowing Owls breed in burrows dug by prairie-dogs (ground-nesting members of the squirrel family).

The steppes of Eurasia lie between the forests to the north and the deserts to the south, stretching in a band of highly variable width from the plains of eastern Europe right across Asia to Mongolia in the east. The steppe was formerly grazed by wild horses, asses, and Saiga Antelope and also by Bactrian Camels in Mongolia—all either rare today or present only in certain protected areas. Typical steppe birds include larks, sandgrouse, eagles, and bustards such as the huge Great Bustard, now classed as Vulnerable largely as a result of habitat loss.

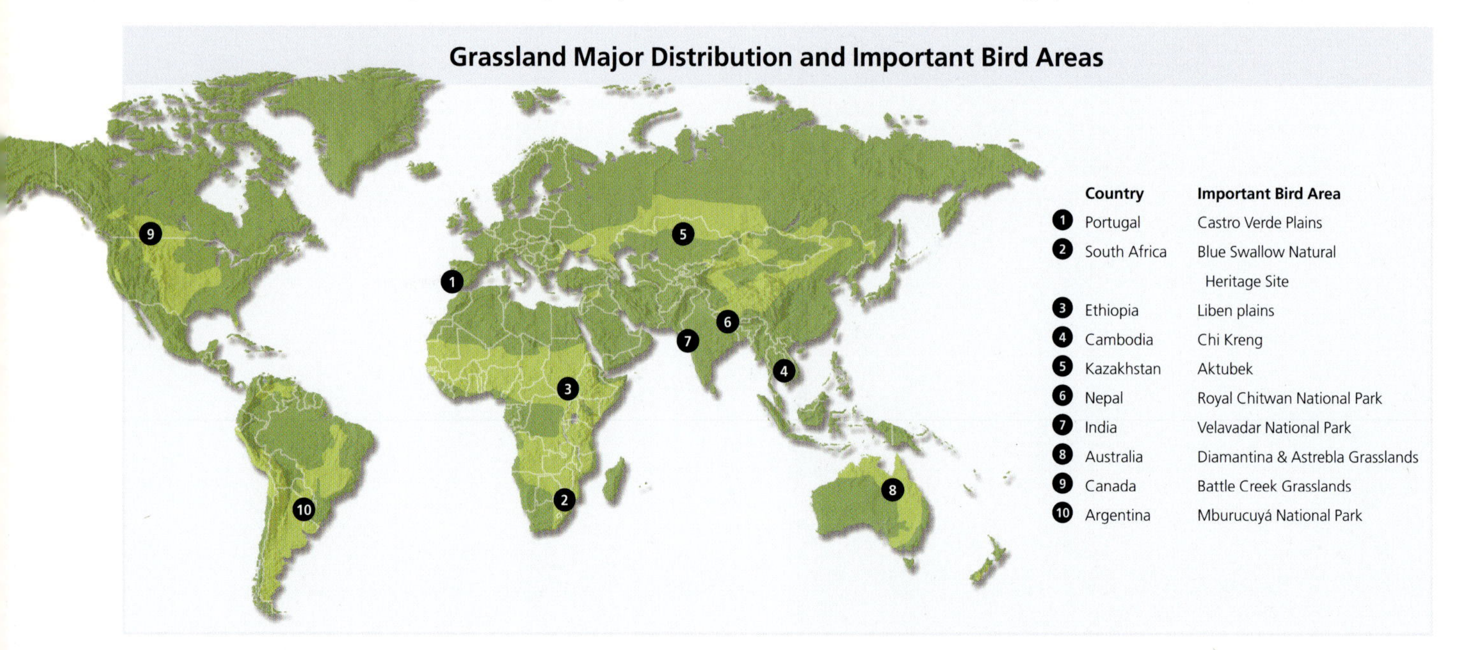

The pampas of South America are found mainly in northern central Argentina, Uruguay, and southern Brazil. There are many types of grassland, influenced mainly by humidity and rainfall and, as elsewhere, much of the land is used for livestock grazing and crops. Most of the pampas region is warmer and moister than the grasslands of the Northern Hemisphere, with some areas receiving as much as 40 in (1,000 mm) of rain annually. The large flightless rheas are characteristic pampas birds, as are the partridgelike tinamous. Both Greater and Lesser Rheas are found in the region, and both are classed as Near Threatened. To the east of the mighty Parana River in northeastern Argentina lie the Argentine Mesopotamian Grasslands, extending into part of Uruguay and southern Brazil. This EBA consists mainly of grassland that is subject to periods of flooding and it is the habitat of three threatened restricted-range birds—the Chestnut, Hooded, and Marsh Seedeaters, respectively Vulnerable, Critically Endangered (possibly Extinct), and Endangered. These small, colorful songbirds are threatened by habitat change, pollution, and trapping as cagebirds. Other threatened species found in the region are the Crowned Eagle (Endangered) and Strange-tailed Tyrant (Vulnerable), and the Yellow Cardinal and Pampas Meadowlark (both Endangered). In Patagonia, southern Argentina, there is a different type of temperate grassland developed under cool, humid conditions. This tends to be dominated by tough, rather tussocky grasses intermingled with herbs and shrubs. Similar grassland is found in parts of Tierra del Fuego and on the islands of the region, including the Falkland Islands (Islas Malvinas), the habitat of the endemic and Vulnerable Cobb's Wren.

▽ *Two male Great Bustards displaying. Bustards are birds of open grassland habitats and despite their large size, they can be hard to spot among tall grasses. This species is one of the world's heaviest flying birds.*

◁ *Listed as Vulnerable, the White-tailed Swallow is endemic to Ethiopia where it is found only in a small area in the south of the country. This dainty swallow hunts over semi-arid grassland and thorn scrub.*

SUBTROPICAL AND TROPICAL GRASSLANDS

Either side of the equator, lying roughly between 10° and 25° N and S, is a zone in which the climate is affected by seasonal monsoon rains. Many of the habitats here are dominated to a varying extent by grassland, maintained by a combination of climate, grazing pressure, and fire. Most of the rain falls in the summer but is often unpredictable, and annual rainfall varies between 20 in (500 mm) and 70 in (1,800 mm). Habitat types range from open grassland to savanna with scattered trees and shrubs presenting a parklike appearance.

South America has numerous types of grassland, with many intermediates. Some of the regularly flooded grassland habitats, such as the pantanal of Brazil, are better considered wetlands, while others grade into open woodland of various types. In Colombia and Venezuela, there are grassland habitats known locally as *llanos*. Found mainly in the floodplains of rivers, they are subject to periodic inundation. The grasslands of Brazil and Paraguay range from open grassland (*campo limpo*) through many intermediates to open savanna (*campo cerrado*).

Savanna grasslands are highly characteristic of the monsoon belt of Africa, and probably the best known are those of East Africa, notably in Tanzania and Kenya. These are protected in several important national parks, such as the Amboseli, Tsavo, and Masai Mara in Kenya, and the Serengeti and Selous in Tanzania. Herds of zebra, wildebeest, and gazelles graze these grasslands, and the birds are also very diverse.

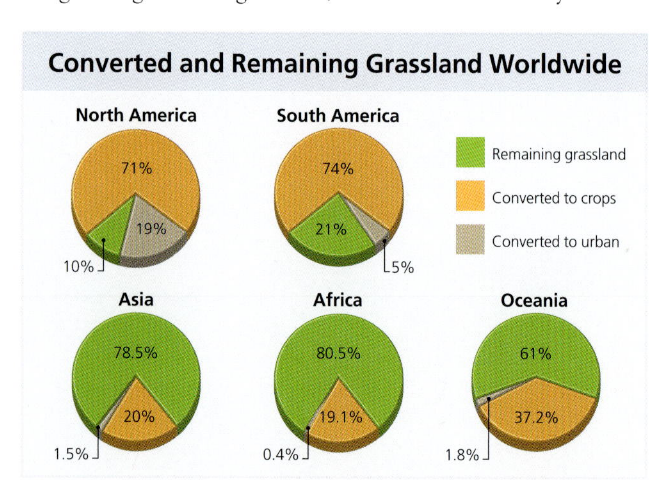

Flocks of vultures pick clean the bones of dead mammals and these savannas are home to one of the world's largest eagles, the magnificent Martial Eagle. This splendid raptor is Near Threatened as it has declined steeply through direct persecution and indirect poisoning. At the other end of the size scale, the dainty White-tailed Swallow is a Vulnerable savanna species with a very small range in Ethiopia.

Many savanna grasslands have evolved with the help of occasional fires from lightning strikes or being set deliberately. Ash increases the soil fertility and the grasses quickly grow back providing pasture for grazing animals. At the same time, fire (and grazing pressures) inhibit colonization of the grassland by shrubs and trees.

In southern Africa, the high veldt is an important type of grassland, and patches of this habitat are found, notably in the hills and highlands of central and eastern South Africa. Much of the grassland is overgrazed by domestic livestock or planted with crops, but remnants remain and are home to threatened birds such as Botha's Lark (Endangered), Rudd's Lark and Yellow-breasted Pipit (both Vulnerable), and also to the strange Southern Bald Ibis, another Vulnerable species. Further north, from Angola across to Mozambique, there are also regions dominated by a habitat known as the bush veldt, in a transition zone between the semi-desert and desert to the south and the "miombo" (*Brachystegia*) woodland to the north, a habitat that is favored by birds such as hornbills, shrikes, and weavers.

While open, savanna-like woodland is found in many parts of Asia, from the Indian subcontinent to Thailand, treeless grass-dominated habitats are less common. In India, patches of tropical grassland are found for example on the Deccan Plateau, and north to the desert edges in Rajasthan. The Assam Plains in northeastern India and neighboring Bangladesh also have pockets of grassland, often marshy with areas of Elephant Grass (*Pennisetum purpureum*) that can grow to 20 ft (6 m). Threatened birds in the Assam grasslands include the Black-breasted Parrotbill, Manipur Bush-quail, and Marsh Babbler (all Vulnerable), and the Bengal Florican (Critically Endangered).

In Australia, tropical grassland of the savanna type can be found especially in Queensland and in the Northern Territory, and the balance of grassland to trees varies with local conditions, grazing pressures, fire frequency, and rainfall. Fires are particularly frequent in Australia, and the seeds of many grass species are stimulated to germinate by exposure to the high soil temperature resulting from fires. The most extensive of these are the Mitchell Grasslands, which stretch from southeastern Queensland into the center of the Northern Territory. These consist of fairly short grasses, including Mitchell Grass (*Astrebla*). The Letter-winged Kite (Near Threatened) still hunts over some of these grasslands, and three notable Endangered birds are found in Australian grasslands and savanna grasslands—the Buff-breasted Buttonquail, Plains-wanderer, and beautiful Gouldian Finch.

▷ *Grassland dominates on these slopes of the Drakensberg Amphitheatre in Royal Natal National Park in KwaZulu-Natal, South Africa. Bearded Vultures and Verreaux's Eagles soar over the cliffs and hunt in the surrounding country.*

Serengeti Plains/Masai Mara

STRADDLING THE BORDER BETWEEN KENYA AND TANZANIA, TO THE WEST OF THE GREAT RIFT VALLEY, LIE THE LINKED PROTECTED AREAS OF THE SERENGETI NATIONAL PARK IN TANZANIA AND THE MASAI MARA RESERVE IN KENYA.

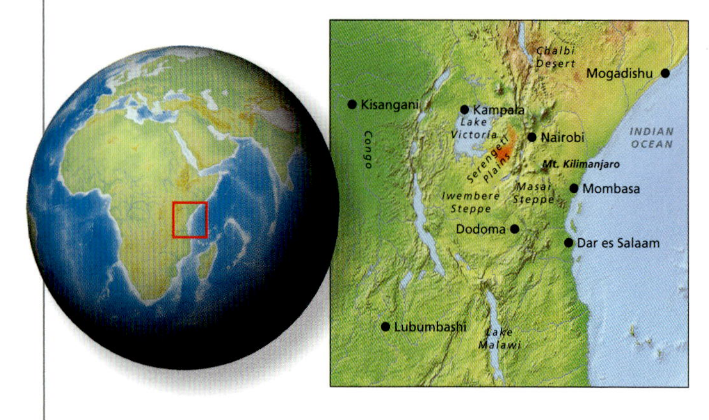

These two spectacular reserves are among the most intensively studied of all and hold a genuinely awesome array of wildlife. Best known for their large mammals (herbivores such as wildebeest, gazelles, zebras, African buffalo, African elephant, and giraffe, and carnivores including lions, leopards, cheetahs, hyenas, and jackals), they are also home to an impressive diversity of birds, among them several local and threatened species. The Serengeti supports the largest numbers of grazing mammals and predators in Africa and is one of the places most visited by naturalists and ecotourists. Another popular destination for ecotourists is the Ngorongoro Conservation Area, centered on the massive volcanic crater on the eastern edge of the Serengeti Plain. This crater is a natural amphitheater containing mainly grassland habitats and it provides one of the world's finest natural wildlife spectacles.

The habitats of the Serengeti–Mara complex are varied, but in most areas they are dominated by grassland and wooded savanna, with the balance between the two responding to soil type, rainfall pattern, and grazing pressures. Much of the soil, especially in southeastern Serengeti, is volcanic in origin, deriving partly from ash spewed from volcanoes in the Ngorongoro region (some of which are still active). This soil is highly fertile and supports grass growth that is particularly rich in nutrients, one reason for the remarkable annual migrations of wildebeest and zebras in a cycle also driven by rainfall patterns. The herds rumble across the plains each year in a more or less predictable roughly circular route, following the best quality grazing. The numbers are huge, with over a million wildebeest, and hundreds of thousands of Thomson's Gazelle and zebras. Calving is usually timed to take place on the lush, fresh growth on the fertile volcanic soils of the southeast. As in many grassland habitats, another factor in maintaining the open grassland is fire, either started deliberately or begun by natural causes such as volcanic events or lightning strikes.

The birds include more than 30 species of raptor and six vultures, and the Serengeti Plains are recognized as an EBA by BirdLife, based on the numbers of restricted-range birds they contain, with the Serengeti National Park listed as one of Tanzania's Important Bird Areas (IBAs). The number of bird species in

Some Local Birds of the Serengeti Plains

	SCIENTIFIC NAME	COMMON NAME	STATUS	WORLD POPULATION
■	*Apalis karamojae*	Karamoja Apalis	VU	10,000–20,000
■	*Gyps rueppellii*	Rüppell's Vulture	NT	30,000
■	*Glareola nordmanni*	Black-winged Pratincole	NT	100,000–500,000
■	*Phoeniconaias (Phoenicopterus) minor*	Lesser Flamingo	NT	2,200,000–3,240,000
■	*Agapornis fischeri*	Fischer's Lovebird	NT	290,000-1,002,000
■	*Prionops poliolophus*	Grey-crested Helmet-shrike	NT	Unknown
■	*Euplectes jacksoni*	Jackson's Widowbird	NT	Unknown
■	*Francolinus rufopictus*	Grey-breasted Spurfowl	LC	Unknown
■	*Trachyphonus usambiro*	Usambiro Barbet	LC	Unknown
■	*Histurgops ruficaudus*	Rufous-tailed Weaver	LC	Unknown
■	*Crex crex*	Corncrake	LC	5–10 million
■	*Parus fringillinus*	Red-throated Tit	LC	Unknown

RED LIST ■ CR = Critically Endangered ■ EN = Endangered ■ VU = Vulnerable ■ NT = Near Threatened ■ LC = Least Concern

Key Features of Serengeti Plains/Masai Mara

Location: Kenya, Tanzania
Area: 62,000 sq. miles (160,000 sq. km)
Altitude: 3,300–7,200 ft (1,000–2,200 m)
Habitat: Grassland, riverine forest, *Acacia* woodland, swamps and lakes, thickets and scrub, boulder-strewn escarpments
Threats: Spread of agriculture and overstocking by pastoralists, collection of fuelwood, uncontrolled tourism and poaching

the Serengeti National Park have been estimated at about 500.

Notable species here include the Rufous-tailed Weaver, Gray-breasted Spurfowl, and the pretty Fischer's Lovebird, the latter listed as Near Threatened partly through its capture for the cagebird trade. The Karamoja Apalis (Vulnerable) has a small range in the southern Serengeti, with most of its population occurring in the National Park.

Other rare birds that are found in this area include the Gray-crested Helmet-shrike and Black-winged Pratincole. Rüppell's Vulture is often seen feeding at carcasses, but the species is declining in part due to poisoning through the use of diclofenac. This drug is given to cattle to treat fever, but is lethal to vultures and has been responsible for the catastrophic declines in the vulture population in India and Pakistan. Efforts are being made to promote the use of an alternative drug—meloxicam—which has a similar therapeutic effect on cattle, but is not thought to be lethal to vultures.

△ *Fischer's Lovebird is a small, attractive parrot that is found in grassland and savanna in north–central Tanzania, most notably in the Serengeti National Park. Popular as a cagebird, this colorful bird suffers from being captured to supply this trade.*

▽ *Grassland is subject to occasional burning, either occurring naturally (mainly from lightning) or deliberately started to promote re-growth. Many birds, here White Storks in Kenya's Masai Mara, take advantage of such events, feeding on animals driven out by the heat and flames.*

MEDITERRANEAN SCRUB AND HEATHLAND

FOUND IN MANY DIFFERENT TEMPERATURE ZONES AND ALTITUDES, HEATH AND SCRUB HABITATS ARE OFTEN TEMPORARY AND CREATED BY PEOPLE. HOWEVER, THEY ARE SAFE HAVENS TO A WIDE RANGE OF BIRDS DUE PARTLY TO THEIR RICH SUPPLY OF FOOD.

Heath and scrub habitats are found in many parts of the world, and in temperate regions they are almost always the result of human interference. Grazing by sheep, cattle, or goats, or deliberate burning, often prevent areas that were cleared of trees from reverting via scrub to woodland. Heath and scrub may also be part of a natural succession, and therefore temporary in nature. Scrub habitats can be found at the latitudinal and altitudinal limits to tree growth, for example at the northern edge of the boreal taiga where conditions are too cold and dry to support forest. Here, in the tundra, there is a complex patchwork of vegetation, with drier, bushier elements interspersed with pools and mires. In mountain sites, a relatively narrow belt of scrub often develops above the tree-line and below the true alpine zone. Perhaps the most important of the heath and scrub habitats are those found in regions with a Mediterranean climate (hot, dry summers; cool, wet winters) and these support many rare and local bird species. Whether semi-natural or temporary, heath and scrub habitats are important refuges for many birds, combining secure nesting sites with a range of microhabitats offering usually abundant prey, notably insects and other invertebrates.

◁ *Rocky hillsides in the Valle della Luna, on the island of Sardinia, Italy. Much of this land once carried evergreen broadleaved woodland, but generations of grazing and cultivation have removed most of the original vegetation, leaving pockets of* maquis *and other types of scrub.*

Types of Scrub and Heathland

TUNDRA

Huge areas of the northern landmasses, stretching from northern North America through the far north of Europe and across northern Asia, are covered by tundra. Most of the tundra lies north of the Arctic Circle, though in continental central Canada it dips down further south around Hudson Bay. There are pockets of tundra in the Southern Hemisphere, but these are mainly restricted to a handful of islands in the Southern Ocean such as the Falkland Islands (Islas Malvinas), South Georgia, and Kerguelen, and to parts of Tierra del Fuego and nearby islands. Antarctica itself is effectively an ice desert.

In the Arctic tundra, permafrost (permanently frozen soil) is a feature, with only the surface layers melting in the summer, creating many pools and mires among the drier hillocks. The annual precipitation over much of the tundra is low, but the cool conditions mean that evaporation is also low so the soils are moist enough to support vegetation. Low-growing shrubs are a feature of many tundra sites—typically willow and birch—with sedges, mosses, and lichens. Tundra provides important breeding grounds for many birds, notably waders and wildfowl, long-distance migrants that take advantage of the short yet highly productive summers with continuous daylight. One of the world's rarest waders, the Bristle-thighed Curlew (Vulnerable) breeds on the tundra of western Alaska, wintering as far south as the Pacific islands. Another tundra nester is the beautiful dainty Red-breasted Goose, an Endangered species nesting in the Russian high Arctic. A few species are resident in and around the tundra, for example the Rock Ptarmigan, Snowy Owl, and Gyr Falcon. The Snowy Owl has a wide distribution, but seems to be declining markedly over most of its range.

△ *The Red Grouse (*Lagopus lagopus scoticus*), or Willow Ptarmigan, is an endemic subspecies of northern Britain. It favors upland heather moorland, a habitat maintained to support its population as a gamebird.*

HEATHLAND

In many parts of northwestern Europe, most notably in Ireland, Scotland, England, Denmark, Sweden, and north Germany heathland replaces temperate forests. This is very much a habitat created by people and prevented from reverting through natural succession to something like the original mixed deciduous woodland. The chief factors that prevent such reversion are grazing (mainly by sheep) and deliberate burning. In Scotland much of the upland heath, more often referred to as moorland, is maintained mainly to encourage the shooting of Red Grouse (*Lagopus lagopus scoticus*), a subspecies of Willow Ptarmigan. The dominant shrubs of heathland are members of the heather family (for example, genera

Mediterranean Scrubland and Tundra Major Distribution and Important Bird Areas

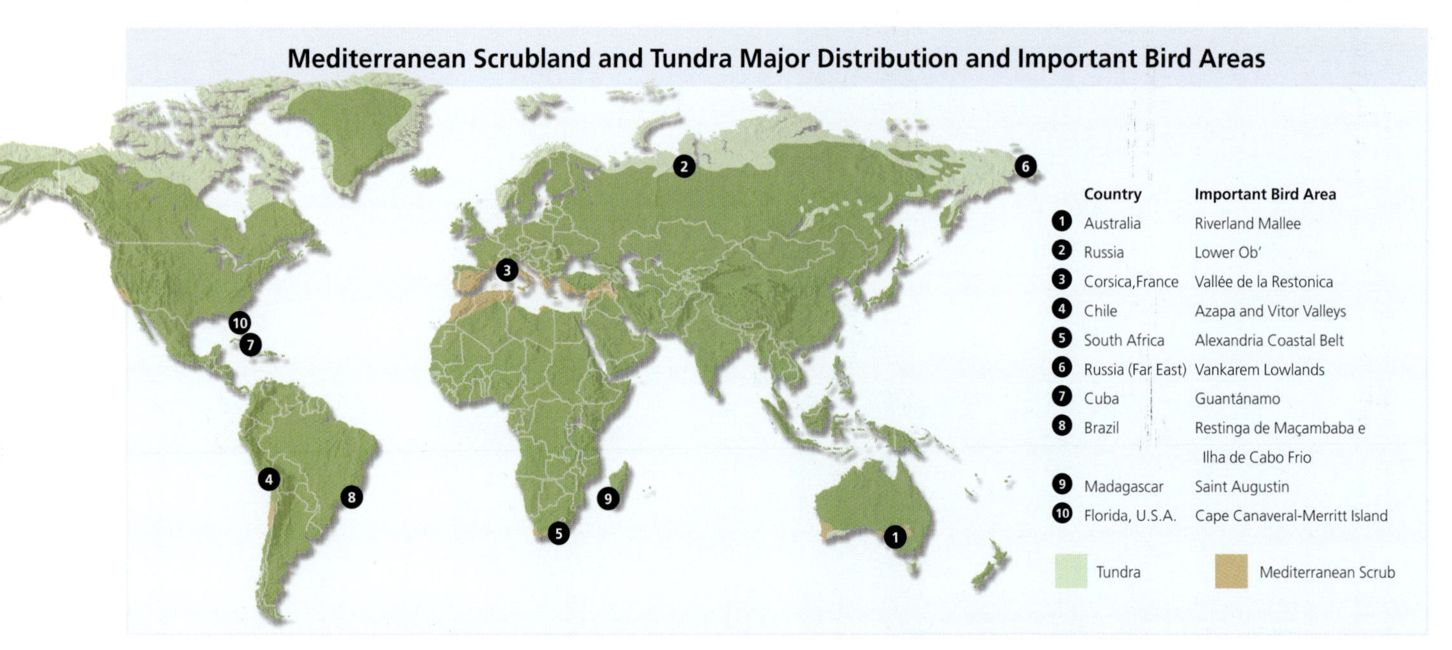

△ *Fire is a frequent hazard in scrub habitats, especially during hot, dry weather in warm regions such as here in southern France, and also in California and southern Australia. This is a controlled burn creating a firebreak. Heath and scrub will quickly regenerate in such areas.*

Erica and *Calluna*), and pea family (for example, genera *Ulex* and *Cytisus*). In the most oceanic and high rainfall regions, such as in western Britain and Ireland, heather moorlands grade into peat mires (bogs or fenland habitats, depending on the acidity); these are best considered as wetland habitats. In global terms, this kind of heath covers only a small area and few rare birds are associated with it. However, the Red Grouse is an endemic subspecies of Willow Ptarmigan found only in Britain. A charismatic raptor of heathland and heather moor is the Hen (Northern) Harrier (*Circus cyaneus*), which though not threatened as a species globally, has suffered severe persecution as a perceived threat to the Red Grouse and has declined markedly in Britain. The plight of this beautiful raptor offers an example of how a species may be under severe pressure locally.

MEDITERRANEAN HABITATS

Heath and scrub communities typically dominate Mediterranean habitats. Just as the heaths and moorland of northwestern Europe replace woodland, so the heathlike habitats here replace mixed woodland, in this area featuring evergreen broadleaved trees such as evergreen oaks. Rainfall in the typical Mediterranean climate is between 12 in (300 mm) and 32 in (800 mm) annually, mainly falling from the fall months through to spring. Frosts are rare and many plants flower in winter, early spring, and fall. The summers are normally hot and dry, with parched vegetation and increased risk of fires.

In addition to the region surrounding the Mediterranean Sea, such climates and habitats are found in southern California, in central Chile, in the southern Cape of South Africa, and in parts of southern Australia. The main factors that create and maintain such habitats are forest clearance, grazing of livestock, and fire (natural and unnatural). A variety of local terms are used to describe such habitats. In France *maquis* (*macchia* in Italy) refers to denser heath and scrub and *garrigue* to more open habitats. In Greece, the open scrub is called *phrygana* and in Spain, *tomillares*. Elsewhere, local terms for similar vegetation types are chaparral (California); *matorral* (Chile); *fynbos* (South Africa); *kwongan* and mallee (Australia), though each has its own special characteristics. Though most of these habitats are mainly managed or created by people, they are among the richest of all in terms of interesting and threatened wildlife and are themselves under threat from tourism, housing development, agriculture, and the increased frequency of fires.

In southern Europe the vegetation is essentially degraded woodland dominated by evergreen trees such as Holm Oak (*Quercus ilex*) and Cork Oak (*Q. suber*), with shrubs—many of them aromatic—such as juniper, lavender, thyme, and rosemary.

A number of rare birds breed or feed in these *maquis* habitats, for example the Spanish Imperial Eagle (Vulnerable), the locally rare Bonelli's Eagle and the Black-winged Kite, and the European Roller (Near Threatened).

California's chapparal is derived partly from original oak woodland with tree species such as *Quercus agrifolia* and *Q. engelmannii*. The shrubs include the blue-flowered *Ceanothus*, often grown as a garden plant and species of sage (*Salvia*). This is one of North America's most endangered habitats, and patches can be found from southern coastal California south to northwestern Mexico. Interesting birds of the chapparal include the California Gnatcatcher, Cactus Wren, and Nuttall's Woodpecker. The Tricolored Blackbird is an Endangered species that breeds colonially in marshy sites in the region. The re-introduced and Critically Endangered California Condor may occasionally be spotted hunting over the coastal chapparal. This precious habitat and the rare plants and animals it supports are under great threat from housing development, introduced plants and animals, overgrazing, pollution, and increased frequency of fire.

In Chile the *matorral* is found between the Andean foothills and the Pacific Ocean in central Chile between the temperate forest region in the south and the Atacama Desert to the north. In common with other similar habitats elsewhere, it is very species-rich and holds many endemics. Though none of these is threatened, notable birds here are the Chilean Mockingbird, Chilean Tinamou, and both Ocher-flanked and White-throated Tapaculo.

In terms of plant diversity, the *fynbos* region of South Africa has developed in relative isolation and is the richest of all Mediterranean scrub habitats. What is more, nearly 70 percent of the *fynbos* plant species are endemic. There are few trees, and the habitat is dominated by low shrubs with hard, evergreen foliage. There are many species of heathers (*Erica*), proteas (*Protea*), and curious rushlike restios (*Restio*). "*Fynbos*" is Afrikaans for "fine bush" and refers to the narrow, needlelike leaves of many of the shrubs that dominate. This is an exposed, windy landscape, and the pruning effects of the wind encourage the dominance of low-growing shrubs. In the southwest of the *fynbos*, the climate is mainly cool and wet in winter, with about half the annual rainfall in May, June, and July, while the summers are warm and dry. This region is home to many species of the genus *Pelargonium*, so popular with gardeners, as well as over 200 species of orchid. *Fynbos* also boasts 660 species in the iris family (Iridaceae), 485 of which are endemic. The proteas have large showy flowers that bloom mainly in the winter, providing welcome nectar for sunbirds and the endemic Cape Sugarbird. Other fascinating endemic birds here include the following: Cape Rock-jumper, Cape Siskin, Orange-breasted Sunbird, Protea Canary, and the elusive Victorin's Scrub-warbler.

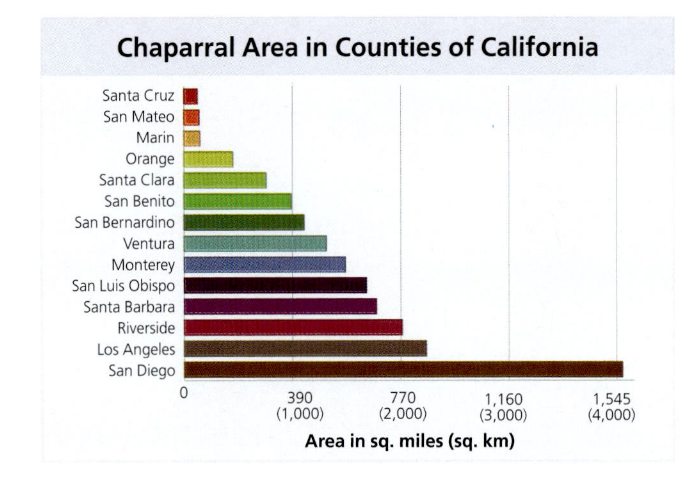

There are two main areas in Australia that feature scrub as a major vegetation type: parts of the Murray and Darling river catchments of South Australia, New South Wales and Victoria in the southeast, and the Mediterranean climate region of Western Australia in the southwest. Both these regions have a mixture of dry woodland and mallee scrub and many interesting and local birds feature in these habitats, including the Mallee Emuwren (Endangered) in the southeast, and Baudin's Black-cockatoo (Endangered), the Noisy Scrub-bird (Vulnerable) and Western Bristlebird (Vulnerable) in the southwest. The fascinating turkeylike Malleefowl is common to both regions and has suffered rapid declines, due mainly to loss of habitat. The Malleefowl is also listed as Vulnerable.

▷ *Endemic to South Africa, the elegant Cape Sugarbird, here perched on a protea flower, belongs to a family with only one other species. It uses its long, curved bill to probe for nectar.*

▽ *The* fynbos *(pronounced "fainboss") of South Africa's Southern Cape has the richest of all scrub habitats, and is home to many garden plants such as this agapanthus. It also has a high proportion of endemic plants.*

Australian Mallee

The scrub habitats of Australia are comparable with similar vegetation types found in Mediterranean-type climates elsewhere, though they do have their own special features.

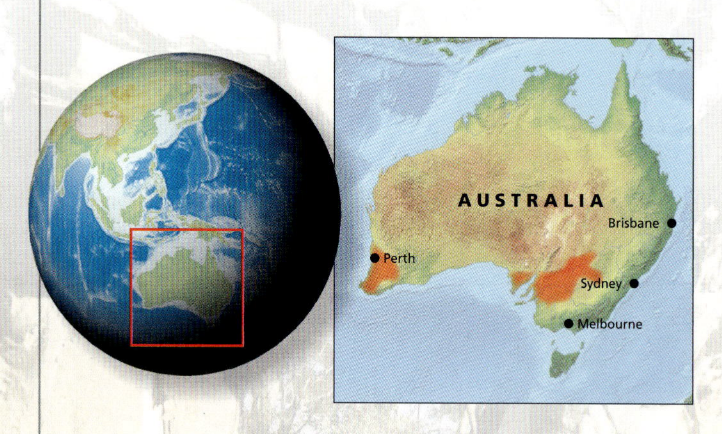

A notable feature of mallee is the presence of many species of eucalyptus. The term "mallee" refers to a growth form shown by certain species of eucalypt, in which many stems arise from a large, woody base—a swollen root crown or lignotuber. This contains buds and a store of carbohydrates that help the plant regenerate after fires. Along with the eucalypts, species of *Acacia* and *Casuarina* are also typical of such habitats. Though mallee is the most common term used, in southwestern Australia especially the term "*kwongan*" is used to refer to a wide range of Mediterranean shrublands, of which mallee is but one form. The *kwongan* is very rich in plant species and like the *fynbos* of South Africa it also contains many endemics. Large areas of these scrublands have been cleared for agriculture and are also threatened by housing developments. Though many of the plants are adapted to survive fire, and may even depend upon a regime of occasional burning, fires have become increasingly frequent, threatening the integrity of these habitats and their birdlife and other animal life.

One of the most famous of all mallee birds is the extraordinary Malleefowl, a Vulnerable species that is found in both of the main scrub regions of southern Australia. This hen-sized, ground-dwelling bird has a highly unusual breeding method. The male excavates a large hole about 16 ft (5 m) wide and 3 ft (1 m) deep and fills this with leaves and twigs, covering it with sand and soil, eventually forming a mound over 3 ft (1 m) high. Much of the year is spent maintaining what is in effect a giant compost heap. The eggs are laid into this mound, and the heat generated from decomposition of the vegetable matter incubates the eggs. The parent birds carefully monitor the temperature so that it stays at an optimum of 90–93°F (32–34°C). Remarkably, they do this using their bills as thermometers, probing into the mound at regular intervals to take "readings." They make any necessary adjustments by rearranging the amount of soil covering the eggs. When they hatch, the chicks

▽ *The scrub habitats of Australia are mainly dominated by species of eucalypt, as here in Hattah-Kulkyne National Park in Victoria, a protected area with large areas of mallee. Malleefowl still breed in the park, which is also home to many other species, including cockatoos and other parrots.*

dig their way out of the mound and are entirely independent, dispersing from the area and fending for themselves. Juvenile mortality is very high, made worse by introduced predators such as the Red Fox (*Vulpes vulpes*) and domestic dogs. Unfortunately, the numbers of Malleefowl have declined quite steeply over the last few decades from a combination of habitat loss and introduced predators.

Important reserves protecting this habitat in Western Australia are Fitzgerald River National Park, Stirling Range National Park, and Two Peoples Bay Nature Reserve. One of the rare birds affected by fires is the Western Bristlebird, a Vulnerable species with a small and decreasing population, mainly in the Fitzgerald River National Park, which is the largest protected area on the south coast of Western Australia. Though the Western Bristlebird seems able to survive the occasional fire, frequent burning has led to local extinctions.

In the east, mallee is well represented in Murray–Sunset National Park, Wyperfield National Park, Hattah–Kulkyne National Park, Billiat Conservation Park, and Ngarkat Conservation Park. Two notable Endangered birds are found in the mallee of this region: the Black-eared Miner and Mallee Emuwren. The Black-eared Miner is a pretty honeyeater endemic to this region of South Australia, Victoria, and New South Wales, with a small and decreasing population estimated at fewer than 1,000 birds. The Mallee Emuwren is a charming bird with a red crown, long threadlike tail, and a tiny plump body, orange below; the male has a bright blue bib. It occurs mainly in tussock grassland in and around the mallee, in heath and scrub. Frequent, large-scale fires are a major threat, as is livestock grazing and agriculture.

△ *The Mallee Emuwren is an endemic species found only in an area to the south and east of the Murray River in South Australia. With a small, decreasing population it is listed as Endangered.*

Threatened and Local Birds of the Australian Mallee

SCIENTIFIC NAME	COMMON NAME	STATUS	POPULATION ESTIMATE
Botaurus poiciloptilus	Australasian Bittern	EN	1,000–2,500
Stipiturus mallee	Mallee Emuwren	EN	1,500–2,800
Manorina melanotis	Black-eared Miner	EN	250–1,000
Calyptorhynchus latirostris	Carnaby's Black-cockatoo	EN	40,000
Leipoa ocellata	Malleefowl	VU	100,000
Atrichornis clamosus	Noisy Scrub-bird	VU	1,000–1,500
Dasyornis longirostris	Western Bristlebird	VU	1,000
Pachycephala rufogularis	Red-lored Whistler	NT	10,000
Psophodes nigrogularis	Western Whipbird	NT	20,000
Polytelis anthopeplus	Regent Parrot	LC	21,500
Lichenostomus cratitius	Purple-gaped Honeyeater	LC	Unknown
Purpureicephalus spurius	Red-capped Parrot	LC	Unknown
Platycercus icterotis	Western Rosella	LC	Unknown
Stagonopleura oculata	Red-eared Firetail	LC	Unknown

RED LIST CR = Critically Endangered EN = Endangered VU = Vulnerable NT = Near Threatened LC = Least Concern

Wetland and Coast

THE WETLAND HABITATS OF THE WORLD ARE VERY SPECIAL PLACES FOR BIRDS, PROVIDING THEM WITH VITAL FOOD RESOURCES THAT ARE OFTEN IN ABUNDANT SUPPLY, AND MANY BIRDS HAVE PARTICULAR ADAPTATIONS TO EXPLOIT WETLANDS TO THE FULL. THE TERM "WETLAND" IS BROAD AND IMPRECISE, BUT HERE IT INCLUDES A WIDE RANGE OF BOTH FRESHWATER AND SALTWATER HABITATS, SUCH AS RIVERS, LAKES, ESTUARIES, SALTMARSHES, AND COASTAL WATERS.

Most birds that feed over the open seas breed on offshore or oceanic islands, and these are mainly included in the Island chapter (see pp. 102–115). Waders and wildfowl in particular are highly dependent upon wetlands for their survival and this is especially true in the case of those species that migrate from polar and subpolar regions to overwinter at wetland sites in warmer latitudes.

Wetlands are highly sensitive habitats that are very easily damaged by human activities, including drainage for building developments, and pollution from industrial waste and sewage. Chemicals and other pollutants are readily transported in the water, or they dissolve, contaminating wetlands, sometimes with catastrophic consequences.

In agricultural and heavily populated regions, ponds and lakes often suffer from excess eutrophication, resulting especially from increased amounts of phosphates reaching the water, mainly from fertilizers, sewage, and detergents. This process may result in occasional algal blooms, which are sometimes toxic to animal life. The algae also tend to cut out light from underwater plant life and alter the whole ecology of the water.

Many coastal sites have suffered the effects of oil spilled from stricken tankers, damaged oil wells, or because the oil has been deliberately released. Oil spills kill thousands of wetland birds by rendering them unable to fly or feed, as well as ruining the habitats. Continuous surf action eventually disperses and degrades even heavy crude oil, but the immediate effects are often devastating and the residues may persist for months or even years.

◁ *Parallel curving lines of forest surrounded by Amazonian floodwater in the Anavilhanas Archipelago on Brazil's Rio Negro. This unique wetland habitat has a very rich diversity of bird species.*

WETLANDS OF THE WORLD

Inland wetland habitats include rivers, streams, and freshwater and saline lakes, as well as land that is permanently waterlogged, habitats known collectively as mires. The vegetation of mires is typically rich in mosses and sedges and develops on peaty soils that are derived from semidecayed plant remains. Acid mires are often referred to as bogs, while neutral or alkaline mires are more accurately referred to as fens. The soil beneath shallow lakes and mires is usually very suitable for growing crops and therefore many such wetlands have been drained deliberately to provide more land for agriculture.

The wetland habitats of the coasts are very varied, ranging from estuaries, lagoons, mudflats, and saltmarshes to mangrove swamps, rocky shores, and sea cliffs. While some wetlands are permanent, many are seasonal, with the water levels varying markedly, some even drying up completely for periods. Where water levels are low, and the soil conditions suitable, emergent plants may be able to grow, even in some cases trees, as for example in habitats such as the Swamp-cypress (*Taxodium distichum*) forests of southeastern U.S.A., or the seasonally flooded tropical forests of the Amazon Basin, known in Brazil—and often by extension elsewhere—as *varzea*. Some wetlands are the direct result of human activity, such as the rice fields of southern China and Southeast Asia, or lakes formed from flooded gravel diggings and reservoirs created to supply drinking water or hydroelectricity. Though these are artificial, they can provide good habitats for wildfowl and other birds.

Estuaries and coastal waters are important feeding grounds for waders and wildfowl, many of which migrate hundreds or

thousands of miles each year to take advantage of the riches offered by mudflats, creeks, and saltmarshes. So too are seasonally flooded meadows such as the Ouse Washes in eastern England, which are visited every year by about 9,000 wild swans that breed in the Arctic. Whooper Swans breed on wet tundra, mainly in Iceland, and migrate mostly to Britain and Ireland to winter at lakes and flooded grassland. Tundra Swans (the European race of which is called Bewick's Swan) are also highly migratory, breeding in the wet tundra of Arctic Siberia, Alaska, and Canada, and wintering at wetlands in western Europe, southern China and the wetlands of the Pacific and Atlantic coasts of the U.S.A. Most of the North American Tundra Swans winter at Chesapeake Bay on the Atlantic coast, a site that attracts about one-third of all the waterfowl on this coast.

△ *The Okavango River floods to form a complex inland delta with a wide range of habitats including reedbeds and papyrus swamps, as well as areas of open water and flowing channels, and occasional islands of higher ground. It is a haven for birds and other wildlife.*

▽ *An African Fish-eagle plucks a fish from the water in the Okavango Delta of Botswana. The loud, musical calls of this eagle are a characteristic sound of Africa's lakes and rivers. This species has a wide distribution ranging over most of continental Africa south of the Sahara Desert. Its population is in excess of 300,000 and the species is evaluated as Least Concern.*

ENDANGERED CRANES

LARGE AND GRACEFUL, CRANES (FAMILY GRUIDAE) ARE SOME OF THE MOST IMPRESSIVE OF ALL WETLAND BIRDS. MOST SPECIES MAKE REGULAR MIGRATIONS TO AND FROM THEIR BREEDING GROUNDS, OFTEN IN LARGE, STRAGGLING FLOCKS. WITH THEIR ARRESTING PLUMAGE AND LOUD BUGLING CALLS, THESE CHARISMATIC BIRDS HAVE ENTERED FIRMLY INTO HUMAN FOLKLORE IN MANY PARTS OF THE WORLD.

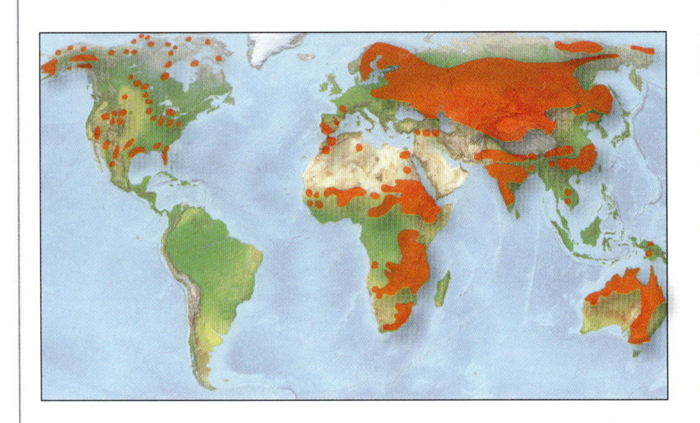

Of the 15 living species of crane, only four are listed as of Least Concern with populations that are not currently considered to be at risk; the remaining 11 species are all threatened to some degree. The Whooping Crane of North America and Red-crowned Crane from the Far East, are Endangered, while the magnificent Siberian Crane is Critically Endangered. It may be no coincidence that the Siberian Crane is also the most reliant on wetland habitats, both in its breeding range and at its wintering sites. All of the remaining eight species are listed as Vulnerable.

The Endangered Whooping Crane of North America almost became extinct in the 1930s, mainly through hunting. Though its population remains very small, between 50 and 250, it is slowly recovering, responding to protection of its breeding grounds in Canada's Wood Buffalo National Park and reintroduced flocks to Florida. The Canadian birds breed in wet prairie and marshes and winter in Texas, at or close to the Aransas National Wildlife Refuge.

In Africa, there are three threatened cranes: the Gray Crowned-, Blue, and Wattled cranes, and all three are classed as Vulnerable. Though its population stands at about 50,000, with a range from Uganda and Kenya south to Mozambique, the Gray Crowned-crane is decreasing partly through loss of wetland habitats, but also due to hunting. The closely related Black Crowned-crane is also in retreat, confronted by habitat loss, hunting, and drought. Originally listed as Near Threatened, it was recently uplisted to Vulnerable. The small Blue Crane, with a declining population of some 26,000, is almost restricted to South Africa, with a small population in Namibia. It has suffered badly from poisoning. Much larger is the Wattled Crane, found scattered through southern, central, and eastern Africa. A favored site for this species is the Okavango Delta in Botswana.

Asia is blessed with no fewer than six crane species. The large

▽ *Juvenile Siberian Cranes in flight. This Critically Endangered species is under pressure at its breeding sites in Arctic Russia and western Siberia and at its wintering grounds, mainly in China's lower Yangtze River valley.*

Sarus Crane (Vulnerable) is mainly pale gray, with naked, red flesh on the head. It has an unusual distribution with three widely separated populations—one mainly in India and Nepal (8,000–10,000 birds); one in Cambodia, Laos, Vietnam, and Myanmar (about 1,800 birds), and another in northern Australia (about 8,000 birds). Drainage of wetlands and pesticide poisoning affect this species, and in India especially fatal collisions with power lines have taken their toll.

White-naped and Hooded Cranes are two Vulnerable species with largely gray plumage. Both breed in the wetlands, wooded bogs, and steppe grasslands of southeastern Russia and northeastern China where the main threats are from habitat loss and occasional fires. As with many migrant species, threats to their wintering grounds also cause significant losses. Both species winter partly on lakes in the lower reaches of the Yangtze River in China, habitats that may be adversely affected by the Three Gorges Dam. The Red-crowned Crane, a mainly white species with black neck, black legs, and tail, has a similar range, with a small population also in Japan, and winters mainly to the delta of China's Yellow River. Somewhat smaller with rather similar plumage, though grayer, the Black-necked Crane breeds mainly at wetlands on the high plateaux of Tibet and Qinghai in China, wintering at lower altitudes, notably to lakes in Yunnan. It is feared that climate change may be causing some high altitude lakes to dry out, reducing the availability of suitable wetlands. Most at risk is the magnificent Siberian Crane. With a dwindling population of a little over 3,000, it is at serious risk, mainly at or on the way to its few chosen wintering grounds. The major wintering site is at Poyang Hu, a lake in the lower Yangtze valley in China, with other lakes used as staging posts en route.

△ *A pair of Red-crowned Cranes calling as part of bonding behavior. This Endangered species breeds in southeastern Russia, northeastern China, and Japan, and winters to Korea and the coast of central and northeastern China.*

Threatened Species of Cranes

	SCIENTIFIC NAME	COMMON NAME	STATUS	WORLD POPULATION
■	*Balearica pavonina*	Black Crowned-crane	VU	43,000–70,000
■	*Balearica regulorum*	Gray Crowned-crane	VU	47,000–59,000
■	*Grus leucogeranus*	Siberian Crane	CR	3,750
■	*Grus antigone*	Sarus Crane	VU	19,000–21,800
■	*Grus rubicunda*	Brolga	LC	26,000–100,000
■	*Grus vipio*	White-naped Crane	VU	6,500
■	*Grus canadensis*	Sandhill Crane	LC	520,000–530,000
■	*Grus virgo*	Demoiselle Crane	LC	230,000–280,000
■	*Grus paradisea*	Blue Crane	VU	26,000
■	*Grus carunculatus*	Wattled Crane	VU	6,000–8,000
■	*Grus grus*	Common Crane	LC	360,000–370,000
■	*Grus monacha*	Hooded Crane	VU	2,500–10,000
■	*Grus americana*	Whooping Crane	EN	50–250
■	*Grus nigricollis*	Black-necked Crane	VU	8,800
■	*Grus japonensis*	Red-crowned Crane	EN	1,700

RED LIST ■ CR = Critically Endangered ■ EN = Endangered ■ VU = Vulnerable ■ NT = Near Threatened ■ LC = Least Concern

△ *A Red-throated Loon takes off from a lake in the boreal forests of Sweden. This is a migratory species that breeds in the Arctic regions of the Northern Hemisphere and winters along the Pacific and Atlantic coastlines.*

Chesapeake Bay is the largest estuary in the U.S.A. and one of North America's prime coastal wetlands.

Also breeding on the tundra and near rivers in northern Siberia, the beautiful Red-breasted Goose has a small population, estimated at 37,000. Its wintering grounds are wetlands of the western Black Sea, mainly in Bulgaria and Romania. Hunting, both at its wintering grounds and during migration stopovers, is a major threat to this Endangered goose.

TROPICAL LAKES

Some of the most unusual wetlands are shallow lakes in warm regions in situations where high evaporation has concentrated natural salts to the point where only the most specialized life forms can survive. Examples of such salt or soda lakes can be found in several deserts, high in the Andes, and also in East Africa along the Rift Valley. Crustaceans such as brine shrimps and cyanobacteria (blue-green algae) dominate in such waters and are exploited by flamingos that often gather in very large flocks to sift out this nutritious soup. The cyanobacteria contain a red pigment that

▷ *The Rospuda Valley in northeastern Poland is a wetland area of open water and bogs, inhabited by several protected bird species including the Common Crane. It has been reprieved from the proposed building of a major road.*

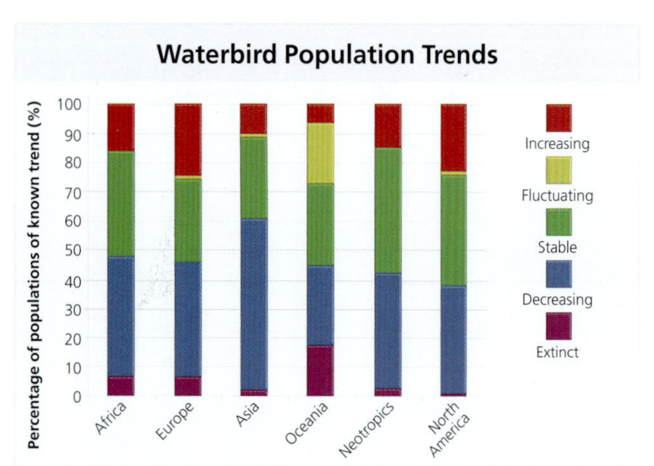

colors the water and also gives the flamingos their pink tint. Lesser and Greater Flamingos breed at Lake Natron in Tanzania and Lake Magadi in Kenya, and may also be seen in huge flocks at other alkaline lakes such as Lakes Nakuru and Bogoria (both in Kenya). The value and vulnerability of wetlands is underscored by their importance to birds and other wildlife though freshwater habitats cover less than one percent of the Earth's surface. It is also perhaps no surprise that many wetland birds are classed as threatened. For example, no fewer than 33 species of rails and their allies are listed as Vulnerable, Endangered, or Critically Endangered, as are 28 species of ducks, geese, and swans, and 20 species of wader (plovers, and sandpipers and allies). The figures for the other main wetland families are: gulls and terns (11 threatened species); kingfishers (12 threatened species); cormorants (10 threatened species); cranes (11 threatened species); herons and egrets (8 threatened species); ibises and spoonbills (8 threatened species); storks (5 threatened species).

IMPORTANT WETLAND SITES

Throughout the world, wetlands support large numbers of birds and it is no surprise that many have become major attractions to ornithologists and ecotourists. The Convention on Wetlands—commonly called the Ramsar Convention as it was held in Ramsar, Iran in 1971—recognizes important wetland sites throughout the world. These include many of the most biodiverse wetlands, some of which are also recognized by UNESCO on their World Heritage List. All three of the sites mentioned below are both World Heritage and Ramsar listed.

The **Everglades National Park** in Florida is one of North America's finest wildlife reserves, home to more than 350 species of bird. Fed largely by the waters of the Kissimee River, which flows into Lake Okeechobee, this marvellous wetland is a complex mixture of habitats, including wet "rivers of grass," river channels, coastal marshes, and one of the world's largest systems of mangroves. Set among these wet habitats are raised areas with trees, including cypresses clad in bromeliads and other epiphytes. The trees are welcome breeding sites for egrets and the Anhinga, Roseate Spoonbill, and Wood Stork, as well as the Red-cockaded Woodpecker, a Vulnerable and decreasing endemic in the U.S.A. that makes use of old pine trees for nesting. Though not threatened internationally, the Wood Stork is a local rarity, as is the remarkable Snail Kite, here at its only known North American site and with a population of just a few hundred. The latter has a long, decurved upper mandible, an adaptation used for extracting the meat from the apple snails that are its main prey.

▽ *Dalmation Pelicans feeding on Lake Kerkini in Greece. This artificial reservoir is considered a Wetland of International Importance according to the Ramsar Convention, and is an important site for this Vulnerable species.*

KAKADU NATIONAL PARK

ONE OF AUSTRALIA'S TOP WILDLIFE SITES, THE PARK IS FAMOUS AS A REFUGE FOR OVER 280 SPECIES OF BIRDS—ABOUT A THIRD OF THE TOTAL BIRD FAUNA OF AUSTRALIA. IT COVERS NEARLY 7,700 SQ. MILES (20,000 SQ. KM), STRETCHING INLAND FROM THE COAST OF THE TROPICAL NORTHERN TERRITORY, EAST OF DARWIN.

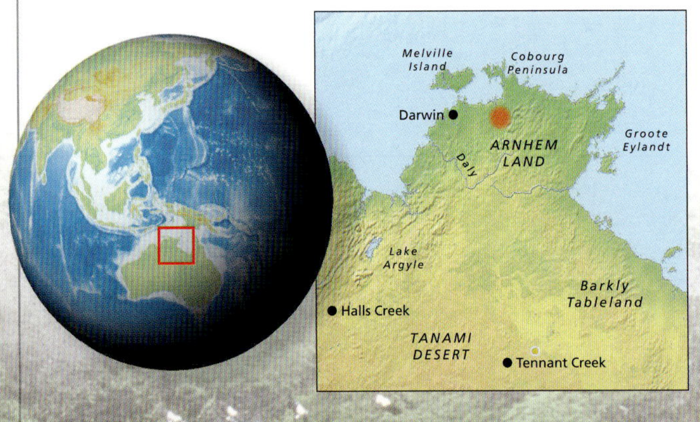

Ramsar-listed and designated by UNESCO as a World Heritage Site, this park has been home to Aboriginal people for over 40,000 years and therefore has deep cultural significance in addition to its value as a haven for wildlife. Much of the park is owned by native Australian Aboriginal people, and leased by them as a national park.

Though about 80 percent of the area is covered by woodland of various sorts, the park also includes some of the world's greatest wetland habitats. There are about 200 sq. miles (500 sq. km)

▽ *The floodplains and wetlands of the Kakadu National Park, such as the Magela Wetlands below, are important refuges and feeding grounds, especially in the dry season, for many Australian waterbirds, such as egrets and geese.*

of coastal wetland, with estuaries, tidal flats, and large areas of mangroves, as well as seasonally flooded plains and billabongs (small oxbow lakes). It includes four major rivers, the Wildman River and the West, East, and South Alligator rivers, the latter named for their crocodiles (wrongly identified as alligators by early explorers). Both types of Australian crocodile, the Freshwater (*Crocodylus johnsoni*) and the Saltwater (*C. porosus*) are found in the park,and the latter is capable of growing to a length of 20 ft (6 m).

The climate is tropical and affected by monsoons, with distinct wet and dry seasons. The dry season lasts typically from April to September when southerly and easterly winds are dominant. The wet season starts roughly in November, with the main rains falling from January through March. During this period, the summer monsoon storms bring regular floodwaters to the low lying plains, transforming the landscape into gigantic shallow lakes, ideal for hundreds of waterfowl and other wetland birds. Then as the cycle of drier conditions returns, the waters gradually recede, leaving billabongs and other pools at which the water birds concentrate.

The large White-bellied Sea-eagle is a common sight over these wetlands, as is the Whistling Kite, while egrets and Black-necked Storks patrol the water margins. Comb-crested Jacanas use their long toes to spread their weight as they walk on floating vegetation, and Chestnut Rails skulk in the waterside vegetation.

Notable waterbirds of the Kakadu wetlands are the striking Magpie Goose, the Radjah Shelduck, Australian Pelican, Green Pygmy-goose, and Plumed Whistling-duck. Though none of these species is listed as threatened, they all suffer elsewhere in the world from habitat loss, and Kakadu offers them a welcome safe haven.

Threatened birds of the park endemic to Australia include the Red Goshawk, a Vulnerable species, and the beautiful Endangered Gouldian Finch. The latter is only found scattered in the north of Australia, where it is sometimes spotted drinking at waterholes; its Kakadu population has been estimated at only 50–150 birds.

△ *The Comb-crested Jacana is also known as the Lilytrotter for its habit of running across floating vegetation with its long toes spread out. This species occurs in wetlands from Borneo to northern and eastern Australia.*

One of Kakadu's best birding sites is the Mamukala Wetlands, especially toward the end of the dry season when wildfowl, including thousands of Magpie Geese, flock to the water to feed. Another good site is Yellow Water, a billabong connected to the South Alligator River system. Fascinating birds such as the Azure and Little kingfishers, Brolga (a species of crane), Royal Spoonbill, White-necked Heron, and Rufous Night-herons are regular highlights here.

Selected Birds of Kakadu National Park

SCIENTIFIC NAME	COMMON NAME	STATUS	WORLD POPULATION
Erythrura gouldiae	Gouldian Finch	EN	2,000–10,000
Erythrotriorchis radiatus	Red Goshawk	VU	660–1,000
Turnix castanotus	Chestnut-backed Buttonquail	NT	50,000
Geophaps smithii	Partridge Pigeon	NT	20,000
Petrophassa rufipennis	Chestnut-quilled Rock-pigeon	LC	Unknown
Psitteuteles versicolor	Varied Lorikeet	LC	Unknown
Platycercus venustus	Northern Rosella	LC	Unknown
Psephotus dissimilis	Hooded Parrot	LC	20,000
Pitta iris	Rainbow Pitta	LC	Unknown
Meliphaga albilineata	White-lined Honeyeater	LC	Unknown
Poephila acuticauda	Long-tailed Finch	LC	Unknown

RED LIST CR = Critically Endangered EN = Endangered VU = Vulnerable NT = Near Threatened LC = Least Concern

△ *The Eurasian Curlew is a widely distributed wader, breeding across Europe. It is listed as Near Threatened since the global population is thought to have fallen 20–30 percent in the past 15 years or three generations.*

The **Danube Delta** in eastern Romania is one of the largest wetlands of Europe and provides a welcome habitat for millions of birds, both those breeding there and the flocks that make stopovers to re-fuel on their migration. The mighty Danube River spreads out into a fan of channels and lagoons as it slows and merges with the waters of the Black Sea, creating an array of interconnecting habitats and the largest area of compact reedbeds on the planet. The mosaic of lakes, streams, marsh, reeds, and wet woodland are a haven for birds and other wildlife. This Special Protection Area supports over 320 species of bird, including the Great White Pelican, Vulnerable Dalmatian Pelican, and Endangered Red-breasted Goose, along with many other attractive species such as the Collared Pratincole, the Western Marsh-harrier, Montagu's and Pallid harriers, and the Red-footed Falcon. Many other birds of prey are also attracted to the delta with its abundant food sources, including four Vulnerable species—the Eastern Imperial and Greater Spotted eagles, the Lesser Kestrel, and the Saker Falcon.

In a scenario reflected in so many deltas and other wetland sites worldwide, all is not well in the Danube Delta. Development projects impinge on these productive wetlands: unrestrained tourism, wind farms, road building, fishing, and hunting continue to impact on the birdlife and these vital habitats. One area at specific risk from new roads and building developments is that of the splendid coastal dunes of the Chituc Levee. Almost 7½ miles (12 km) of the coast here could be turned into a tourist complex, greatly diminishing its wildlife value.

The **Okavango Delta** in Botswana is the most important wetland in southern Africa and one of the world's greatest sites for wetland birds and other wildlife. More than 450 bird species have been recorded here. The Okavango River enters northwest Botswana, emptying into what is an otherwise mainly arid semi-desert landscape, and fanning out into an unusual inland delta, watering the soils and saturating a wide flood-plain on the northern edge of the Kalahari Desert. This close proximity of arid habitats and wetlands is very unusual and creates a magnet for wildlife in an otherwise harsh and unproductive environment. The wetland habitats here are dynamic, changing with seasonal variations in water flow. They include reed (*Phragmites*) and Papyrus (*Cyperus papyrus*) swamps, open water, river channels, lagoons, grassland, and areas of woodland. In all, 18 species of heron are found in the delta, including the Vulnerable Slaty Egret, for which this is the most important breeding site. Another Vulnerable wetland species, both resident and wintering here, is the Wattled Crane. Most of the Okavango Delta is protected, one of the major conservation areas being included in the Moremi Game Reserve.

◁ *Lake Bogoria, an alkaline lake in Kenya's Rift Valley, is a critical refuge for hundreds of thousands, and occasionally millions, of Lesser Flamingos; and more than 300 other bird species have been seen here. It is listed in the Ramsar Convention's Directory of Wetlands of International Importance.*

▽ *Lake Maracaibo, Venezuela, was once was a pristine sanctuary for flamingos and other birds. Oil discovery in the 1920s encouraged industrial growth, which has caused pollution that can cause dense algal blooms.*

Rare Birds of the Andean Lakes

The Andes Mountains contain a number of special high-altitude lakes, unusual wetlands that are home to several fascinating and rare waterbirds.

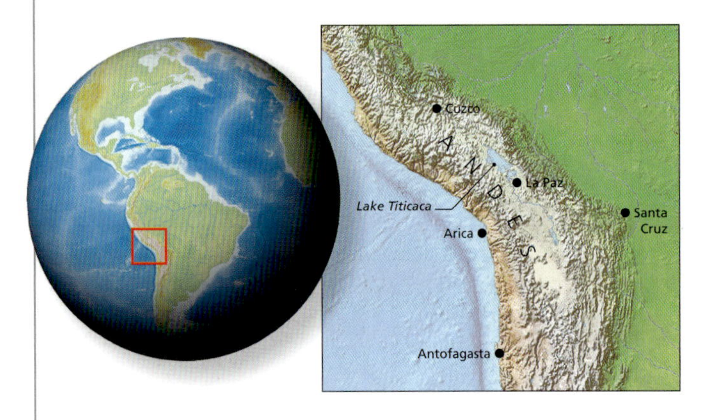

▽ *Reedbeds around Lake Junín in the Andes are the breeding site of several threatened birds. Water level changes and pollution are affecting this habitat.*

In central Peru, in the Department of Junín to the northeast of the capital Lima, lies the Lago de Junín at an altitude of 13,400 ft (4,080 m) in an area dominated by the dry tussocky grassland known as *puna*. Recognized as an EBA due to restricted-range species, its star bird is the Critically Endangered Junín Grebe, a flightless species with a population of just two or three hundred that is found only in this area. The lake is shallow and fringed by marshes and flooded reedbeds, the latter the haunt of the small fish and invertebrates that form the diet of the grebes. The grebes also breed in the reedbeds. Mining in the area has polluted the water with iron oxides, and another threat comes from extreme variation in water level, partly because of extraction for a hydroelectric plant and also through the increasingly unstable climate. The Junín Grebe has been adopted as a symbol of wetland conservation in the high Andes, and ecotourism is being fostered at the lake,

helping to raise awareness of the plight of the Junín Grebe.

Also restricted to this lake is the Junín Rail, with a total population estimated at no more than 2,500 birds and listed as Endangered. In common with many rails, it is secretive and also very small, easily hiding undetected among the reeds and rushes of the lakeside marshes. It, too, is threatened by pollution and water-level changes, and the increasingly dry conditions in the marshes.

Further south in Peru, straddling the border with Bolivia at 12,500 ft (3,800 m), is the huge Lake Titicaca, the largest by volume of all South America's lakes and one of the world's highest altitude navigable lakes. This lake is the haunt of another flightless waterbird, the Endangered Titicaca Grebe. Assessed as having a population of about 1,600, it is decreasing and may soon be reclassified as Critically Endangered. Its small range includes a handful of smaller nearby lakes and in adjacent Bolivia.

Titicaca Grebes breed in reedy marshes dominated by Tule-rush (*Schoenoplectus tatora*), from which they emerge to feed on the open waters of the lake. They rely almost entirely on a particular genus of small fish (*Orestias*), itself threatened by the introduction of exotic fish. Fishing with gill-nets is another problem here and grebes often become entangled and drowned as a bycatch. One study found an average of almost three grebes per fisherman each month, which when extrapolated indicates a very real threat to a species with such a small population. Tule-rush beds are sometimes burned, partly in order to gain grazing land for cattle, thus reducing the prime breeding habitat of the grebe. In addition to this, increased tourist activity, especially from boats, puts pressures on this fascinating wetland bird that are indeed substantial. Titicaca National Reserve protects part of the grebe's habitat, but the threats still remain.

Further south in the region, another Endangered species—the Hooded Grebe—inhabits a small number of lakes on the dry Patagonian steppes in southwestern Argentina, and in adjacent Chile. Numbers of this bird fluctuate, but they are estimated in the range 250–2,500; the trend is definitely downward, a situation that is hampered by the slow reproductive rate of the species.

△ *The flightless Titicaca Grebe is listed as Endangered because of its recent very rapid population decline, which if it continues may result in the species being listed as Critically Endangered.*

Selected Birds of Andean Lakes

	SCIENTIFIC NAME	COMMON NAME	STATUS	WORLD POPULATION
Lago de Junín				
■	*Podiceps taczanowskii*	Junín Grebe	CR	50–250
■	*Laterallus tuerosi*	Junín Rail	EN	1,000–2,500
■	*Oreotrochilus melanogaster*	Black-breasted Hillstar	LC	Unknown
■	*Geositta saxicolina*	Dark-winged Miner	LC	Unknown
■	*Cinclodes palliatus*	White-bellied Cinclodes	CR	50–250
■	*Asthenes virgata*	Junín Canastero	LC	Unknown
Lake Titicaca				
■	*Rollandia microptera*	Titicaca Grebe	EN	1,600
■	*Phoenicopterus chilensis*	Chilean Flamingo	NT	200,000
■	*Plegadis ridgwayi*	Puna Ibis	LC	10,000–15,000
■	*Fulica gigantea*	Giant Coot	LC	10,000–100,000
Patagonian Steppes lakes				
■	*Podiceps gallardoi*	Hooded Grebe	EN	250–2,500

RED LIST ■ CR = Critically Endangered ■ EN = Endangered ■ VU = Vulnerable ■ NT = Near Threatened ■ LC = Least Concern

ISLAND

THE OCEANS OF THE WORLD ARE DOTTED WITH ISLANDS. THE MAJORITY OF OCEANIC ISLANDS ARE FOUND IN SUBTROPICAL AND TROPICAL LATITUDES, WITH SOME ALSO IN THE SOUTHERN OCEAN. THE DEFINITION OF "ISLAND" IS SOMEWHAT ARBITRARY, BUT HERE WE ARE LOOKING MAINLY AT ISOLATED AND RELATIVELY SMALL OCEANIC ISLANDS THAT ARE NOT INDEPENDENT NATIONS. THE ISLANDS OF THE CARIBBEAN, FOR EXAMPLE, ARE NOT COVERED, NOR IS MADAGASCAR, AN ISLAND BLESSED WITH A VERY RICH WILDLIFE.

Islands are attractive to birds for a number of reasons. Seabirds find that islands offer safe breeding sites from which they can easily forage and feed in the coastal waters and open oceans. Many oceanic islands are rather inaccessible to people and have as a result remained relatively undisturbed. Birds that have evolved on islands have been able to exploit habitats that have less competition from other animals such as mammals, or indeed from other birds.

As a result of their separation from mainland continents and also in many cases from other islands, island birds are often rather unusual, and islands are well known as being miniature centers of avian endemism. Darwin's famous studies in the Galápagos Archipelago drew attention to the speed at which evolution produces separate species from a common ancestor, through isolation on adjacent islands. On these and on many other oceanic islands, birds have evolved into the endemic species we see today. As many islands are small in size, endemic birds often have naturally small populations and are therefore of particular concern to conservationists. In summary, oceanic islands have more than their "fair" share of rare and endemic species.

◁ *A view of the small islands off the northern coast of Viti Levu, the main island of Fiji. Listed as an Endemic Bird Area of the World (EBA), Fiji boasts numerous unique species, including the Critically Endangered Red-throated Lorikeet and Endangered Long-legged Thicketbird.*

THE THREATS TO ISLAND BIRDS

Island birds are especially vulnerable to threats posed by people and introduced alien animals and plants. Birds, with their power of flight, disperse and colonize islands more effectively than mammals, and many islands have few natural mammalian predators. The native birds have often not evolved defensive strategies against such predators. As a result, some have even lost the power of flight altogether, and many nest in the open, on or near the ground with no protection from predators. A major problem for island birds is posed by the introduction of domestic or farmed animals—whether they have arrived by accident or brought in deliberately—and also alien wild animals (mainly mammals). Cats and dogs find easy pickings among the local birds, while sheep, pigs, and especially goats browse the native vegetation, destroying or altering the landscape to the detriment of the local birdlife.

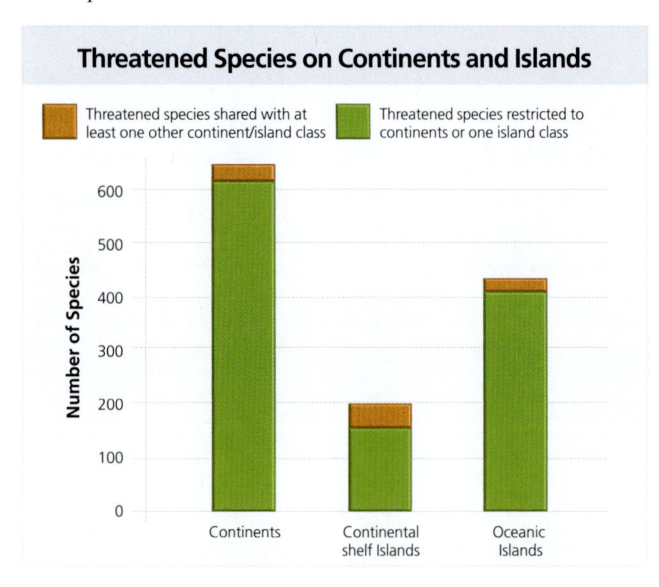

The threats faced by island birds are many, the major ones being predation or competition from alien species, and habitat loss due to agriculture, logging, or clearance for building. Many island birds seem unusually tame, unfamiliar as they are to attack, and alien carnivores find them easy prey. Introduced mammals are the major threat, followed by alien plants, introduced diseases, and alien birds. On many island sites the threats are multiple. On Hawaii for example, no fewer than eight Critically Endangered birds are threatened by a combination of invasive species, habitat degradation, and the effects of climate change, a pattern repeated all too commonly elsewhere. On the Pacific island of New Caledonia a major threat comes from open-cast mining.

The main mammals responsible for reducing numbers of island birds are Brown (*Rattus norvegicus*) and Black (*R. rattus*) rats, the domestic cat, and domestic dog, while farm animals such as goats, sheep, pigs, and cattle have major impacts on the natural vegetation through grazing and trampling. Alien birds that compete with or predate native birds include the Common Mynah, Red-whiskered Bulbul, Barn Owl, and crows. Other alien carnivores include the Mona Monkey (*Cercopithecus mona*), Indian Mongoose (*Herpestes javanicus*), African Civet (*Civettictis civetta*), weasels and stoats (*Mustela* species), and mice. Efforts are being made to control alien or invasive species on many oceanic islands.

With only about 17 percent of bird species restricted to islands, these places have disproportionately higher numbers of threatened species. Almost equal numbers of threatened birds are found on continents and islands (see chart left).

The importance of islands to rare birds may be seen clearly by looking at the figures for those threatened species that are Critically Endangered. Out of a global total of 190 species currently listed as Critically Endangered, no fewer than 92 are island birds. Of the 20 rarest of these, the majority are restricted to island sites.

The Puerto Rican Amazon, a bright green parrot numbering only 30–35, clings on in the forests of eastern Puerto Rico. Most of its natural habitat has been destroyed and it has also suffered through collection, predation by the Black Rat and Indian Mongoose, and also from hurricanes. The Bali Starling is a beautiful, almost pure white bird that has declined almost to the point of extinction, mainly through illegal trapping for the cagebird trade. On the Galápagos Islands, the rarest bird species is the Floreana Mockingbird. Though it became extinct on the island that gave it its name, it survives with a population of about 50 on two of the other islands. Extinction on Floreana seems to have been caused by Black Rats.

About 45 of the world's Critically Endangered species are so severely threatened or have such small (or badly known) populations that sadly some of these may already be extinct. On Christmas Island, south of Java, an alien snake, the Wolf Snake (*Lycodon aulicus*), was introduced via shipping cargo in 1987 and may be a threat to the birdlife. Another snake, the Brown Tree Snake (*Boiga irregularis*), was accidentally introduced to the island of Guam in the Mariana Islands in the 1940s or 1950s. This specialist bird predator devastated populations of the local forest birds such as the Mariana Fruit-dove, Rufous Fantail, and Guam Flycatcher (the latter now extinct).

Though island birds, with their small ranges, face such a large number of threats, careful research, followed by targeted conservation measures, have proved successful in some cases, and several species have been saved from probable extinction. Four classic examples are the Black Robin (Chatham Islands, New Zealand), Mauritius Parakeet, Rarotonga Monarch (Cook Islands), and Seychelles Magpie-robin. All four of these birds were reduced to a mere handful of individuals, yet they have all been rescued, and now enjoy healthier populations that have increased steadily since about the mid-1980s.

▷ *At one point in the 1960s, fewer than 15 Seychelles Magpie-robins survived in the Seychelles Islands, their population decimated by introduced predators such as feral cats and by habitat loss. A recovery program was initiated in the 1990s, and by 2006 the population had risen to 178 birds.*

Island Conservation Measures

SCIENTIFIC NAME	COMMON NAME	ISLAND	MAIN ACTION
Papasula abbotti	Abbott's Booby	Christmas Island	Control of invasive ants
Pterodroma baraui	Barau's Petrel	Réunion	Ban on hunting
Ninox natalis	Christmas Island Hawk-owl	Christmas Island	Control of invasive ants
Ducula whartoni	Christmas Island Imperial-pigeon	Christmas Island	Control of invasive ants
Zosterops natalis	Christmas Island White-eye	Christmas Island	Control of invasive ants
Ducula galeata	Marquesan Imperial-pigeon	Marquesas	Translocation
Falco punctatus	Mauritius Kestrel	Mauritius	Captive breeding/release, nest box provision, predator control
Cyanoramphus cookii	Norfolk Island Parakeet	Norfolk Island	Nest protection, rat and cat control
Nesoenas mayeri	Pink Pigeon	Mauritius	Habitat restoration, predator control, feeding, brood fostering
Megapodius pritchardii	Polynesian Megapode	Tonga	Translocation
Acrocephalus rodericanus	Rodrigues Warbler	Mauritius	Habitat protection, reforestation

SUCCESS IN THE SEYCHELLES

SET IN THE MIDDLE OF THE INDIAN OCEAN, THE CLUSTER OF ISLANDS THAT MAKE UP THE SEYCHELLES ARE A FAVORITE DESTINATION FOR SUN-SEEKING TOURISTS AND ECOTOURISTS ALIKE. THE MAIN ROCKY ISLANDS ARE COMPOSED MAINLY OF ANCIENT GRANITE, UNLIKE MOST OCEANIC ISLANDS, WHICH TEND TO BE RELATIVELY RECENT AND VOLCANIC OR CORALLINE IN NATURE.

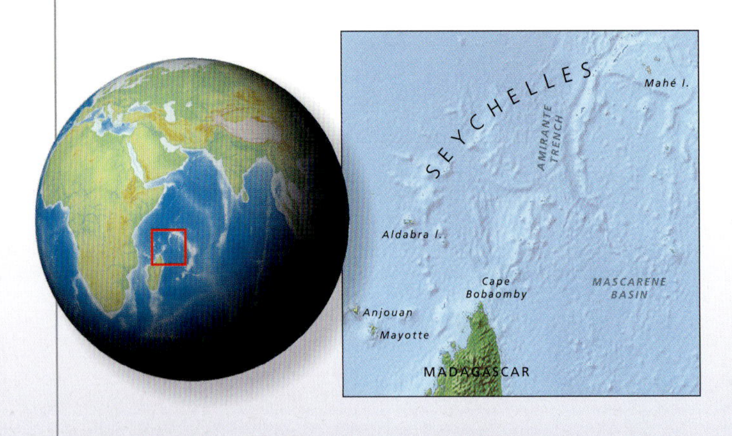

These beautiful islands are home to a splendid set of fascinating birds, including rare, local endemic species, some of which are threatened. On the main island of Mahé it is still possible to see (or more likely hear) the Endangered Seychelles Scops-owl, whose rasping call has been likened to that of a frog. This small owl lives in the misty montane forests of central and northern Mahé, a habitat largely included in the Morne Seychellois National Park. Introduced Black Rats, domestic cats, and Barn Owls are thought to predate both adults and nests, and the numbers of this charming owl are alarmingly low, between 250 and 300.

Thanks to some far-sighted conservation initiatives, some of the Seychelles' rare and threatened species have been brought back from the brink, most notably the Seychelles Warbler (Vulnerable), Seychelles Magpie-robin (Endangered), and Seychelles Paradise-flycatcher (Critically Endangered). The latter two are most attractive and approachable species that have become local symbols of successful bird conservation, enchanting ecotourists and inspiring local naturalists alike.

The charming island of La Digue, northeast of Mahé, is the haunt of the Seychelles Paradise-flycatcher, the male of which is jet black, with very long tail streamers. These beautiful birds nest in damp forest of mature trees, where they number only about 250. The invasive Water Lettuce (*Pistia stratiotes*) introduced to La Digue's marshes may have reduced the supply of the flycatcher's insect prey and it is now routinely removed. Following protection in a dedicated reserve, they seem to be gradually increasing. In late 2008, 23 birds were moved by helicopter to the isolated island of Denis, with two nests and eggs recorded the following year. The Seychelles Warbler, Seychelles Magpie-robin, and Seychelles Fody have also been successfully translocated to Denis, providing valuable backup to their former ranges.

Once much more numerous, the Seychelles Warbler used to occur on several of the islands, but by 1968 it was down to fewer than 30 birds at just one site, the small island of Cousin, north of Mahé. Since then it has been successfully translocated to the islands of Aride, Cousine, and Denis, and it has a stable population of about 2,500. The Seychelles are fortunate in consisting of numerous small islands, several of which remain free of introduced predators, or from which such alien carnivores can be eliminated or controlled.

▽ *The Special Reserve of Cousin Island, Seychelles, covers just 67 acres (27 hectares), but is the home to five of the land birds endemic to the Seychelles and to large breeding colonies of several seabirds.*

The thrushlike, black-and-white Seychelles Magpie-robin has a small but increasing population of about 200. Like the warbler, its population plummeted to a handful on one island, in this case Frégate. Following better protection on Frégate and successful translocations to Aride, Cousin, and Cousine, the numbers of this species are now increasing. Other threatened birds of these islands include the Endangered Seychelles White-eye, and the Seychelles Swiftlet and Seychelles Kestrel (both Vulnerable).

Aldabra, the largest coral atoll in the world, is part of the Seychelles, but lies about 680 miles (1,100 km) southwest of the main group, between the African mainland and the northern tip of Madagascar. Aldabra is a very different place from the granite islands of the main Seychelles group. A huge coral atoll encircling a shallow lagoon fringed with mangroves, it has areas of rocky scrub, and is the haunt of the Aldabra Giant Tortoise (*Geochelone gigantea*) that reaches a length of over 3 ft (1 m). Recognized as a World Heritage Site, Aldabra is also home to the Aldabra subspecies of the White-throated Rail. This bird differs from the (nominate) subspecies found on Madagascar in that it is flightless—the only remaining flightless bird on the western Indian Ocean islands. With no natural enemies, this rail gradually lost the ability or need to fly, in common with a number of other island birds. Sadly the Aldabra Warbler is now listed as extinct, having been last seen in 1983, a victim probably of rat predation and habitat degradation by goats. However, the Near Threatened Aldabra Drongo still survives, breeding in the scrub and mangroves.

△ *The population of the Vulnerable Seychelles Kestrel recently numbered around 370 pairs on Mahé and a further 10–15 pairs on its satellite islands.*

◁ *The preserved palm forest of the Vallée de Mai, Praslin is home to the unique Coco de Mer palm, as well as to an endemic subspecies of the Black Parrot.*

Key Features of Aldabra

Land area: 59 sq. miles (153 sq.km)
Total area, including lagoon: 134 sq. miles (346 sq. km)

Threatened Bird Species:

Comoro Blue-pigeon	*Alectroenas sganzini*	LC
Red-headed Fody	*Foudia eminentissima*	LC
Aldabra Drongo	*Dicrurus aldabranus*	NT
Aldabra Warbler	*Nesillas aldabrana*	EX

Ten bird subspecies are endemic

Main threats: introduced predators and plants; loss of mangrove habitat

Oceanic Islands

It is convenient to consider the islands of the main oceans separately. Out of the many thousands of oceanic islands, especially in the complex Pacific archipelagos, certain islands or groups of islands stand out as of particular significance for their threatened and endemic birds. Below is a selection of some of the more important islands in each of the major oceans, providing a snapshot of their value as havens for unusual endemic and threatened birds. As they harbor so many endemic or restricted-range species, it is no surprise that so many of these islands or groups (or parts of them) have been defined as Endemic Bird Areas (EBAs).

ATLANTIC OCEAN ISLANDS

Though the Atlantic Ocean has relatively few islands, there are a number of individual islands or island clusters that are of particular interest for their threatened birds. The Atlantic islands are mostly of volcanic origin and many of them have steep terrain and mountains, supporting a range of habitats.

The Azores are the most northerly, sited due west of Portugal, with Madeira and the Canary Islands to the southeast, off the coast of Morocco. The Azores hold colonies of Cory's Shearwater (half the global population), three-quarters of Europe's Roseate Terns, and Fea's Petrel may also breed there. All three of these island groups have the restricted-range Island Canary, while Madeira is home to Fea's Petrel and the Endangered endemic Zino's Petrel, threatened by Black Rats. The laurel forests of Madeira and the Canaries are the habitat of three local species: the Madeira, Dark-tailed, and White-tailed Laurel Pigeons, the former endemic to Madeira and the latter two to the Canary Islands. The White-tailed Laurel Pigeon is Endangered; the other two are Near Threatened. Pockets of these unique damp forests remain, but were once much more extensive. Black Rats are one of the main threats to ground-nesting birds and these adaptable rodents have been unwittingly spread by ships to many of the islands in the North Atlantic.

Two more island groups lie further south off the west coast of Africa: the Cape Verde Islands, and São Tomé and Príncipe. São Tomé was once well wooded and many of its special birds such as the Critically Endangered São Tomé Grosbeak are forest species. Though much of the original forest was cleared, many old plantations have reverted to secondary woodland to the benefit of the birds. The forests of southwestern and central São Tomé are some of the best refuges for threatened birds in the whole of tropical Africa.

Much further south lie Tristan da Cunha and Gough Island. The vegetation is typically tussock grassland and scrub, becoming tundralike at about 5,000 ft (1,500 m). These islands are breeding sites for rare seabirds, including Endangered Atlantic and Vulnerable White-chinned Petrels, and the Wandering Albatross (Vulnerable). Another species, the Endangered Black-browed Albatross, nests on the Falkland Islands and on South Georgia, off the southern tip of South America. One of the main threats to albatrosses is as a bycatch in longline fisheries that kill the birds as they forage at sea. The chicks also suffer when they ingest hooks from the fish brought back by the parent birds. The Black-browed Albatross is one of the seabirds most frequently killed by longline fishing, especially in the waters of the south Atlantic.

INDIAN OCEAN ISLANDS

These islands are some of the richest in terms of the rare and endemic birds that they support. This is especially true of the island clusters in the tropical seas around the huge island country of Madagascar. To the north lie the Seychelles with their suite of endemic birds, several of them threatened, though responding well to conservation efforts (see pp. 106–107).

The Seychelles are a mecca for tourism, some of which now promotes the wildlife as much as the beaches. Here, too, introduced carnivorous mammals are a problem, though several islands are predator free and translocation of certain threatened species to such islands is proving effective.

To the east of Madagascar, Mauritius, Réunion, and Rodrigues, known collectively as the Mascarene Islands, are home to a fascinating group of threatened birds, mainly found in the remnants of tropical forest that once clothed much of the islands. Mauritius

▷ *The Seychelles Paradise-flycatcher is Critically Endangered. There are no more than 250 individuals, most of which are on the island of La Digue.*

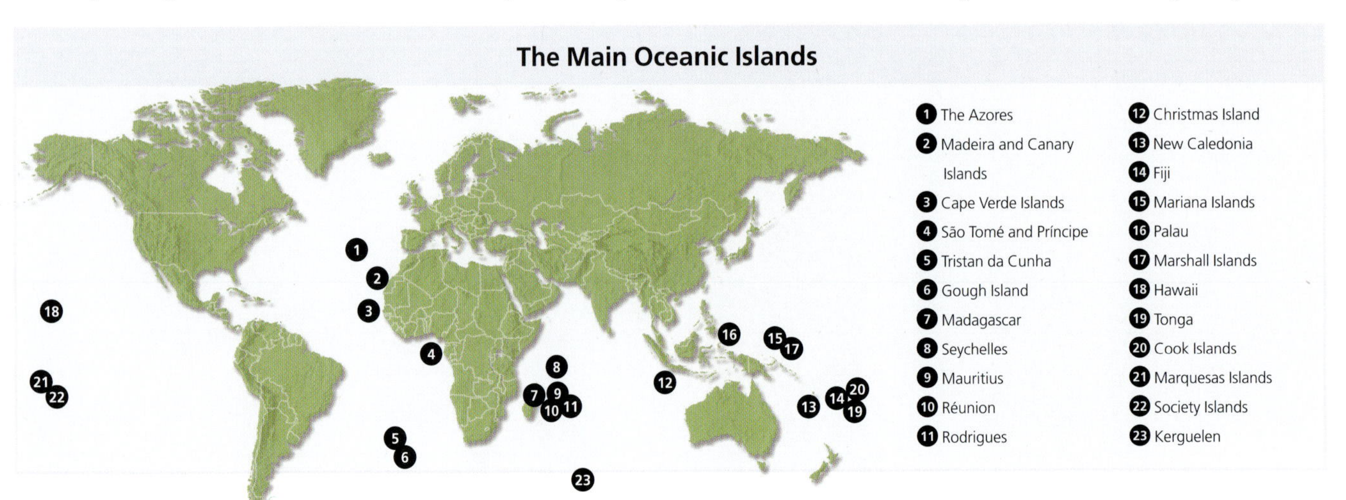

△ *The Society Islands in French Polynesia are listed as an EBA by BirdLife International with a number of threatened forest species and endemics. They are among the most devastated island groups in the whole of the Pacific.*

is famous as the haunt of the extraordinary Dodo, a giant flightless pigeon that was hunted to extinction in the 1660s, giving rise to the expression "dead as a Dodo." With luck, and targeted conservation initiatives, the remaining threatened species will not suffer the same fate. The Pink Pigeon, Mauritius Parakeet, and Mauritius Fody are all Endangered, while the Mauritius Olive White-eye is listed as Critically Endangered. Most of the forest on Mauritius has been converted to plantations of tea, sugar-cane, and conifers. Introduced deer, pigs, and monkeys, along with both Black and Brown rats, also cause problems for the native birds. The White-eye also suffers nest-robbing by the introduced Red-whiskered Bulbul. The Black River National Park protects some 27 sq. miles (70 sq. km) of forest habitat, including tracts considered the most important for threatened birds in the whole of tropical Africa. One consequence of deforestation has been to render the islands more susceptible to cyclones, which are becoming more frequent due to climate change.

On the eastern side of the Indian Ocean, to the south of Java, Christmas Island is the breeding site for the Endangered Abbott's Booby and the Critically Endangered Christmas Frigatebird. Forest loss, mainly through clearance for phosphate mining, has caused declines in both species, but most nest sites are now included in the Christmas Island National Park.

Amsterdam and St Paul Islands lie right in the center of the southern Indian Ocean. The endemic Amsterdam Albatross breeds only on the upland plateau of Amsterdam Island. Of a population of about 80, only about 25 pairs breed each year, each pair breeding every two years. The threats come from disturbance from introduced cattle, predation by feral cats, longline fishing, and possibly diseases such as avian cholera.

PACIFIC OCEAN ISLANDS

The Pacific Ocean is studded with thousands of islands, many of them being extremely remote. Most are in archipelagos, following submerged ridges, and situated in the warm waters of the tropics. There are three main groups:

Melanesia: includes New Caledonia, Vanuatu, Fiji, and the Solomon Islands.

▷ *The Endangered Pink Pigeon, endemic to Mauritius, was brought back from near extinction in the early 1990s, but still numbered only 380 in 2007.*

◁ *The Vulnerable Iiwi is among the most extraordinary of Hawaii's honeycreepers with its bright scarlet plumage and decurved bill. Though still relatively abundant in some areas, there is a marked decline in the population.*

△ *The Hawaiian Goose or Néné was reduced to some 30 individuals in the 1930s, but with intensive conservation and captive breeding the wild population has increased to 1,000 with much the same number in captivity.*

△ *The Critically Endangered Kakapo was once widespread on New Zealand. Its highly protected population of about 125 birds is located on the predator-free Codfish, Chalky, Anchor, and Maud islands off the coast of New Zealand.*

▷ *The distinctive volcanic landscape of Isabela Island is typical of the Galápagos Islands, which combined with their isolation has produced a unique flora and fauna, with numerous endemic and threatened species.*

Micronesia: includes the Mariana Islands, Palau, and the Marshall Islands.

Polynesia: includes Hawaii, Samoa, Tonga, Cook Islands, and French Polynesia (the Marquesas Islands and the Society Islands).

Some of the most important areas for threatened birds are the Hawaiian Islands, New Caledonia (see pp.114–115), and the Galápagos Islands.

The volcanic Hawaiian Islands lie in the central waters of the north Pacific and have a very distinctive set of birds, with several endemic species, including five genera found nowhere else. Some of these species have only local language common names. Notable Endangered species are Akepa, Akiapolaau, Hawaiian Duck, and Hawaii Creeper. Worryingly, the following are all listed as Critically Endangered: the Akekee, Akikiki, Akohekohe, Maui Parrotbill, Nukupuu, Oahu Alauahio, Olomao, Ou, Palila, Po'o-uli, and Puaiohi.

These islands have suffered more extinctions than any other similar area. The main threats are forest clearance for agriculture, grazing by farm animals, hunting, alien predators, competition from non-native birds, and also diseases carried by introduced mosquitoes. An introduced tree—the invasive Bush Currant (*Miconia calvescens*)—penetrates and alters native forest. Native to Central and South America, it has severely altered native habitats both here and elsewhere, notably on Tahiti. On Laysan Island, the Laysan Duck and Millerbird are Critically Endangered, and the island is an important wintering site for the Vulnerable Bristle-thighed Curlew.

The Galápagos Islands, off the coast of Ecuador in the eastern Pacific, are famous mainly through the studies made there by Charles Darwin, helping him to arrive at his theory of evolution by natural selection. Darwin studied the finches in particular, noticing that similar, related forms showed highly distinctive differences on different islands, deducing that they had almost certainly evolved into separate species from a common ancestor after generations of separation. As well as the 13 species of "Darwin's" finch, there are nine other endemic birds here including Vulnerable species such as the Galápagos Hawk, Galápagos Martin, Galápagos Rail, and Lava Gull; the Endangered Galápagos Penguin and Flightless Cormorant; and the Critically Endangered Floreana Mockingbird, Galápagos Petrel, and Waved Albatross. The whole archipelago is a National Park, Biosphere Reserve, and World Heritage Site. Some threats persist, notably predation from alien carnivores and overgrazing.

SOUTHERN OCEAN ISLANDS

The islands of the Southern Ocean lie south of latitude 40°S, with a major cluster to the south and east of New Zealand's South Island. The prevailing winds create the so-called Antarctic Circumpolar Current, causing the surface waters to circulate from west to east around Antarctica. Birds such as petrels, shearwaters, and albatrosses exploit these exposed, windy seas and travel long distances for food. Many of them breed on isolated oceanic islands. During the long days of the Antarctic summer, the seas become very productive as phytoplankton increase, using sunlight, carbon dioxide, and nutrients brought to the surface by upwellings of deep water. This rising water is rich in nitrates and phosphates derived from the decayed bodies of marine plants and animals on the ocean floor. The phytoplankton bloom feeds zooplankton including crustaceans such as krill, which in turn support fish, squid, and large numbers of marine mammals, penguins, and birds.

One of the major threats to seabirds, especially to albatrosses, is longline fishing, practiced on a large, commercial scale. The huge Wandering Albatross, listed as Vulnerable, breeds mainly on the Prince Edward Islands, Crozet Islands, and Kerguelen Islands, South Georgia and Macquarie Island, and longline fishing is the main threat to this majestic ocean bird as it is to many albatrosses.

New Caledonia

Very large by Pacific island standards, the French territory of New Caledonia is a tropical island group some 930 miles (1,500 km) east of Australia. The main island, Grande Terre, is mountainous, reaching a peak altitude of 5,341 ft (1,628 m). It is continental in origin, once being part of the great southern landmass of Gondwanaland, and has been separated from Australia since about 65 million years ago.

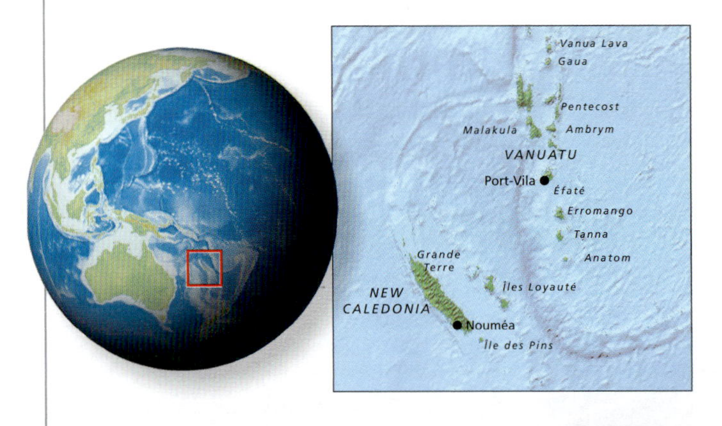

Such long separation from the mainland has resulted in the evolution of endemic species, including a large number of special birds, several of them threatened; indeed it has been likened to a veritable "Noah's Ark", preserving such special wildlife. The flora of Grande Terre is special too: 80 percent endemic and including no fewer than five endemic plant families. The habitats range from reef-fringed shorelines to lowland rain forest, scrub, and montane cloud-forest, with remnants of dry forest and savanna.

Most of the restricted-range bird species that define the New Caledonia EBA are found in the forests of the hills and lowlands. However, most of these lowland forests have been cleared and used for grazing or for plantations. Fortunately for the local economy, though not for the wildlife, Grande Terre also has 50 percent of the world's known deposits of nickel, and logging, followed frequently by opencast mining, has destroyed large areas of montane forest. Such clearance has also resulted in hugely damaging erosion as rainstorms fall directly onto the bare soil. Another recurrent threat is fire—often started deliberately to improve grazing, but frequently spreading to adjacent forest and scrub.

Two of New Caledonia's star birds are the extraordinary Kagu, and the New Caledonian Imperial-pigeon. The Endangered Kagu is so unusual that it is classified in its own family (Rhynochetidae). It is a medium-sized flightless, crested bird with dove-gray plumage, a long bill, and long red legs, looking rather like a cross between a small heron and a pigeon. Ghostlike, the Kagu stalks the dark forest floor searching for invertebrates and small lizards. Unfortunately for this bizarre bird, settlers introduced rats and domestic cats and dogs, and many fall prey to these alien predators. Its population is estimated at between 250 and 1,000, and seems to be stable, probably as dogs are controlled within the main protected areas. The New Caledonian

◁ *A pair of Endangered Kagu, with the male displaying by raising his crest.*

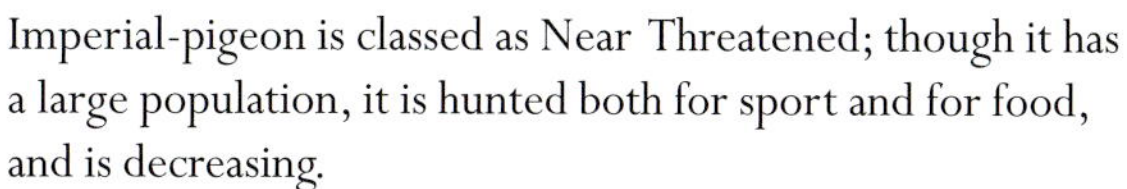

Imperial-pigeon is classed as Near Threatened; though it has a large population, it is hunted both for sport and for food, and is decreasing.

Four Critically Endangered birds are found here: the Crow Honeyeater, New Caledonian Lorikeet, New Caledonian Owlet-nightjar, and New Caledonian Rail. All four are teetering on the brink of extinction, with populations as low as 50 or fewer. The rail, being flightless, has certainly suffered at the jaws of introduced carnivores and may already be extinct. There are no recent records of the owlet-nightjar, nor of the pretty, bright green lorikeet, and both may also have vanished. The Crow Honeyeater is also precariously placed and seems to have suffered from predation of its young by introduced rats.

Vital Signs

Original extent of vegetation: 7,325 sq. miles (18,972 sq. km)
Vegetation remaining: 1,978 sq. miles (5,122 sq. km)
Endemic plant species: 2,432
Endemic threatened birds: 7
Endemic threatened mammals: 3
Endemic threatened amphibians: 0
Extinct species: 1
Human population density: 29 people/sq. mile (11 people/sq. km)
Total area protected: 1,619 sq. miles (4,192 sq. km)
Nature reserves: 192 sq. miles (497 sq. km)

△ *The New Caledonian Imperial-pigeon is classified as Near Threatened with an estimated population of 100,000.*

▽ *Open-caste nickel mines have devastated the landscape on Grande Terre, New Caledonia. The montane forest that supported a variety of bird species has all but disappeared in many parts of the island.*

ENDANGERED BIRDS

THE MODERN ERA HAS SEEN SPECIES OF BIRDS BECOMING EXTINCT AT AN ALARMINGLY HIGH RATE—MANY THOUSANDS OF TIMES HIGHER THAN WHAT IS CONSIDERED TO BE THE "NATURAL BACKGROUND RATE."

In the few hundred years since 1500, some 150 bird species have disappeared, and one in eight species is now threatened with extinction. Those species with small ranges are very susceptible to even minor changes, and there have also been steep and worrying declines in many species that were once common and widespread. The number officially listed as threatened now stands at 1,240, a figure that is under constant review and one that continues to have a distinctly upward trend.

While figures like these are worrying, they do at least highlight the problem and also provide researchers and conservation practitioners with the tools to tackle the issues and attempt to reverse these trends. However, it is only when concentrating on the individual species that the reality of the pressures that face wild birds throughout the world really begins to sink in.

In the following pages, individual bird species speak for themselves, telling varied stories about their struggles for survival against a multitude of threats, many of which are the result of human activities. In among the bad news stories, there are some successes, and the heartening fact is that, with the targeted conservation efforts and education programs taking place around the world, birds can be brought back from the brink and saved for future generations. And we as individuals can each do our bit to help in such endeavors.

▽ *Now listed as Least Concern, the bald eagle was once threatened by intense hunting, unintentional poisoning, and habitat destruction.*

▷ *The Black Crowned-crane from sub-Saharan West Africa and the Sudan is listed as Vulnerable, having suffered population declines in the recent past.*

The Bird Profiles

The birds in this section of the book are grouped by family, with the families following a generally accepted systematic sequence. The general characteristics of each family are described, with an indication of the numbers of species that are threatened. For each family or group of families there is also a table that lists the species that are threatened, with an indication of their Red List status. The species are listed alphabetically by genus, and within each genus alphabetically by species name. The common names are also given. Selected species are profiled, with information about their range, characteristics, and threats.

The Profiles Section lists all the world's threatened species of bird—assigned to the Red List categories (see chart below).

The numbers in these categories change as species move from one category to another, sometimes becoming less threatened, but all the more often suffering population declines, or even becoming extinct. Occasionally, new species are discovered and added to the list, or taxonomic revision may result in a new species being recognized.

The latest information reports the extinction of the Alaotra Grebe, a waterbird found only in a tiny area of eastern Madagascar and previously listed as Critically Endangered. It was driven to extinction by a combination of the introduction of carnivorous fish and through being trapped in nylon gill-nets used by the local fishermen. The Zapata Rail from Cuba is another wetland bird that has suffered from introductions, in this case from introduced catfish and mongooses. Its status has changed from Endangered to Critically Endangered. White-bellied Cinclodes of Peru and

▽ *From only 27 surviving birds in 1987, the population of the California Condor was approaching 400 by 2010 through massive conservation efforts; despite this progress, the species remains Critically Endangered.*

The Red List Categories

CATEGORY		DESCRIPTION	NUMBER
■ Extinct	EX	There is no reasonable doubt that the last individual has died.	132
■ Extinct in the Wild	EW	Known only to survive in captivity.	4
■ Critically Endangered	CR	Facing an extremely high risk of extinction in the wild.	190
■ Endangered	EN	Facing a very high risk of extinction in the wild.	372
■ Vulnerable	VU	Facing a high risk of extinction in the wild.	678
■ Near Threatened	NT	Close to qualifying for or is likely to qualify for a threatened category in the near future.	838
■ Least Concern	LC	Widespread and abundant.	7,751
□ Data Deficient	DD	There is inadequate information to make a direct, or indirect, assessment of its risk of extinction.	62
□ Not Evaluated	NE	Not yet evaluated.	Not known

Those species listed as Vulnerable, Endangered or Critically Endangered are collectively described as threatened, and it is principally the species in these three categories, along with those Extinct in the Wild, that are listed in the profiles section of this book.

Black-winged Starling of Indonesia have also been uplisted to Critically Endangered. Both the Great Knot and the Far Eastern Curlew have been uplisted to Vulnerable as populations of both of these once common waders seem to have crashed, mainly because of drainage and pollution of coastal wetlands. The loss of certain mudflats at a migratory stopover site in South Korea resulted in a 20 percent decline in numbers of the Great Knot.

On a more positive note, the Azores Bullfinch has been recently downlisted from Critically Endangered to Endangered. This follows carefully targeted conservation work to restore natural vegetation on its island home. There is good news, too, for the Yellow-eared Parrot of Colombia, now downlisted to Endangered following nest protection and education of local communities. The Chatham Albatross, has been downlisted from Critically Endangered to Vulnerable as its population is increasing, though it breeds on only one rock stack in New Zealand's Chatham Islands.

THE RED LIST

The IUCN Red List of Threatened Species, usually referred to as the IUCN Red List—or simply the Red List—is the most comprehensive information source on the global conservation status of plant and animal species. IUCN—the International Union for Conservation of Nature—with headquarters in Gland, Switzerland, is the world's oldest and largest global environmental network. BirdLife International, a member of IUCN, is the Red List Authority for birds and gathers and holds data on the world's birds, including their Red List status, as recorded in this book. Currently, BirdLife recognizes 10,027 bird species, a number that changes in the light of taxonomic revisions and the occasional discovery of new species, such as the recent discovery of the Olive-backed Forest Robin (*Stiphrornis pyrrholaemus*) in Gabon.

△ *The Bali Starling qualifies as Critically Endangered because it has an extremely small range and a tiny population, which is still suffering from illegal poaching for the cagebird trade.*

▽ *The Chatham Albatross, which breeds only on The Pyramid, a large rock stack in the Chatham Islands, New Zealand, is listed as Vulnerable, but its population is stable, and even increasing.*

Kiwis, tinamous, and cassowaries

Kiwis (family Apterygidae) are some of the strangest of all birds. They are flightless and their feathers are narrow and hairlike. The four species of kiwi (three of which are threatened) are nocturnal forest birds found in New Zealand. They lay very large eggs, relative to their body size, which are incubated by the male. Kiwis use their long bill to probe in damp soil for invertebrates such as earthworms. Introduced predators, including domestic cats, dogs, and stoats are the major threat to these highly unusual and fascinating birds.

Tinamous (family Tinamidae) are ground birds found in Central and South America, in scrub, forest, and grassland. They are rather dumpy in build and partridgelike in appearance. They fly reluctantly and clumsily, preferring to scuttle through the undergrowth. There are about 40 species, five of which are threatened, mainly through habitat loss, disturbance, and hunting.

Cassowaries (Casuariidae) are very tall, bulky flightless forest birds from New Guinea and Australia. With their powerful legs and very large, sharp claws—the middle claw is daggerlike and 4¾ in (12 cm) long—they forage mainly for fruit in rain forests and savanna habitats. There are three species of cassowary, two of which are threatened. In New Guinea cassowaries are hunted for food, feathers, and claws (used to make spear-tips), which has caused local extinctions of the bird.

Southern Brown Kiwi

Apteryx australis

This kiwi survives as three separate populations in New Zealand: 20,000 on Stewart Island, 7,000 in Fjordland, and in another small isolated population. A medium-sized flightless bird, it is a ground feeder, using its long decurved bill to eat invertebrates, as well as some fruit and other plant material. It survives in various habitats, from coastal areas to grasslands. Plumage color varies: Fjordland birds are dark grayish-brown; those on Stewart Island have dark brown feathers with a reddish-brown streak along their length. Both sexes call noisily in the early hours of darkness. The male whistles piercingly, while the female emits a low, gruff call. Their life span is up to 20 years, with females laying a single egg per brood. The young birds are vulnerable to introduced stoats and domestic cats, while Brush-tailed possums and stoats also take the eggs. Adults and juveniles are predated by dogs and ferrets. Though the population on Stewart Island is relatively large, the species is in overall decline and classed as Vulnerable. Conservation measures for the smaller populations include the possible relocation of birds, new legislation, controlling predator numbers, and a community involvement program.

Family Apterygidae
Size 15¾ in (40 cm)
Status Vulnerable (decreasing)
Population 27,000
Range New Zealand (endemic)
Habitat Coastal sand dunes, forest, subalpine scrub, tussock grassland
Main threats Introduced predators

Southern Cassowary

Casuarius casuarius

This large flightless bird lives a solitary and elusive existence in a variety of tree-dominated habitats in Northern Australia and New Guinea. The adult is mainly black, with distinctive neck markings. Most of the neck is bright blue with a red nape and a red double wattle dangling from its neck. Cassowaries communicate using a variety of hissing and rumbling calls, as well as with booming sounds when displaying. They feed mainly on fallen fruit. Numbers of the Southern Cassowary have decreased rapidly over the last 30 years and there have been local extinctions in parts of New Guinea. In Australia the main threats are road kills, predation of young by dogs, and loss or fragmentation of habitat, now mainly halted. In Indonesia and Papua New Guinea this species suffers heavy losses through hunting, mainly for meat, as well as habitat loss.

Family Casuariidae
Size 6 ft (1.8 m)
Status Vulnerable (decreasing)
Population 10,000–19,000
Range Australia (Queensland), New Guinea, Papua New Guinea
Habitat Rain forest, savanna forests, mangroves, fruit plantations
Main threats Habitat fragmentation and loss, hunting, logging

THREATENED KIWIS, TINAMOUS, AND CASSOWARIES

FAMILY		COMMON NAME	SCIENTIFIC NAME	STATUS
APTERYGIDAE Kiwis	■	Southern Brown Kiwi	*Apteryx australis*	VU
	■	Great Spotted Kiwi	*Apteryx haastii*	VU
	■	Northern Brown Kiwi	*Apteryx mantelli*	EN
TINAMIDAE Tinamous	■	Choco Tinamou	*Crypturellus kerriae*	VU
	■	Taczanowski's Tinamou	*Nothoprocta taczanowskii*	VU

RED LIST ■ EX = Extinct ■ EW = Extinct in the Wild ■ CR = Critically Endangered ■ EN = Endangered ■ VU = Vulnerable

THREATENED KIWIS, TINAMOUS, AND CASSOWARIES *Continued*			
FAMILY	**COMMON NAME**	**SCIENTIFIC NAME**	**STATUS**
TINAMIDAE Tinamous	Lesser Nothura	*Nothura minor*	VU
	Dwarf Tinamou	*Taoniscus nanus*	VU
	Black Tinamou	*Tinamus osgoodi*	VU
CASUARIIDAE Cassowaries	Southern Cassowary	*Casuarius casuarius*	VU
	Northern Cassowary	*Casuarius unappendiculatus*	VU

RED LIST ■ EX = Extinct ■ EW = Extinct in the Wild ■ CR = Critically Endangered ■ EN = Endangered ■ VU = Vulnerable

PENGUINS

Penguins (family Spheniscidae) are unmistakable flightless marine birds. Dumpy and rather comical, they stand upright on land but are streamlined and efficient swimmers, both on and below the sea surface. Their wings are short and stiff and used not for flight but as paddles to provide propulsion when swimming. Their plumage is black or blue-gray above and white below. Young penguins are covered in thick soft down. Nearly all penguins are restricted to the southern hemisphere, though the cold currents along the west coast of South America allow some to survive as far north as the Gálapagos Islands. The highest diversity is reached in the South Atlantic, and penguins have been known to breed on the Antarctic continent. Of the 18 existing species, no fewer than 11 are now considered threatened, by factors including ensnarement in fishing nets, scarcity of prey caused partly by overfishing, and in some cases through predation by introduced carnivores. Like many seabirds, penguins are also badly affected by marine oil spills and many die as a result of such incidents. The Galápagos Penguin suffers periodic declines as a result of the El Niño phenomenon, when changes in the ocean currents cause a reduction of fish stocks. These effects may increase in frequency as a result of climate change.

Galápagos Penguin
Spheniscus mendiculus

This small penguin breeds further north than any other, and is endemic to the Galápagos Archipelago on the equator. Most are found on the large western islands of Isabela and Fernandina, with the remainder in clusters on a few smaller islands. They feed on shoaling fish in cool, nutrient-rich waters close to shore, and breed throughout the year in burrows or rock crevices near sea level. Gray-black above and white below, they have a black head with a white stripe running from behind the eye and around the ear coverts to the chin. There are two black bands across the breast, the lower one running down the flanks. The bill is dark with pink on the lower mandible. Juveniles have pale cheeks and lack breast bands. With such a restricted distribution and complete dependence on fish supplies, the population is vulnerable to fluctuations in the ocean currents that affect their food, particularly warming of the seas. In recent decades numbers have plummeted by over 60 or 70 percent as a result of El Niño Southern Oscillation. Recovery is slow. The main population has also been decimated by introduced domestic cats, and the birds are killed by a variety of fishing methods. Conservation will rely on tighter management of fishing within the Galápagos Marine Reserve, limiting disturbance at breeding sites, and continued protection from introduced predators.

Family	Spheniscidae
Size	20¾ in (53 cm)
Status	Endangered (decreasing)
Population	1,800
Range	Galápagos Islands, Ecuador (endemic)
Habitat	Cool, oceanic waters close to rocky shores
Main Threats	Fluctuations in ocean currents, introduced predators, inshore fisheries

Yellow-eyed Penguin
Megadyptes antipodes

This medium-sized penguin does indeed have pale yellow eyes. It also has a pale yellow head with a brighter yellow band extending around the back of the head from the eyes. The pale feathers have contrasting black feather shafts. The young birds are uniformly gray in color. The calls are rather more musical than those of most penguins. The Yellow-eyed Penguin has a small breeding range and there are small populations on the southeast coast of New Zealand's South Island, in forest and scrub. Though there are signs that it is benefiting from conservation measures, it is listed as Endangered. It feeds mainly on sprat, red cod, opal fish, and squid taken from coastal waters. Disturbance close to the nesting sites from livestock or people reduces their breeding success, a problem that is being dealt with by fencing off certain areas. Sadly, the chicks are also very sensitive to outbreaks of fatal diseases including blood parasites and possibly also avian malaria, seriously depleting populations in some years. Nesting birds are also eaten by introduced mammals including stoats, ferrets, and cats, and efforts are underway to control these alien predators. A natural predator, Hooker's sea-lion (*Phocarctos hookeri*), occasionally feeds on this species, though losses are usually small.

Family	Spheniscidae
Size	25½ in (65 cm)
Status	Endangered (decreasing)
Population	5,000
Range	New Zealand (endemic)
Habitat	Forest, scrub
Main threats	Introduced predators, disturbance, reduction in habitat quality

THREATENED PENGUINS				
FAMILY		COMMON NAME	SCIENTIFIC NAME	STATUS
SPHENISCIDAE Penguins	■	Southern Rockhopper Penguin	*Eudyptes chrysocome*	VU
	■	Macaroni Penguin	*Eudyptes chrysolophus*	VU
	■	Northern Rockhopper Penguin	*Eudyptes moseleyi*	EN
	■	Fiordland Penguin	*Eudyptes pachyrhynchus*	VU
	■	Snares Penguin	*Eudyptes robustus*	VU
	■	Royal Penguin	*Eudyptes schlegeli*	VU
	■	Erect-crested Penguin	*Eudyptes sclateri*	EN
	■	Yellow-eyed Penguin	*Megadyptes antipodes*	EN
	■	African Penguin	*Spheniscus demersus*	EN
	■	Humboldt Penguin	*Spheniscus humboldti*	VU
	■	Galápagos Penguin	*Spheniscus mendiculus*	EN

RED LIST ■ EX = Extinct ■ EW = Extinct in the Wild ■ CR = Critically Endangered ■ EN = Endangered ■ VU = Vulnerable

Grebes

Grebes (family Podicipedidae) are wonderfully adapted to their aquatic life. Their slender, streamlined bodies lie low in the water and their lobed feet are set far back, acting as paddles on and under the water. When they dive, they expel air trapped in their feathers, thus reducing buoyancy and facilitating underwater agility. They are expert at chasing and catching prey—mainly aquatic invertebrates and fish—using their large feet to steer. The 19 species of grebe are found across the globe, mainly in lakes and rivers. Of the five threatened species, two are Endangered, and one, the Junin Grebe, is Critically Endangered. Grebes build floating nests woven from water plants such as reeds and waterweeds. Many grebes have elaborate courtship behavior, involving synchronized pattering along the water surface and repeated head bobbing and shaking, displaying head ruffs and crests. Some species are threatened by the introduction of exotic fish (such as *Tilapia*) that reduce the food availability for the grebes, or of carnivorous fish that include small grebes in their diet. Changes in the levels and quality of the water in lakes is also a factor, sometimes as a result of climate change, as for example in the case of the Hooded Grebe of Argentina. Another problem faced by grebes is posed by the use of gill-nets, which trap and kill the birds as well as the fish for which they are intended.

In May 2010, the extinction of the Alaotra Grebe (*Tachybaptus rufolavatus*) was announced. Restricted to a tiny area of eastern Madagascar, this species declined rapidly after carnivorous fish were introduced to the lakes in which it lived. This along with the use of nylon gill nets by fishermen that caught and drowned birds, has driven this species into the abyss.

Titicaca Grebe
Rollandia microptera

Dark-brown and flightless, this grebe has a yellow bill and white chin, throat and upper neck, shading to reddish-brown on the lower neck. Nonbreeding adults lack a crest, and have duller plumage, while the young are gray with a striped head and white neck and breast. The Titicaca Grebe lives only in certain lakes on the altiplano of Peru and Bolivia, including the famous Lake Titicaca itself, but also some others such as Lake Arapa and Lake Umavo. These grebes feed mainly on a particular species of local fish (genus *Orestias*, endemic to the altiplano, and one of which has already become extinct), rendering them susceptible to changes that affect this fish, notably the introduction of exotic fish species. The main breeding habitats are in broad marshes dominated by the rush *Schoenoplectus tatora* or on floating masses of waterweeds. Though it breeds throughout the year, numbers are dramatically affected by various problems throughout their range. These include unintentional capture in fishing nets, water pollution from nearby settlements and industry, and destruction of the reed habitat for grazing livestock. The birds are also hunted by locals for their eggs. Boat trips for tourists on Lake Titicaca may also be a factor affecting their breeding success.

Family Podicipedidae
Size 11–17¾ in (28–45 cm)
Status Endangered (decreasing)
Population 1,600
Range Peru and Bolivia
Habitat Freshwater lakes
Main threats Accidental trapping in fishing nets, pollution, natural water level fluctuations, competition for habitat from cattle rearing, introduction of exotic fish species, tourism (disturbance by boats), egg collection for food

Hooded Grebe

Podiceps gallardoi

This endangered grebe has a small range and seems to be in rapid decline, not helped by its low rate of reproduction. Medium-sized, its body is white with a dark gray back and lower neck. Its head is black with a peaked reddish crown and a white forehead. The birds spend winters in saltwater habitats on the Atlantic coast of Santa Cruz Province in southern Argentina. From October to March they relocate to freshwater lakes inland on the dry steppes of Patagonia. Here they build floating nests from aquatic plants, and breed colonially. They feed mainly on aquatic invertebrates. Kelp Gulls predate the grebes on some lakes, but their main threat is salmon and trout stock in private lakes. Changes in water levels attributed to climate change also affect their habitats.

Family	Podicipedidae
Size	12½ in (32 cm)
Status	Endangered (decreasing)
Population	250–2,500
Range	Southern Chile, Southwest Argentina
Habitat	Coastal estuaries and saltwater lakes (winter), upland freshwater lakes
Main threats	Climate change, introduced fish species

THREATENED GREBES

FAMILY		COMMON NAME	SCIENTIFIC NAME	STATUS
PODICIPEDIDAE Grebes	■	Hooded Grebe	*Podiceps gallardoi*	EN
	■	Junin Grebe	*Podiceps taczanowskii*	CR
	■	New Zealand Grebe	*Poliocephalus rufopectus*	VU
	■	Titicaca Grebe	*Rollandia microptera*	EN
	■	Madagascar Grebe	*Tachybaptus pelzelnii*	VU

RED LIST ■ EX = Extinct ■ EW = Extinct in the Wild ■ CR = Critically Endangered ■ EN = Endangered ■ VU = Vulnerable

Albatrosses

Albatrosses (family Diomedeidae) are supreme seabirds, covering long distances gliding close to the waves as they scour the oceans for food. They feed mainly on fish, crustaceans, and squid, usually plucked from the surface of the sea. They have very long, narrow wings that they use to catch wind currents deflected from the waves, a technique that enables them to incorporate long periods of gliding and thus reduce the need for energy-expensive, flapping flight. The largest species of albatrosses have the greatest wingspan of any living bird—up to about 11½ ft (3.5 m). The bill of an albatross is long and heavy, with a hooked tip and tubular nostrils at the base. Most species are found in the Southern Ocean, breeding typically on oceanic islands that are devoid of natural mammalian predators. Satellite tracking has revealed that they often travel huge distances, ranging widely, some as far as the northern oceans and some circumnavigating the globe. Of the 22 species of albatross, all but five are threatened, six are listed as Endangered and three Critically Endangered. Drift netting and longline fishing pose major threats, along with introduced predators, notably rats, mice, and cats, and marine pollution such as from oil spills. Some species have suffered high chick mortality through diseases such as avian cholera.

Wandering Albatross

Diomedea exulans

This huge seabird spends much time soaring over the Southern Ocean in search of food—fish, squid, and waste from fishing boats. The adult birds resemble a long-winged gull with mainly white feathers and black markings on their outer tail feathers and wing tips. Juveniles are darker, with a variety of chocolate brown and gray markings. Adult birds have flesh-colored legs and a pink bill. They nest on coastal hills and ridges on islands in the Southern Ocean. Many albatross habitats are now protected but the "incidental" capture of birds by longline fishing remains a problem. The fishing industry is being tackled to try to modify fishing techniques.

Family	Diomedeidae
Size	3¾ ft (115 cm)
Status	Vulnerable (decreasing)
Population	26,000
Range	Prince Edward Islands, Crozet and Kerguelen Islands, South Georgia, Macquarie Island
Habitat	Forages at sea, nests on exposed coastal sites
Main threats	Longline fishing, introduced predators such as cats

Waved Albatross

Phoebastria irrorata

This medium-sized, long-winged albatross is mainly brown with a contrasting white face and neck, grading to yellow-buff on the back of head and neck. The large bill is yellow, and the feet bluish. At close quarters most of the dark body feathers are seen to be finely barred. The breeding population is largely confined to just one island, Española in the Galápagos, where they nest in colonies on rather bare, flat areas of rock-strewn lava, or sometimes among scrub. Intermittent breeding has occurred on Isla de la Plata, off the coast of mainland Ecuador, and small numbers of nonbreeding birds visit other islands in the Galápagos. Outside the breeding season, they forage the seas off Ecuador and Peru. They feed on fish, squid, and crustaceans, and while some die as an incidental fisheries bycatch, others are taken deliberately for food or feathers. The young rely on being fed by both parents, and do not reach breeding age for five or six years, so rates of loss are critical to a species with such a restricted range.

Family Diomedeidae
Size 35½ in (90 cm)
Status Critically Endangered (decreasing)
Population 35,000
Range Galápagos Islands and Isla de la Plata, Ecuador (endemic)
Habitat Forages at sea, nests on rocky areas with little vegetation or scrub
Main Threats Trapping by fishing lines, nest predation by people, domestic cats and rats on Isla de la Plata

Sooty Albatross

Phoebetria fusca

The Sooty and the Waved Albatross differ from most albatrosses in having brown plumage, in this case a dark sooty brown. Other distinguishing features include a white crescent above and behind the eye, slightly darker brown coloring on the sides of the head, and a diamond-shaped tail. The bill is black with a yellow or orange marking on the lower mandible. Unusually for albatrosses, immature birds are similar in appearance to the adults. The adult makes a distinctive two-syllable call at the nest, which it builds on steep coastal slopes or cliffs in small colonies of 50–60, on islands in the Indian and South Atlantic Oceans. The main colonies are on Gough Island, Tristan da Cunha, Prince Edward and Marion Islands, Crozet Islands, and Amsterdam Island. Sooty Albatrosses forage over subtropical seas for fish, squid, and crustaceans, sometimes scavenging edible refuse from fishing boats. Longline fishing has affected population numbers, along with past problems of introduced predators on some islands.

Family Diomedeidae
Size 33½ in (85 cm)
Status Endangered (decreasing)
Population 42,000
Range Islands in the Southern Atlantic and Indian Oceans: Prince Edward and Marion Islands, Crozet Islands, Kerguelen Island, Amsterdam Island, Gough Island, Tristan da Cunha
Habitat Forages in subtropical southern oceans, steep coastal nesting sites
Main threats Longline fishing, introduced predators

THREATENED ALBATROSSES

FAMILY		COMMON NAME	SCIENTIFIC NAME	STATUS
DIOMEDEIDAE Albatrosses	■	Amsterdam Albatross	*Diomedea amsterdamensis*	CR
	■	Antipodean Albatross	*Diomedea antipodensis*	VU
	■	Tristan Albatross	*Diomedea dabbenena*	CR
	■	Southern Royal Albatross	*Diomedea epomophora*	VU
	■	Wandering Albatross	*Diomedea exulans*	VU
	■	Northern Royal Albatross	*Diomedea sanfordi*	EN
	■	Short-tailed Albatross	*Phoebastria albatrus*	VU
	■	Waved Albatross	*Phoebastria irrorata*	CR
	■	Black-footed Albatross	*Phoebastria nigripes*	EN
	■	Sooty Albatross	*Phoebetria fusca*	EN
	■	Indian Yellow-nosed Albatross	*Thalassarche carteri*	EN
	■	Atlantic Yellow-nosed Albatross	*Thalassarche chlororhynchos*	EN
	■	Gray-headed Albatross	*Thalassarche chrysostoma*	VU
	■	Chatham Albatross	*Thalassarche eremita*	VU
	■	Campbell Albatross	*Thalassarche impavida*	VU
	■	Black-browed Albatross	*Thalassarche melanophrys*	EN
	■	Salvin's Albatross	*Thalassarche salvini*	VU

RED LIST ■ EX = Extinct ■ EW = Extinct in the Wild ■ CR = Critically Endangered ■ EN = Endangered ■ VU = Vulnerable

PETRELS AND SHEARWATERS, STORM-PETRELS, AND DIVING-PETRELS

Petrels and shearwaters (family Procellariidae) are rather like smaller versions of albatrosses in overall shape. Like albatrosses they are mostly birds of the open seas and they use a similar flight technique, using their long, narrow wings to take advantage of the variations in wind speed close to and above the ocean waves. They use long glides between wing flaps to conserve energy as they scour the surface waters for food. Of the 80 species, 35 are listed as threatened.

Storm-petrels (family Hydrobatidae) are dainty ocean birds that flutter above the waves, plucking food such as plankton and small fish, from the surface of the sea or just below, sometimes following ships to feed on scraps. Storm-petrels are colonial breeders, usually on isolated islands that are free from ground predators. There are 22 species, four of which are considered in danger.

Diving-petrels (family Pelecanoididae) comprise a small family of only five species. One of these, the Peruvian Diving-petrel, is classified as Endangered. Dumpy, compact, and auklike, these small birds have the remarkable habit of flying straight into and through ocean waves and diving to catch their prey—which consist mainly of invertebrates such as krill. They nest in colonies, in burrows excavated in the soil.

Townsend's Shearwater
Puffinus auricularis

This medium-sized shearwater is endemic to Mexico where it now breeds only on a single island, Socorro, off the Pacific coast. Its coloration is mostly black above and white underneath, with a black half collar, under tail coverts and thighs, and white patches on the flanks. The face markings shade from black at the neck to a contrasting white area. The underside of the wing has black markings on the leading and trailing borders. The birds nest in rocky burrows at the edges of scrub and forest. They suffer predation from introduced cats and possibly also from rats, while sheep also trample and disturb their habitat in some places. The birds are also easily disturbed by light pollution and require very specific conservation measures to encourage them to nest. Careful habitat management is being developed, which includes the limiting of farming activities and controlling the numbers of predators on the island.

Family Procellariidae
Size 13 in (33 cm)
Status Critically Endangered (decreasing)
Population 250–950
Range Socorro (Revillagigedo Islands), Mexico (endemic)
Habitat Burrows in grassy slopes or rocky forest edge
Main threats Introduced predators (cats, rats), habitat destruction by pigs and sheep

Ashy Storm-petrel
Oceanodroma homochroa

A dainty, dark seabird with few distinguishing features, making it quite hard to identify in the field. The upper tail coverts have pale gray edges and the underwing displays a pale bar in flight. This storm-petrel breeds in burrows situated on a few rocky islands off the coasts of California and Mexico, with the main colonies on the South Farallon Islands and Channel Islands (California). At sea they often gather in large flocks—up to 6,000 have been noted in Monterey Bay in the fall. Their food consists mainly of plankton and small fish. In common with many other seabirds, they are vulnerable to oil spills and chemical pollution when feeding over the ocean. They are also easily disturbed at their nesting sites, and in some places are confused by lights from boats as they return to their colonies at dusk. Native predators include Burrowing and Tawny Owls and Western Gulls, all three posing threats in some colonies.

Family Hydrobatidae
Size 7¾ in (20 cm)
Status Endangered (decreasing)
Population 5,200–10,000
Range Coast of California, U.S.A., and Mexico
Habitat Nests in burrows on rocky islands
Main threats Introduced predators (rats), increase in native predators, pollution, disturbance to nesting sites

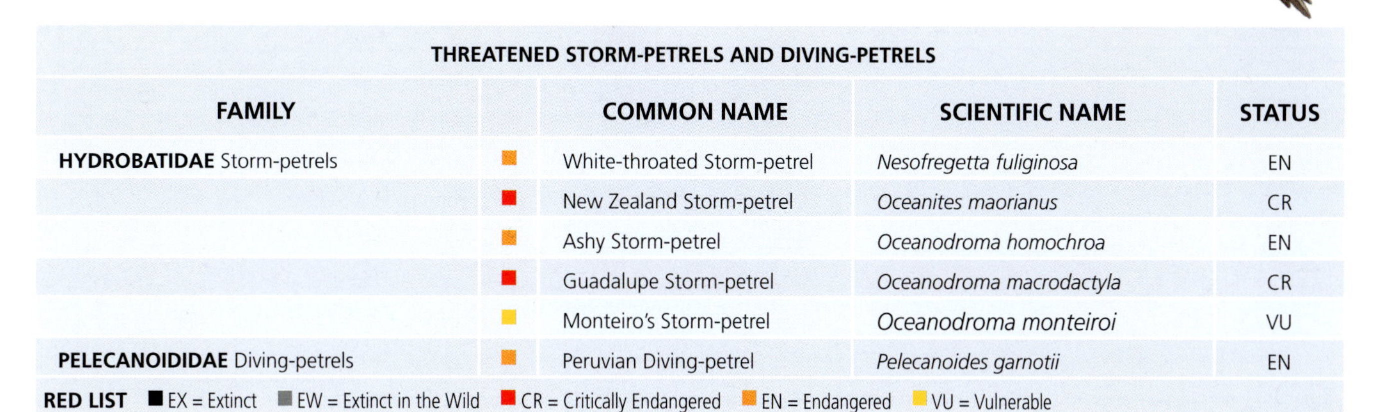

THREATENED STORM-PETRELS AND DIVING-PETRELS

FAMILY		COMMON NAME	SCIENTIFIC NAME	STATUS
HYDROBATIDAE Storm-petrels	■	White-throated Storm-petrel	*Nesofregetta fuliginosa*	EN
	■	New Zealand Storm-petrel	*Oceanites maorianus*	CR
	■	Ashy Storm-petrel	*Oceanodroma homochroa*	EN
	■	Guadalupe Storm-petrel	*Oceanodroma macrodactyla*	CR
	■	Monteiro's Storm-petrel	*Oceanodroma monteiroi*	VU
PELECANOIDIDAE Diving-petrels	■	Peruvian Diving-petrel	*Pelecanoides garnotii*	EN

RED LIST ■ EX = Extinct ■ EW = Extinct in the Wild ■ CR = Critically Endangered ■ EN = Endangered ■ VU = Vulnerable

Peruvian Diving-petrel

Pelecanoides garnotii

This small, dumpy seabird is black above and paler and dusky underneath with a brown face and sides of neck. Rather auklike in shape and behavior, diving-petrels fly fast and low over the sea with rapidly beating short wings. The Peruvian Diving-petrel lives on just four small islands off the coast of Chile and Peru, where it nests in deep burrows dug in guano, or sometimes in sandy soil or among large rock fissures. The burrows are used for nesting almost year round with probably two broods being reared. They feed on a range of prey including fish larvae, crustaceans, and small fish such as anchovies. Though the total population is quite high, it is listed as Endangered as the range is small and it seems to be in fairly rapid decline. It suffers predation from introduced animals including dogs, cats, rats, and foxes. The habitat is also threatened by harvesting the guano. They are sometimes caught in fishing nets, and overfishing reduces their food supply. The effects of El Niño on ocean currents cause periodic food scarcity.

Family Pelecanoididae
Size 8¾ in (22 cm)
Status Endangered (decreasing)
Population 25,000–28,000
Range Four small islands off Peru and Chile
Habitat Sand, rock fissures or guano burrows
Main threats Introduced predators (foxes, rats, cats, and dogs), guano collection, overfishing, accidental capture by fishing

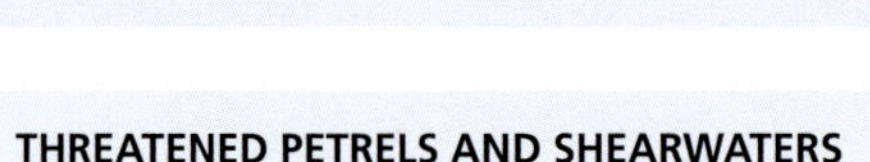

THREATENED PETRELS AND SHEARWATERS

FAMILY	COMMON NAME	SCIENTIFIC NAME	STATUS
PROCELLARIIDAE Petrels and shearwaters	White-chinned Petrel	*Procellaria aequinoctialis*	VU
	Spectacled Petrel	*Procellaria conspicillata*	VU
	Parkinson's Petrel	*Procellaria parkinsoni*	VU
	Westland Petrel	*Procellaria westlandica*	VU
	Mascarene Petrel	*Pseudobulweria aterrima*	CR
	Beck's Petrel	*Pseudobulweria becki*	CR
	Fiji Petrel	*Pseudobulweria macgillivrayi*	CR
	Phoenix Petrel	*Pterodroma alba*	EN
	Trindade Petrel	*Pterodroma arminjoniana*	VU
	Henderson Petrel	*Pterodroma atrata*	EN
	Chatham Petrel	*Pterodroma axillaris*	EN
	Barau's Petrel	*Pterodroma baraui*	EN
	Bermuda Petrel	*Pterodroma cahow*	EN
	Jamaica Petrel	*Pterodroma caribbaea*	CR
	White-necked Petrel	*Pterodroma cervicalis*	VU
	Cook's Petrel	*Pterodroma cookii*	VU
	De Filippi's Petrel	*Pterodroma defilippiana*	VU
	Juan Fernandez Petrel	*Pterodroma externa*	VU
	Black-capped Petrel	*Pterodroma hasitata*	EN
	Atlantic Petrel	*Pterodroma incerta*	EN
	Gould's Petrel	*Pterodroma leucoptera*	VU
	Stejneger's Petrel	*Pterodroma longirostris*	VU
	Zino's Petrel	*Pterodroma madeira*	EN
	Magenta Petrel	*Pterodroma magentae*	CR
	Galapagos Petrel	*Pterodroma phaeopygia*	CR

RED LIST EX = Extinct · EW = Extinct in the Wild · CR = Critically Endangered · EN = Endangered · VU = Vulnerable

THREATENED PETRELS AND SHEARWATERS *Continued*				
FAMILY		COMMON NAME	SCIENTIFIC NAME	STATUS
PROCELLARIIDAE Petrels and shearwaters	■	Pycroft's Petrel	*Pterodroma pycrofti*	VU
	■	Hawaiian Petrel	*Pterodroma sandwichensis*	VU
	■	Providence Petrel	*Pterodroma solandri*	VU
	■	Townsend's Shearwater	*Puffinus auricularis*	CR
	■	Buller's Shearwater	*Puffinus bulleri*	VU
	■	Pink-footed Shearwater	*Puffinus creatopus*	VU
	■	Heinroth's Shearwater	*Puffinus heinrothi*	VU
	■	Hutton's Shearwater	*Puffinus huttoni*	EN
	■	Balearic Shearwater	*Puffinus mauretanicus*	CR
	■	Newell's Shearwater	*Puffinus newelli*	EN

RED LIST ■ EX = Extinct ■ EW = Extinct in the Wild ■ CR = Critically Endangered ■ EN = Endangered ■ VU = Vulnerable

PELICANS, GANNETS AND BOOBIES, CORMORANTS, AND FRIGATEBIRDS

Pelicans (family Pelecanidae) are large waterbirds with broad, powerful wings and a long, straight bill with an unusual inflatable pouch that is attached to the underside of the lower mandible. Their amusing and unusual appearance has made them well known in folklore and fable. The elastic pouch helps them scoop up fish, retaining the catch until the water is drained out before swallowing. Most pelicans fish from the surface but one species, the Brown Pelican, dives for its prey. There are eight species, of which one, the Dalmatian Pelican, is listed as threatened (Vulnerable).

Gannets and boobies (family Sulidae) are elegant, powerful seabirds with long, daggerlike bills and streamlined bodies. They nest in (sometimes huge) colonies and soar over the sea searching for shoals of fish that they catch by diving. Of the ten species, Cape Gannet and Abbott's Booby are threatened.

Cormorants (family Phalacrocoracidae) are long-necked fish-eating birds with mainly dark plumage. They catch their prey by diving and swimming under water. They often stand with their wings stretched out to dry. There are 33 species of cormorant, of which ten are listed as threatened.

Frigatebirds (family Fregatidae) are impressive seabirds with long, angled wings and a long, forked tail. They are powerful and highly maneuverable in flight and often remain airborne for days at a time, sometimes using thermals to soar to a great height. Of the five species of frigatebird, two are threatened.

Abbott's Booby
Papasula abbotti

Distinctive with its black-and-white plumage, this booby is now found only on Christmas Island in the eastern Indian Ocean. It breeds at a slow pace, maturing at eight years, after which it raises one chick every other year and may live to 40 years. The slow reproductive rate may partly explain its dwindling population numbers, when combined with habitat changes. Abbott's Booby nests in tall trees in the rain forest, but patches of this forest have been cleared, mainly to assist mining activities. Alien plants have also escaped and established themselves, as has an introduced species of ant, the Yellow Crazy Ant (*Anoplolepis gracilipes*), which has established huge colonies, damaging trees and disrupting natural food chains by feeding on the local crabs. These ants may also attack young Abbott's Booby chicks, or their presence can cause the adult birds to abandon their nests. Extreme local weather with windy conditions also impacts on nesting birds, an effect enhanced by the removal of patches of forest. The Christmas Island National Park, established in 1980, offers improved protection for most of the birds.

Family Sulidae
Size 31 in (79 cm)
Status Endangered (decreasing)
Population 6,000
Range Christmas Island (endemic)
Habitat Nests in tall rain forest trees, forages at sea
Main threats Habitat destruction, introduced ants, mining activities, extreme local weather conditions

Flightless Cormorant
Phalacrocorax harrisi

The absence of mammalian predators at its remote location has allowed this species to abandon the power of flight, its wings degenerating into small, untidy vestiges. Strangely, it continues to indulge in the typical cormorant habit of standing with its wings outstretched. It is found only around the shores of Fernandina and Isabela in the Galápagos. Adults are blackish-brown above, paler brown below, with bright turquoise eyes; immatures are glossy black with dull brown eyes. They feed on fish, octopus, and eels, and build nests of seaweed on shingle or flat lava outcrops close to the shore. The size of population fluctuates considerably in response to El Niño events, sometimes with a loss of 50 percent; though numbers can recover quickly, the flightless nature of the birds makes them particularly vulnerable to other threats such as oil pollution, fishing, introduced predators, and volcanic eruptions. Though the birds are all within the Galápagos National Park and Marine Reserve there is a need for tighter restrictions and enforcement concerning net-fishing and hunting, and for continued control of cats on Isabela.

Family Phalacrocoracidae
Size 3 ft (95 cm)
Status Endangered (fluctuating)
Population 900
Range Fernandina and Isabela in the Galápagos Islands (endemic)
Habitat Sea and shoreline
Main threats Fishing, introduced predators, marine fluctuations

Christmas Frigatebird
Fregata andrewsi

Large and menacing both in its appearance and some of its feeding habits, the Christmas Frigatebird breeds only in three colonies, on the Australian territory of Christmas Island in the Indian Ocean. It shares the angular wings, deeply forked tail and long, hooked bill of all frigatebirds. The male is mainly iridescent black with a pale wingbar and a small white belly patch, which distinguishes it from other species. It has the family's typical bright red, balloonlike gular pouch which it inflates during courtship displays. The female is dark with a white breast and belly. These birds feed by harassing other seabirds in flight until they regurgitate their catch, or by picking fish and squid from the surface of the sea; they rarely land on the water and the waterproofing of their plumage is poor. Nesting in tall forest trees, each pair can only raise one fledgling every two years, so the recovery of the population after any decline is very slow. The breeding colonies have been reduced through loss of habitat to the phosphate mining industry, and with the majority of nests now located in just a single colony the species is very vulnerable to hurricanes. Away from Christmas Island there are threats to non-breeding birds from tree clearance and hunting on their roost islands.

Family Fregatidae
Size 3 ft (95 cm)
Status Critically Endangered (decreasing)
Population 2,400–4,800
Range Christmas Island (endemic)
Habitat Nests in tall forest trees, forages great distances at sea
Main threats Habitat destruction, fishing; hurricanes

THREATENED PELICANS, GANNETS AND BOOBIES, CORMORANTS, AND FRIGATEBIRDS

FAMILY		COMMON NAME	SCIENTIFIC NAME	STATUS
PELECANIDAE Pelicans	■	Dalmatian Pelican	*Pelecanus crispus*	VU
SULIDAE Gannets and boobies	■	Cape Gannet	*Morus capensis*	VU
	■	Abbott's Booby	*Papasula abbotti*	EN
PHALACROCORACIDAE Cormorants	■	Campbell Island Shag	*Phalacrocorax campbelli*	VU
	■	New Zealand King Shag	*Phalacrocorax carunculatus*	VU
	■	Stewart Island Shag	*Phalacrocorax chalconotus*	VU
	■	Auckland Islands Shag	*Phalacrocorax colensoi*	VU
	■	Pitt Island Shag	*Phalacrocorax featherstoni*	EN
	■	Flightless Cormorant	*Phalacrocorax harrisi*	EN
	■	Bank Cormorant	*Phalacrocorax neglectus*	EN
	■	Socotra Cormorant	*Phalacrocorax nigrogularis*	VU
	■	Chatham Islands Shag	*Phalacrocorax onslowi*	CR
	■	Bounty Islands Shag	*Phalacrocorax ranfurlyi*	VU
FREGATIDAE Frigatebirds	■	Christmas Frigatebird	*Fregata andrewsi*	CR
	■	Ascension Frigatebird	*Fregata aquila*	VU

RED LIST ■ EX = Extinct ■ EW = Extinct in the Wild ■ CR = Critically Endangered ■ EN = Endangered ■ VU = Vulnerable

HERONS AND EGRETS, STORKS, IBISES AND SPOONBILLS, SHOEBILL, AND FLAMINGOS

Herons and egrets (Ardeidae) are graceful waterbirds with a long, narrow neck, long bill, and long legs. Most herons have gray or dark plumage, while most egrets are mainly white. The family also includes the bitterns with their camouflaged plumage. They watch for fish, amphibians, and invertebrates in shallow water, stabbing swiftly when the prey comes within reach. They have broad, powerful wings and fly slowly, with their neck tucked in and long legs stretched out behind. The distribution of the family is global. There are 62 species, of which eight are threatened.

Storks (Ciconiidae) are also long-necked and long-legged, but are generally bulkier-bodied than herons and have a heavier bill. Unlike herons, nearly all stork species fly with their neck outstretched. The family is at its most diverse in tropical and warm temperate regions. Storks are strong fliers and their long, broad wings allow them to make good use of thermals, economizing by gliding for long periods, in the manner of vultures. Of the 19 species, five are threatened.

Ibises and spoonbills (Threskiornithidae) are mostly rather smaller than storks and their bills are either decurved (in ibises) or flattened and expanded at the tip (in spoonbills). Ibises use their long bill to probe for food in soft soil or mud. Spoonbills drag the bill tip from side to side across the water searching for invertebrates or small fish. Out of a total of 34 species, eight are threatened, and four of these are Critically Endangered.

The Shoebill (Balaenicipitidae) is the only member of its family. It is found in the wetlands of central and eastern Africa. It is gray, with a large head and an extraordinary shoe-shaped broad bill. It feeds mainly on fish, small mammals, reptiles, and amphibians. It is listed as Vulnerable.

Flamingos (Phoenicopteridae) are beautifully graceful, with very long legs and a long neck. Highly social birds, flamingos are often seen gathered together in large flocks at coastal lagoons or at inland salt or soda lakes. The flamingo has a complex, highly specialized bill and tongue that is used for filtering small invertebrates from saline water. The upper and lower mandibles of the bill contain fibrous lamellae that sieve food particles from the water pumped through the bill by the powerful muscular tongue. Of the six species, the Andean Flamingo is threatened, listed as Vulnerable.

Madagascar Heron

Ardea humbloti

Typically tall and stately, this heron is mostly shades of gray enlivened by its very large, pale yellow bill, which deepens when breeding to darker yellow or even orange. Once seen throughout Madagascar, it is now confined chiefly to suitable aquatic habitats in Western Madagascar, especially coastal mangroves and also lakes and rivers in the Melaky region. Usually seen alone or in small groups, this species sometimes mixes with Gray Herons and the two may even use the same nests at different times. The Madagascar Heron prefers unpolluted water and medium-sized prey (4–8 in;10–20 cm), especially fish, but also crabs, crayfish, shrimps, and similar crustaceans. However, suitable habitats have been greatly degraded or converted to human use, including shoreline tourist development and inland rice paddies. Other threats are clearance of nesting trees, and eggs and chicks collected by people for food. Conservation measures focus on rigorous protection of designated Ramsar Sites, including Lake Befotaka and Lake Ankerika.

Family	Ardeidae
Size	3¼–3½ ft (100–105 cm)
Status	Endangered (decreasing)
Population	1,400
Range	Madagascar (endemic)
Habitat	Coasts, freshwater wetlands
Main threats	Habitat conversion, human predation of eggs, chicks

Asian Crested Ibis

Nipponia nippon

With white plumage out of the breeding season, and upper parts graying in season, this ibis is named for its extravagant backswept crest; the red face and skin are also distinctive. Known locally as the toki, it has a wide freshwater-based diet including small fish, frogs, crabs, snails, also water beetles, their larvae, and other aquatic insects. Formerly widespread across East Asia, including Japan, the species was thought to be virtually extinct from the 1960s until just seven individuals were located in central China, in Shaanxi Province. With great conservation efforts, including several captive breeding centers in both China and Japan, their numbers have recovered to several hundred birds. However, habitat disturbance is a constant threat, especially drainage of natural wetlands and former rice paddies for dry crop production, and agrochemical water pollution. Despite legal protection, occasional shootings still occur. In 2008, 10 birds were released into its former range in Japan, and further reintroductions are planned.

Family	Threskiornithidae
Size	21¾–22½ in (55–57 cm)
Status	Endangered (increasing)
Population	400–500
Range	China (Japan)
Habitat	Mixed forest and freshwater wetlands
Main threats	Wetland drainage and pollution

Andean Flamingo

Phoenicoparrus andinus

The only flamingo with yellow feet and legs, the Andean Flamingo is otherwise typically pinkish-white, with a yellow black-tipped bill and a small patch of black primaries. In its remote high habitat, around 6,600–14,440 ft (2,000–4,400 m), it relies for food on tiny organisms such as diatoms and cyanobacteria (blue-green algae) that flourish in salt lakes. The flamingo filters these from the water using the comblike sieving apparatus in its deep, narrow bill. Colonial nesting was one factor in its drastic decline during the mid-20th century, with people trekking to the lakes to collect thousands of eggs as food. Birds were also trapped or shot for their flesh and highly decorative feathers. These threats are being tackled by setting aside protected areas such as Salinas and Aguada Blanca Nature Reserve (Peru), Salar de Atacama National Flamingo Reserve (Chile), and Las Chinchillas Provincial Natural Reserve (Argentina). However, a possible increase in mining for minerals such as salt and borax, and general human disturbance, represent potentially increasing threats.

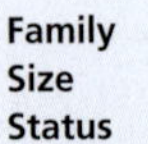

Family	Phoenicopteridae
Size	3¼–3½ ft (100–110 cm)
Status	Vulnerable (decreasing)
Population	32,000–35,000
Range	Bolivia, Peru, Chile, Argentina
Habitat	High mountain salt lakes
Main threats	Water abstraction, mining, hunting

THREATENED HERONS AND EGRETS, STORKS, IBISES AND SPOONBILLS, SHOEBILL, AND FLAMINGOS

FAMILY	COMMON NAME	SCIENTIFIC NAME	STATUS
ARDEIDAE Herons and egrets	Madagascar Heron	*Ardea humbloti*	EN
	White-bellied Heron	*Ardea insignis*	CR
	Madagascar Pond-heron	*Ardeola idae*	EN
	Australasian Bittern	*Botaurus poiciloptilus*	EN
	Chinese Egret	*Egretta eulophotes*	VU
	Slaty Egret	*Egretta vinaceigula*	VU
	Japanese Night-heron	*Gorsachius goisagi*	EN
	White-eared Night-heron	*Gorsachius magnificus*	EN
CICONIIDAE Storks	Oriental Stork	*Ciconia boyciana*	EN
	Storm's Stork	*Ciconia stormi*	EN
	Greater Adjutant	*Leptoptilos dubius*	EN
	Lesser Adjutant	*Leptoptilos javanicus*	VU
	Milky Stork	*Mycteria cinerea*	VU
THRESKIORNITHIDAE Ibises and spoonbills	Dwarf Olive Ibis	*Bostrychia bocagei*	CR
	Southern Bald Ibis	*Geronticus calvus*	VU
	Northern Bald Ibis	*Geronticus eremita*	CR
	Asian Crested Ibis	*Nipponia nippon*	EN
	Black-faced Spoonbill	*Platalea minor*	EN
	White-shouldered Ibis	*Pseudibis davisoni*	CR
	Giant Ibis	*Thaumatibis gigantea*	CR
	Madagascar Sacred Ibis	*Threskiornis bernieri*	EN
BALAENICIPITIDAE Shoebill	Shoebill	*Balaeniceps rex*	VU
PHOENICOPTERIDAE Flamingos	Andean Flamingo	*Phoenicoparrus andinus*	VU

RED LIST EX = Extinct EW = Extinct in the Wild CR = Critically Endangered EN = Endangered VU = Vulnerable

Ducks, geese, and swans

Ducks, geese and, swans (Anatidae), often known as waterfowl or wildfowl, are waterbirds that are found across the globe with the exception of the Antarctic. They number some 150 species, 28 of which are threatened, including six that are Critically Endangered. They are adapted to a range of aquatic habitats, including large lakes, seas, rivers, and other wetlands, and the majority breed on or close to fresh water. The typical member of this family has a buoyant body with waterproof feathers and webbed feet—adaptations for efficient swimming. Many species nest on the ground, which makes them vulnerable to disturbance or predation, and pollution of rivers, lakes, and the sea has also taken its toll. In addition, wildfowl have long been traditionally shot, either for food or for sport, and in many countries this has also had an impact on the rarer species. While some species, such as most swans, are mainly white, most ducks and geese have dark plumage, while some, especially the male birds, have bright breeding plumage. This striking male plumage is shed in most species after breeding when the male looks more like the drabber female in what is known as "eclipse" plumage. Many species are kept in zoos and other collections, partly for their attractive plumage, and some, such as the Mallard, have been domesticated for their eggs or meat. Artificial selection has produced variations in plumage in domesticated forms. Most waterfowl have an elongated neck and a strong bill that is somewhat flattened, adaptations that have evolved for sifting edible matter from the water, or tearing at grasses and other plants. While most ducks and swans feed on or below the water surface, geese typically feed in flocks on grassland or coastal saltmarshes. Dabbling ducks such as wigeons and Mallard, and swans often "up-end" to stretch down to reach plant or animal food in shallow water. Diving ducks such as pochards and Tufted Duck feed mainly by diving, often to a depth of several meters.

Many species in the family Anatidae—most notably geese and swans—are migratory, regularly flying long distances to their winter feeding grounds such as estuaries, mudflats, and coastal grassland, habitats that have often been subject to drainage and conversion to farmland. Waterfowl have muscular, aerodynamically shaped bodies and large, powerful wings, and are capable of sustained, rapid flight. Swans are some of the heaviest of all flying birds and require a long run for take off, usually pattering across the surface of the water before becoming airborne. By contrast, ducks often take off with a single spring or just a few short steps.

Red-breasted Goose

Branta ruficollis

In Russia's northernmost peninsulas (Taimyr, Gydan, and Yamal), this goose arrives at its breeding grounds of sparsely wooded, scrubby, or fully treeless tundra in April and May. Its main foods here are shoots, buds, and leaves. It nests on the ground, often close to nests of predatory birds such as Snowy Owls and Peregrines whose presence deters other predators such as the Arctic Fox (*Valpes lagopus*). From September the geese return to wintering areas, mainly along the western Black Sea, though some venture as far south as Greece and occasionally turn up in Western Europe. Here they feed on saltmarsh and lagoon vegetation and also on cultivated grasses such as wheat and barley. Replacement of these edible crops with others, as well as general tourist development and housing, has greatly affected the species. A small and neatly marked bird, this goose's chestnut breast extends to the front of the neck and sides of the face, bordered by white lines and patches, with a mainly black body, and wings enlivened by white flank and belly flashes.

Family	Anatidae
Size	21¾ in (55 cm)
Status	Endangered (decreasing)
Population	37,000
Range	Russia (breeds), Bulgaria, Romania (winters)
Habitat	Tundra, subarctic scrub, wetlands
Main threats	Changing agriculture, tourism, wildfowling

Hawaiian Goose

Branta sandvicensis

Reduced to some 30 individuals by the 1950s, the Hawaiian Goose—also known by its local name Néné—has responded to intensive conservation efforts. Protected reserves have been set aside in its native Hawaii and it is also captive-bred at several centers around the world. The species is now relatively well established on the islands of Hawaii, Maui, Molokai, and Kauai. Their total wild populations of up to 1,000 are currently complemented and maintained by about the same number in zoos and parks, famously at Slimbridge, Gloucestershire, England, which spearheaded the initial breeding for reintroduction. The goose has an unusual neck pattern of buff and dark brown, looking vaguely like a plowed field; the bill and head are black, the cheeks faintly golden-brown, and the rest of the body "encrusted" with browns and grays. Often trusting and hand-tame even when wild, the Hawaiian Goose has suffered greatly from introduced species such as cats, dogs, pigs and especially the Indian Mongoose (*Herpestes javanicus*).

Family	Anatidae
Size	23½–25½ in (60–65 cm)
Status	Vulnerable (increasing)
Population	500–1,000
Range	Hawaiian Islands (endemic)
Habitat	Grassland, rocky scrub
Main threats	Habitat loss to agriculture, alien predation

Marbled Teal

Marmaronetta angustirostris

The Marbled Teal has a disjunct distribution from the western Mediterranean to West China. It also has unpredictable habits, making assessment of its population notoriously difficult. There is a general fall migration from more northerly parts of the range—for example, Spain and Turkey—southward, perhaps to sub-Saharan Sahel and Egypt, and back again the following spring. But this is complicated by small-scale wanderings as conditions and food opportunities dictate. Breeding sites and success are similarly haphazard. This small dabbling duck is relatively muted, flecked in shades of gray, brown, and cream, with a dark shadowy eye stripe. It has a wide diet, from midges, other flies and other insects in summer to seeds and shoots in winter, especially from marsh plants such as reeds and sedges. Many of its wetland habitats have been affected by water abstraction, conversion to cropland, domestic stock grazing, and water pollution.

Family Anatidae
Size 15¾ in (40 cm)
Status Vulnerable (decreasing)
Population 14,000–26,000
Range Spain, North Africa, Turkey, east to China
Habitat Wetlands, especially brackish
Main threats Drainage, pollution, grazing, wildfowling

THREATENED DUCKS, GEESE, AND SWANS

FAMILY		COMMON NAME	SCIENTIFIC NAME	STATUS
ANATIDAE Ducks, geese, and swans	■	Auckland Islands Teal	*Anas aucklandica*	VU
	■	Madagascar Teal	*Anas bernieri*	EN
	■	Brown Teal	*Anas chlorotis*	EN
	■	Eaton's Pintail	*Anas eatoni*	VU
	■	Baikal Teal	*Anas formosa*	VU
	■	Laysan Duck	*Anas laysanensis*	CR
	■	Philippine Duck	*Anas luzonica*	VU
	■	Meller's Duck	*Anas melleri*	EN
	■	Campbell Islands Teal	*Anas nesiotis*	CR
	■	Hawaiian Duck	*Anas wyvilliana*	EN
	■	Swan Goose	*Anser cygnoides*	VU
	■	Lesser White-fronted Goose	*Anser erythropus*	VU
	■	Baer's Pochard	*Aythya baeri*	EN
	■	Madagascar Pochard	*Aythya innotata*	CR
	■	Red-breasted Goose	*Branta ruficollis*	EN
	■	Hawaiian Goose	*Branta sandvicensis*	VU
	■	White-winged Duck	*Cairina scutulata*	EN
	■	Blue-winged Goose	*Cyanochen cyanoptera*	VU
	■	West Indian Whistling-duck	*Dendrocygna arborea*	VU
	■	Blue Duck	*Hymenolaimus malacorhynchos*	EN
	■	Marbled Teal	*Marmaronetta angustirostris*	VU
	■	Brazilian Merganser	*Mergus octosetaceus*	CR
	■	Scaly-sided Merganser	*Mergus squamatus*	EN
	■	White-headed Duck	*Oxyura leucocephala*	EN

RED LIST ■ EX = Extinct ■ EW = Extinct in the Wild ■ CR = Critically Endangered ■ EN = Endangered ■ VU = Vulnerable

THREATENED DUCKS, GEESE, AND SWANS *Continued*				
FAMILY		**COMMON NAME**	**SCIENTIFIC NAME**	**STATUS**
ANATIDAE Ducks, geese, and swans	■	Steller's Eider	*Polysticta stelleri*	VU
	■	Pink-headed Duck	*Rhodonessa caryophyllacea*	CR
	■	Salvadori's Teal	*Salvadorina waigiuensis*	VU
	■	Crested Shelduck	*Tadorna cristata*	CR

RED LIST ■ EX = Extinct ■ EW = Extinct in the Wild ■ CR = Critically Endangered ■ EN = Endangered ■ VU = Vulnerable

NEW WORLD VULTURES AND OLD WORLD VULTURES

Both the New World and Old World vultures are large birds that feed mainly on carrion, gliding and soaring for long periods on long, broad wings, making full use of thermals to conserve energy as they scour the ground below for food. Superficially similar, the two groups are not, however, closely related, and recent genetic research implies that the former may in fact be closer to storks. The New World vultures belong to a small family (Cathartidae) of just seven species, including the common rather small Turkey Vulture and the huge condors, one of which, the Andean Condor, is Near Threatened, while the California Condor is Critically Endangered. The California Condor has been rescued by a long and painstaking program of captive breeding and release into suitable habitats. The Andean Condor has the largest wings (in area) of any bird. The Turkey Vulture has a wide range in North and South America and is sometimes confusingly referred to as a buzzard.

The Old World vultures are classified with kites, harriers, hawks, and eagles as part of a large family of raptors (Accipitridae). Eight species of Old World vulture are threatened, of which one, the Egyptian Vulture, is Endangered and four are Critically Endangered. A major threat to the Old World vultures, especially in India and Africa, is the use by farmers of the veterinary drugs diclofenac and ketoprofen. These chemicals accumulate in the vultures if they feed on carcasses that have been treated—resulting in lethal results for the birds. In India, vultures have suffered a catastrophic decline from this one threat alone. Efforts are being made to ban these drugs and replace them with less damaging alternatives such as meloxicam (currently the only drug of the same group known to be safe).

California Condor

Gymnogyps californianus

One of the world's largest and rarest birds, this condor is the beneficiary of one of the most expensive conservation projects ever. Rescued from the brink in 1987, when all surviving 27 birds were taken into captivity amid great controversy, by 2010 its population approached 400, with about 187 living wild. A vast bird with a wingspan approaching 10 ft (3 m), it has the vulture's typical almost-naked head and upper neck, brownish-orange in hue, with otherwise mainly black plumage showing white on the underwings. Its chief food is animal carcasses, varying from voles, rabbits, and birds up to deer and sheep, and including marine mammals. Several condors have died from eating rubbish such as gun cartridge casings. An important conservation measure is banning lead shot from its range, since it ingests pellets while scavenging and accumulates this heavy metal toxin over many years. It is one of the longest-lived of all birds—50 plus years is not unknown.

Family	Cathartidae
Size	4–4¼ ft (120–130 cm)
Status	Critically Endangered (increasing)
Population	187
Range	U.S.A., (California), Mexico
Habitat	Rocky upland scrub and forest
Main threats	Lead poisoning from carcasses, trash ingestion

Egyptian Vulture
Neophron percnopterus

Jostling with other scavengers at a dusty carcass often turns this vulture's mainly white plumage to shades of brown or gray. However, the black edges to the wings, lower rear body, rump, and tail are still discernible. The bill is long, hooked, all-yellow with a black tip (in the Indian subspecies the bill is completely yellow), and the naked facial skin is yellow or orange. With a wingspan of 5½ ft (1.7 m), this medium-sized vulture may have to wait its turn at corpses, but it has an infamously wide diet ranging from insects, lizards, and mice to well-rotted flesh and fecal matter, not only of animals but from humans too.

Shotgun lead pellets in carcasses pose a serious threat, as do veterinary antibiotic and anti-inflammatory drugs in farm stock. On a smaller scale, its bones and feathers are used in traditional medicines in North Africa and India. This bird is one of the very few avian tool-users, with the ability to smash other birds' eggs to feed on their contents using a stone held in the beak.

Family	Accipitridae
Size	21¾–25½ in (55–65 cm)
Status	Endangered (decreasing)
Population	21,000–67,000
Range	Western Europe, Africa, east to India
Habitat	Rocky upland, scrub, open forest
Main threats	Poisoning, competition for nest site

THREATENED NEW WORLD AND OLD WORLD VULTURES

FAMILY		COMMON NAME	SCIENTIFIC NAME	STATUS
CATHARTIDAE New World vultures	■	California Condor	*Gymnogyps californianus*	CR
ACCIPITRIDAE Old World vultures	■	White-rumped Vulture	*Gyps bengalensis*	CR
	■	Cape Vulture	*Gyps coprotheres*	VU
	■	Indian Vulture	*Gyps indicus*	CR
	■	Slender-billed Vulture	*Gyps tenuirostris*	CR
	■	Lappet-faced Vulture	*Torgos tracheliotos*	VU
	■	White-headed Vulture	*Trigonoceps occipitalis*	VU
	■	Egyptian Vulture	*Neophron percnopterus*	EN
	■	Red-headed Vulture	*Sarcogyps calvus*	CR

RED LIST ■ EX = Extinct ■ EW = Extinct in the Wild ■ CR = Critically Endangered ■ EN = Endangered ■ VU = Vulnerable

KITES, HARRIERS, HONEY-BUZZARDS, AND FALCONS

Kites, harriers, and honey-buzzards are medium-sized raptors that are rather buoyant in flight (especially in the first two groups). Some species are very specialized feeders. The Snail-kite, for example, feeds largely on particular freshwater snails, while the European Honey-buzzard includes wasp grubs in its diet. Most harriers are associated with open, often marshy, country and take small birds and small rodents. Harriers patrol low over the ground in buoyant, gliding flight with their wings tilted upward in a shallow "V" shape.

The falcon family (Falconidae) also contains the caracaras of Central and South America. None of the ten species of caracaras is threatened, though one, the Striated Caracara of southern Argentina and Chile and the Falkland Islands, is listed as Near Threatened. Of the 38 species of true falcon (genus *Falco*), four species (Lesser, Mauritius, and Seychelles Kestrels and Saker Falcon) are Vulnerable. The falcons also include eight species of falconet (genera *Microhierax*, *Polihierax*, and *Spiziapteryx*), the forest-falcons (seven species of *Micrastur*, one of which is Vulnerable), and the mainly snake-eating Laughing Falcon.

Threats to these groups come mainly from habitat loss. However, falcons in particular suffer from accumulation of poisons when their prey species (which tend to comprise mainly smaller birds) have ingested pesticides contained in their food. DDT and other chlorinated hydrocarbons that concentrated in the bodies of falcons such as the Peregrine caused major population crashes, from which the species has since recovered well following restrictions in the use of these pesticides.

White-collared Kite
Leptodon forbesi

Clearance of Brazil's Atlantic tropical forests, now reduced to less than one-hundredth of their former extent within the range of this species, has devastated many species. These include the celebrated Golden Lion Tamarin (*Leontopithecus rosalia*), a small monkey, and the White-collared Kite, a large yet elusive raptor. This bird is a striking "monochrome" black-and-white bird with white head topped by a faint pearly crown, white body, black back and upper wings, and a white tail with a broad black band near the tip. The white leading edges to the wings are diagnostic. In the air its broad wings are very apparent and it has a butterfly-like display flight. Its feeding and breeding habits are largely unknown, though they may follow those of the Gray-headed Kite, a species that is very similar in appearance, but much more widespread. This bird watches for prey from a high branch and swoops to feed on lizards, snakes and other reptiles, frogs and other amphibians, and arthropods such as crickets, beetles, and other insects, as well as spiders. The White-collared Kite population is reduced to an estimated size of only 50 breeding pairs, and these occur mainly in Pernambuco and Alagoas, northeast Brazil, though it is possible that unsurveyed areas of nearby Brazilian states, such as Paraíba and Sergipe, harbor small groups. Surveys of such areas are a priority if this unusual kite is to be saved. Though some of the forest tracts where this rare species occurs are protected, many of their habitats are affected by small-scale logging and fires.

Family	Accipitridae
Size	20½ in (52 cm)
Status	Critically Endangered (decreasing)
Population	50–250
Range	Brazil (endemic)
Habitat	Moist tropical forest
Main threats	Deforestation, fire

Mauritius Kestrel
Falco punctatus

This small, neat raptor—with mainly chestnut or brown upperparts and white underparts, all with small black patches, blotches, and crescents—came back from virtual extinction, with just four birds left in the mid-1970s. Following a massive captive breeding effort, particularly involving the Durrell Wildlife Conservation Trust (Jersey Zoo), the Mauritius Kestrel now numbers many hundreds. It has re-established not only in Mauritius' few remaining native subtropical forests, but also in more open scrubby, areas. It is found chiefly in the uplands and mountains of the southwest, also less commonly in the north and east. Strenuous and continued conservation efforts involving provision of nest boxes, guarding of nest sites, supplementary feeding, and reintroduction have led to a steadily increasing population, but the island is not thought able to support more than about 1,000 birds. The main prey of this rather dainty kestrel are lizards, including the island's day geckoes, small rodents, birds, and large insects. Significant in its demise was deforestation by early colonizers, followed later by poisoning and egg-thinning due to DDT and similar pesticides used during the 1950s and 1960s. The current populations are self-sustaining, but ongoing threats include alien species, especially black rats, Crab-eating Macaques (*Macaca fuscata*), Indian Mongooses (*Herpestes javaninus*), and domestic cats,that eat the eggs and young. A forest in the southeast, around the Ferney Valley, holds about half the global population.

Family	Falconidae
Size	8¾–10 in (20–26 cm)
Status	Vulnerable (increasing)
Population	800–1,000
Range	Mauritius (endemic)
Habitat	Forests, open woodland
Main threats	Introduced predators

THREATENED KITES, HARRIERS, HONEY-BUZZARDS, AND FALCONS

FAMILY		COMMON NAME	SCIENTIFIC NAME	STATUS
ACCIPITRIDAE Kites and harriers	■	Cuban Kite	*Chondrohierax wilsonii*	CR
	■	White-collared Kite	*Leptodon forbesi*	CR
	■	Madagascar Harrier	*Circus macrosceles*	VU
	■	Reunion Harrier	*Circus maillardi*	EN
	■	Black Harrier	*Circus maurus*	VU
ACCIPITRIDAE Honey-buzzards	■	Black Honey-buzzard	*Henicopernis infuscatus*	VU
FALCONIDAE Falcons	■	Seychelles Kestrel	*Falco araea*	VU
	■	Saker Falcon	*Falco cherrug*	VU
	■	Lesser Kestrel	*Falco naumanni*	VU
	■	Mauritius Kestrel	*Falco punctatus*	VU
	■	Plumbeous Forest-falcon	*Micrastur plumbeus*	VU

RED LIST ■ EX = Extinct ■ EW = Extinct in the Wild ■ CR = Critically Endangered ■ EN = Endangered ■ VU = Vulnerable

Hawks

The true hawks (genus *Accipiter*) number about 45 species, most of which are known as sparrowhawks or goshawks. These are agile hunters, mainly taking other birds in flight. They have fairly broad, rounded wings (compared with falcons) and a long, flexible tail. They often soar in circles before diving and chasing their prey between the trees. They pluck their prey before eating it. A number of other genera are commonly called hawks and these include *Buteo*, also known as "buzzards" in Europe (confusingly, the Turkey Vulture of the Americas is also often referred to as a buzzard). These are heavier-bodied than the true hawks and many feed on larger prey such as rabbits and other mammals, though their diets are very varied. There are ten species of *Leucopternis*—forest hawks of Central and South America. Also Neotropical are the "black-hawks" of the genus *Buteogallus*, with five species.

Most hawks are birds of woodland or forests and they have suffered through loss or degradation of these habitats. Other threats to hawks include poisoning and many are shot as a perceived danger to poultry. Eleven species of hawk are currently listed as threatened. Eight of these are Vulnerable, two (Gundlach's Hawk and Gray-backed Hawk) are Endangered, and one (Ridgway's Hawk) is Critically Endangered.

Gundlach's Hawk

Accipiter gundlachi

Preying on poultry and other domestic birds means this hawk is persecuted by trapping, shooting, and poisoning, despite its precarious situation. Loss of forest habitat for feeding and breeding is another chief threat, as trees are felled for timber and fuel, and land turned to agriculture. Gundlach's Hawk prefers dry-to-moist lowland forests, being found at altitudes up to 2,600 ft (800 m). It is strongly built with a dark cap, narrow white eyebrow stripe extending from the bill base to above the eye, gray-blue upperparts, mostly rusty underside with darker bars, yellow legs and barred, rounded tail. With only five population areas left, its main conservation hopes lie with Cuban National Parks such as Sierra del Cristal and Sierra Maestra, both in the southeast. Further opportunities for protection include the Zapata Biosphere Reserve (Zapata Swamp, Ramsar Site) in the southwest.

Family	Accipitridae
Size	17¾–19¾ in (45–50 cm)
Status	Endangered (decreasing)
Population	300–400
Range	Cuba (endemic)
Habitat	Woods, forests
Main threats	Habitat loss, human persecution

Gray-backed Hawk

Leucopternis occidentalis

This distinctive hawk has a white-streaked gray head and nape, black wings, white underparts, a white tail with a broad band near the tip, and yellow legs. A forest-dweller, preferring damp, evergreen forests, it is also found in drier deciduous woodlands, at elevations from 300 ft (100 m) to 5,000 ft (1,500 m), and rarely up to twice this altitude. Important prey include rodents, small birds, and reptiles such as lizards and snakes, as well as crabs and crayfish. The Gray-backed Hawk's main downfall is the epidemic of deforestation for timber logging, clearance for crops such as oil palm and for livestock grazing, as well as road building, settlements, and mining and industry. Some of this occurs illegally in protected areas such as the part-coastal Machalilla National Park, West Ecuador, which is also experiencing tourist development, and the Mache-Chindul Ecological Reserve in the west-center.

Family	Accipitridae
Size	17¾–19 in (45–48 cm)
Status	Endangered (decreasing)
Population	250–1,000
Range	Ecuador, Peru
Habitat	Forests
Main threats	Deforestation

THREATENED HAWKS

FAMILY		COMMON NAME	SCIENTIFIC NAME	STATUS
ACCIPITRIDAE Hawks	■	New Britain Sparrowhawk	*Accipiter brachyurus*	VU
	■	Nicobar Sparrowhawk	*Accipiter butleri*	VU
	■	Gundlach's Hawk	*Accipiter gundlachi*	EN
	■	Imitator Sparrowhawk	*Accipiter imitator*	VU
	■	Slaty-mantled Sparrowhawk	*Accipiter luteoschistaceus*	VU
	■	New Britain Goshawk	*Accipiter princeps*	VU
	■	Galapagos Hawk	*Buteo galapagoensis*	VU
	■	Ridgway's Hawk	*Buteo ridgwayi*	CR
	■	Red Goshawk	*Erythrotriorchis radiatus*	VU
	■	White-necked Hawk	*Leucopternis lacernulatus*	VU
	■	Gray-backed Hawk	*Leucopternis occidentalis*	EN

RED LIST ■ EX = Extinct ■ EW = Extinct in the Wild ■ CR = Critically Endangered ■ EN = Endangered ■ VU = Vulnerable

Eagles

As with the hawks, a number of different genera are commonly referred to as "eagles." In general these are some of the largest and most powerful members of the family. The true eagles are classified in the genus *Aquila*, of which there are 12 species. These are large raptors that combine effortless soaring with energetic hunting and can tackle large prey. Four *Aquila* species are Vulnerable: Indian Spotted Eagle, Greater Spotted Eagle, and both Spanish and Eastern Imperial Eagles. One of the largest and most widespread of the "true eagles" is the Golden Eagle, with a range from Europe through Asia to North America. Named for the golden sheen to its head and neck feathers, this bold hunter is found in varied habitats, often in mountainous country. Also large, and in some cases even larger, are the "sea-eagles" or "fish-eagles," *Haliaeetus* (eight species) and *Ichthyophaga* (two species). As their names suggest, these eagles soar over the sea and include fish in their diet, though most do not have an exclusive diet and also eat other prey and carrion. Three other large and powerful eagles are found in the tropics. The Harpy Eagle lives in the rain forests of South America. It uses its muscular legs and large talons to catch prey such as sloths, monkeys, and even deer. Africa's most powerful eagle is the magnificent Martial Eagle, that hunts by soaring over savanna and woodland, then plunging onto its prey—mainly birds, and mammals up to the size of small antelopes. Also large and powerful is the Philippine Eagle, a rare forest predator with a preference for flying lemurs, civets, and monkeys.

The "snake-eagles" (*Circaetus*, six species) and "serpent-eagles" (*Spilornis*, six species; *Dryotriorchis* and *Eutriorchis*, one species each) mostly hunt reptiles, especially snakes, caught mainly in open country. After a successful hunt they can often be seen with a snake dangling down, held securely in their talons. The other main group of eagles are the "hawk-eagles." These are mainly small or medium-sized eagles. The main genera are *Spizaetus* (13 species), tropical forest eagles found in Southeast Asia and Indonesia, and *Hieraaetus* (seven species), with a distribution from southern Europe and Africa to Asia. Hawk-eagles mainly live in wooded habitats. They are agile hunters that catch other birds or small mammals in and around forest or scrub.

Of the 67 species of eagle, 13 are listed as Vulnerable, three (Madagascar Serpent-eagle, Crowned Eagle, and Javan Hawk-eagle) are Endangered, and a further three (Madagascar Fish-eagle, Philippine Eagle, and Flores Hawk-eagle) are Critically Endangered.

Spanish Imperial Eagle

Aquila adalberti

Formerly classified as a subspecies of the Eastern Imperial Eagle (also classed as Vulnerable), this majestic raptor has benefited from wide-ranging conservation measures, from altering electricity pylons and lines to prevent collision and electrocution, to supplemental feeding and nest guarding. The species' abundance and range have been greatly affected by major declines in rabbit numbers. This is partly due to myxomatosis and also to changing practices: former rough land has been put to livestock grazing, while hunting estate management has moved toward larger inhabitants such as boar and deer, and both of these entail reducing the rabbit and small gamebird populations that are the eagle's dietary staples. The adult plumage is flecked browns and blacks, with whiter patches on the shoulders and scapulars, and a pale neck. From only 30 breeding pairs in the 1960s, numbers are approaching 200 pairs. Its strongholds are in central and southern Spain, with recent expansion into adjacent Portugal. However, deliberate and accidental poisoning (pesticides, lead shot), and lack of suitable nesting trees, are continuing problems for this bird.

Family	Accipitridae
Size	29½–33 in (75–84 cm)
Status	Vulnerable (increasing)
Population	300–400
Range	Spain; Portugal
Habitat	Varied, coastal marshes to mountains
Main threats	Habitat conversion, poisoning, powerlines

Crowned Eagle

Harpyhaliaetus coronatus

The crest at the back of its head gives this South American eagle its common name. Its plumage is shades of mid-gray, with slightly paler shoulders and forehead, a darker tail with a white band and white tip, and yellowish bill and legs. The Crowned Eagle has a wide habitat tolerance, from the hot, arid *campo cerrado* (mainly grasses and low shrubs, with scattered trees) to swamps and upland *caatinga* (dry, thorny forest) and gallery woodlands. Its diet is similarly varied, from small rodents to medium-sized birds, snakes, and lizards, and fish, crabs, and other aquatic life. Many of its favored lowland areas are rapidly being converted to farm crops or cattle ranches, along with the accompanying roads, settlements, and industries. Some persecution occurs from hunting and poisoning, even though this is illegal. Conservation efforts focus on raising awareness and also establishing "stepping stone" protected areas and corridors between the currently highly fragmented reserves and national parks.

Family	Accipitridae
Size	29½–33¼ in (75–85 cm)
Status	Endangered (decreasing)
Population	250–1,000
Range	Brazil, Bolivia, Paraguay, Argentina
Habitat	Lowland savanna, woodland
Main threats	Habitat loss to agriculture, hunting

Philippine Eagle

Pithecophaga jefferyi

Sometimes called the monkey-eating eagle, this huge bird is certainly capable of plucking smaller monkeys from branches. However, its dietary mainstay in many areas are colugos ("flying lemurs"), but it can adapt to other prey, varying from tree rats, arboreal snakes, and lizards to medium-sized birds, including parrots, hornbills, owls, and other birds of prey. Its streaked pale-and-dark-brown shaggy mane over the head and neck gives it a fierce appearance; the rest of the upperparts are dark brown, with pale to white on the underbody and underwings. This eagle is celebrated as one of the largest of all raptors, with a weight of up to 15½ lb (7 kg) and a 6½ ft (2 m) wingspan. Such record measurements have made the species valuable for its feathers, skins, and as an exotic but illegal captive—problems that continue despite law enforcement. Its favored habitat is primary tropical forest, usually in steep terrain. Loss of forest habitat is the other main threat, through extraction of timber and clearance for shifting cultivation. These massive birds require enormous hunting territories and they also breed slowly, producing just a single chick every two years. Even in an ideal habitat with good protection, repopulation is therefore a slow process. Endemic to the Philippines, this eagle is known to survive only on Mindanao, Samar, Leyte, and Luzon. Mindanao has the bulk of the population with perhaps 340 pairs. A captive breeding and reintroduction program has been established, along with education initiatives aimed at promoting this magnificent eagle, which is the national bird of the Philippines.

Family Accipitridae
Size 33¾–40 in (86–102 cm)
Status Critically Endangered (decreasing)
Population 180–500
Range Philippines (endemic)
Habitat Forests up to 6,600 ft (2,000 m)
Main threats Forest destruction and fragmentation, hunting

THREATENED EAGLES

FAMILY		COMMON NAME	SCIENTIFIC NAME	STATUS
ACCIPITRIDAE Eagles	■	Spanish Imperial Eagle	*Aquila adalberti*	VU
	■	Greater Spotted Eagle	*Aquila clanga*	VU
	■	Indian Spotted Eagle	*Aquila hastata*	VU
	■	Eastern Imperial Eagle	*Aquila heliaca*	VU
	■	Crowned Eagle	*Harpyhaliaetus coronatus*	EN
	■	New Guinea Eagle	*Harpyopsis novaeguineae*	VU
	■	Philippine Eagle	*Pithecophaga jefferyi*	CR
ACCIPITRIDAE Snake-eagles	■	Beaudouin's Snake-eagle	*Circaetus beaudouini*	VU
ACCIPITRIDAE Serpent-eagles	■	Madagascar Serpent-eagle	*Eutriorchis astur*	EN
	■	Mountain Serpent-eagle	*Spilornis kinabaluensis*	VU
ACCIPITRIDAE Fish-eagles	■	Pallas's Fish-eagle	*Haliaeetus leucoryphus*	VU
	■	Steller's Sea-eagle	*Haliaeetus pelagicus*	VU
	■	Sanford's Sea-eagle	*Haliaeetus sanfordi*	VU
	■	Madagascar Fish-eagle	*Haliaeetus vociferoides*	CR
ACCIPITRIDAE Hawk-eagles	■	Javan Hawk-eagle	*Spizaetus bartelsi*	EN
	■	Flores Hawk-eagle	*Spizaetus floris*	CR
	■	Black-and-chestnut Eagle	*Spizaetus isidori*	VU
	■	Wallace's Hawk-eagle	*Spizaetus nanus*	VU
	■	Philippine Hawk-eagle	*Spizaetus philippensis*	VU

RED LIST ■ EX = Extinct ■ EW = Extinct in the Wild ■ CR = Critically Endangered ■ EN = Endangered ■ VU = Vulnerable

PEAFOWL, PHEASANTS, AND TRAGOPANS

These three groups make up part of the large family Phasianidae, the latter containing over 180 species. Males and females differ markedly in plumage, with the males of many species brightly colored, while the females mainly have dull, camouflaged plumage to help them escape notice when incubating eggs. This family exhibits some of the most extreme sexual dimorphism. In peafowl and pheasants, the females have camouflaged plumage while the males tend to be very brightly colored and have long, often dramatic, tail feathers. In addition to habitat loss, threats to these birds come mainly from being hunted for food and also in some species for their beautiful feathers.

Peafowl are large, with elaborate plumage, notably the decorative tail display feathers of the male. The tail feathers of a male peacock, each adorned with eyelike markings, are fanned into a semicircular shape during displays. The Indian Peafowl is often kept as an ornamental bird in parks, large gardens, and country estates. Two peafowl, the Congo Peafowl (Vulnerable) and the Green Peafowl (Endangered) are threatened.

Pheasants are mostly medium-sized with a long tail. Out of 16 threatened species of pheasants, four, the Edwards' Pheasant, Vietnamese Pheasant, and Bornean and Hainan Peacock-pheasants are Endangered, the remainder being listed as Vulnerable.

Tragopans are shorter-tailed and have characteristic brightly colored fleshy erectile "horns" and inflatable fleshy throat patches ("lappets") used in courtship displays; they nest in trees, whereas the members of the other two groups are ground-nesters. Three tragopans are threatened: Blyth's, Cabot's, and Western (all three listed as Vulnerable).

Cheer Pheasant

Catreus wallichi

This is a gray-and-brown pheasant with a buff tail, distinct head crest, and red face. Both sexes have barring on the tail and upperparts. The female is duller in color and much smaller than the male. The birds have rather particular habitat requirements, preferring rocky terrain with scattered trees, with scrub and tall grasses, in parts of the Himalayas at heights of about 3,950–11,000 ft (1,200–3,350 m). Such habitats tend to be successional and depend partly on low-intensity arable activities, for example where areas of woodland have been removed for grazing. Proximity to people has resulted in the birds being hunted. With the decline in traditional agriculture, suitable habitat has also diminished. Conservation measures include reduction of habitat destruction by encouraging maintenance of more traditional low impact seasonal farming methods. The birds are already legally protected in India and Nepal.

Family	Phasianidae
Size	Male 35½–46½ in (90–118 cm), female 24–30 in (61–76 cm)
Status	Vulnerable (decreasing)
Population	4,000–6,000
Range	Himalayas of North Pakistan, India, and Nepal
Habitat	Rocky terrain with low scrub and grassy clearings
Main threats	Increasingly intensive farming/grazing techniques, hunting

Green Peafowl

Pavo muticus

This beautiful species is a close relative of the much more numerous Indian Peafowl (*Pavo cristatus*), which is more commonly known as Peacock, a name that refers only to the male (the female is called Peahen). The male Green Peafowl has metallic green plumage with black scaly markings and a long head crest. The tail is a green-tinged blackish brown while the primaries are caramel-colored. The female is duller in coloration, lacks the crest, and has predominantly dark brown feathers with a few buff markings. Both make loud, distictinctive territorial calls to each other, commonly being heard at dawn and dusk. Green Peafowl are often hunted for their meat, eggs, and spectacular feathers. Sadly, their habitat has become increasingly fragmented. In certain parts of the world, notably Thailand and China, they are also considered to be a crop pest and the birds are sometimes poisoned by farmers. Though some have been able to adapt to reduced habitat they definitely need additional protection because their numbers are in rapid decline. Among the conservation measures that have been imposed are captive breeding, legislation on trading of birds, and restrictions on the use of poisons and pesticides. Access to a source of water close to dry forest, seems to be a requirement.

Family	Phasianidae
Size	Male 8 ft (244 cm), female 3¼–3½ ft (100–110 cm)
Status	Endangered (decreasing)
Population	10,000–20,000
Range	Vietnam (west central), Cambodia, Myanmar, Thailand (west and north), China (Yunnan), Laos (south), Indonesia (Java)
Habitat	Dry deciduous forest close to wetlands or rivers
Main threats	Habitat loss, hunting for feathers and meat, poisoning

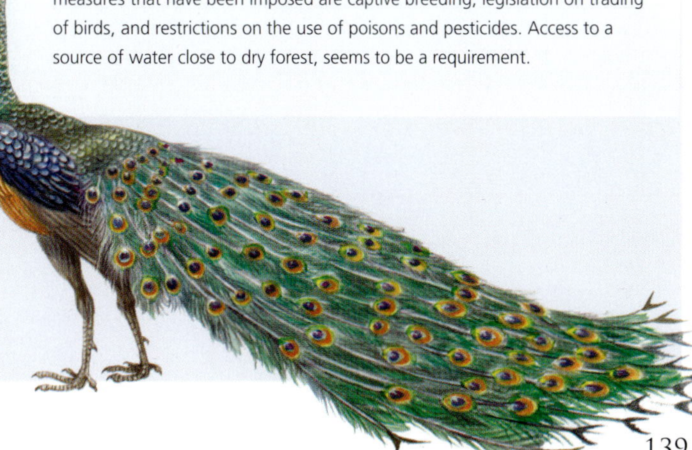

Blyth's Tragopan
Tragopan blythii

This forest-dwelling bird is camouflaged with beautiful disruptive markings, indicating a life spent largely on the dappled forest floor though it also roosts in the trees. It has a distinctive gray belly, and breast and the male has yellow facial skin and a bright red neck and chest. Males emit a loud groaning territorial call that carries a long distance through the forest. They live at an altitude of 4,600–10,800 ft (1,400–3,300 m), in subtropical and temperate forests, moving to the lower areas during winter months. They tend to move about in small groups on steep slopes, and feed mainly on shoots and invertebrates. The main threats come from habitat loss due to timber extraction and forest clearance, including grazing, and slash-and-burn agriculture. The birds are also hunted for food, particularly in northern India. The range of this species includes northeast India (Arunachal Pradesh to Manipur), north Myanmar, and also China (southeast Tibet and northwest Yunnan). This species is legally protected in all countries, but still seems to be declining. Deforestation is one of the main threats, mainly from shifting cultivation, and extraction of timber and fuel wood. In some regions it is also trapped for food along with other gamebirds. Careful monitoring is required to establish more information on numbers. Conservation awareness programs are planned to reduce further exploitation of their forest habitat. On a positive note, the species enjoys some protection in a number of reserves, notably in Blue Mountain National Park in Mizoram, northeast India, Thrumsing La National Park, Bhutan, and the Gaoligongshan National Park in China.

Family Phasianidae
Size Male 25½–27½ in (65–70 cm), female 22¾–23 in (58–59 cm)
Status Vulnerable (decreasing)
Population 2,500–10,000
Range Northeast India, north Myanmar, northwest China (southeast Tibet, northwest Yunnan)
Habitat Subtropical and temperate forest, in rocky terrain
Main threats Deforestation, timber extraction, hunting

THREATENED PEAFOWL, PHEASANTS, AND TRAGOPANS

FAMILY		COMMON NAME	SCIENTIFIC NAME	STATUS
PHASIANIDAE Peafowl	■	Congo Peafowl	*Afropavo congensis*	VU
	■	Green Peafowl	*Pavo muticus*	EN
PHASIANIDAE Pheasants	■	Cheer Pheasant	*Catreus wallichi*	VU
	■	Brown Eared-pheasant	*Crossoptilon mantchuricum*	VU
	■	Chinese Monal	*Lophophorus lhuysii*	VU
	■	Sclater's Monal	*Lophophorus sclateri*	VU
	■	Wattled Pheasant	*Lophura bulweri*	VU
	■	Edwards's Pheasant	*Lophura edwardsi*	EN
	■	Crestless Fireback	*Lophura erythrophthalma*	VU
	■	Vietnamese Pheasant	*Lophura hatinhensis*	EN
	■	Aceh Pheasant	*Lophura hoogerwerfi*	VU
	■	Salvadori's Pheasant	*Lophura inornata*	VU
	■	Reeves's Pheasant	*Syrmaticus reevesii*	VU
	■	Mountain Peacock-pheasant	*Polyplectron inopinatum*	VU
	■	Hainan Peacock-pheasant	*Polyplectron katsumatae*	EN
	■	Malayan Peacock-pheasant	*Polyplectron malacense*	VU
	■	Palawan Peacock-pheasant	*Polyplectron napoleonis*	VU
	■	Bornean Peacock-pheasant	*Polyplectron schleiermacheri*	EN
PHASIANIDAE Tragopans	■	Blyth's Tragopan	*Tragopan blythii*	VU
	■	Cabot's Tragopan	*Tragopan caboti*	VU
	■	Western Tragopan	*Tragopan melanocephalus*	VU

RED LIST ■ EX = Extinct ■ EW = Extinct in the Wild ■ CR = Critically Endangered ■ EN = Endangered ■ VU = Vulnerable

PARTRIDGES, FRANCOLINS, QUAILS, AND ALLIES

Partridges, francolins, and quails are medium-sized or small members of the pheasant family (Phasianidae). They have plump, rather rotund bodies, short but powerful wings, and sturdy legs. Partridges are medium-sized with strong legs. While they run well and spend much of their time on the ground, when necessary they can also burst into sudden, whirring flight to escape danger. Mostly birds of grassland or scrub, they are essentially ground-feeding seedeaters. Seven species of partridge are listed as threatened, five of these Vulnerable and two, Sichuan Partridge and Udzungwa Forest-partridge, are Endangered.

Francolins bear a close resemblance to partridges in their general build, but they are distinguished from them partly by having bare patches of skin on the head and neck. Of the six threatened francolins, two are Vulnerable, three, Mount Cameroon, Swierstra's, and Nahan's Francolin are Endangered, and one, Djibouti Francolin, is Critically Endangered.

Quails include some of the smallest of all gamebirds. Most species are ground-nesters and also forage on the ground for food, usually seeds and grain. With their camouflaged plumage and secretive habits, quails can be very hard to spot as they scurry about in an almost mouselike manner in the undergrowth. Quails produce monotonous, rather rasping or mechanical calls, and are more often heard than seen. Perhaps surprisingly for such small birds, some quails, notably Common Quail and Japanese Quail, migrate over long distances. Two species of quail are threatened—the Critically Endangered Himalayan Quail and the Vulnerable Manipur Bush-quail.

In their habitat on the grasslands of North America both Greater and Lesser Prairie-chickens are listed as Vulnerable, while Gunnison Sage-grouse is Endangered.

Also included in the pheasant family are the prairie-chickens and sage-grouse of North America. These are well-known for the remarkable displays of the males at their lekking grounds, involving puffed up feathers accompanied by strange calls.

The New World quails are similar in shape and size to the quails, but they belong to a separate family (Odontophoridae). Like the Old World quails most are also small, but they have brighter plumage, and their bills are strong and serrated. This family, of about 30 species, includes the California Quail, bobwhites, and wood-partridges as well as 15 species of wood-quails. Five species are listed as Vulnerable, and one, Gorgeted Wood-quail of Colombia, is Endangered.

Sichuan Partridge
Arborophila rufipectus

This partridge is endemic to China, being found only in south central Sichuan and possibly also in northeast Yunnan. The male is gray-brown, and distinguished from similar species by the chestnut breast band and head pattern—brown crown, with rufous ear coverts. The throat is white with black streaks. The female's plumage is similar but duller. The pair make repetitive territorial calls with loud rising whistles, often as a duet. Sichuan Partridges forage in temperate cloud forest at altitudes between 3,300 ft (1,000 m) and 7,400 ft (2,250 m), preferring moist leaf-litter and sparse bamboo ground flora. It is also found in well-established broadleaf plantations. Until the late 1990s, habitat destruction from clear-felling was the major threat, but logging has now been largely stopped in the region. However, risks remain from hunting and also from illegal logging. The birds are also disturbed by grazing animals and by local gathering of medicinal plants and bamboo shoots, causing a problem especially during the breeding season.

Family	Phasianidae
Size	11–12 in (28–30.5 cm)
Status	Endangered (decreasing)
Population	1,000–2,500
Range	China (mainly Sichuan) (endemic)
Habitat	Temperate broadleaf cloud forest
Main threats	Hunting and habitat loss

Greater Prairie-chicken
Tympanuchus cupido

This strongly barred brown-and-buff grouse of the North American prairie was once common, but has declined as their original habitat has been turned into cropland or grazing pasture. Both sexes have a dark eyestripe and unusually long, adapted neck feathers known as pinnae (longer in the males). During lekking displays the males raise these above the head like long ears and inflate yellow air sacs above the eyes and neck. At the end of the display, they also utter very loud, trisyllabic booming calls that may be heard from a great distance. This species is now found mainly in the mid-western states of the U.S.A.. The ideal habitat is native prairie with scattered woodland, and the birds usually select areas of short grass for leks. Fragmentation of the habitat has produced isolated populations that may suffer reduced fitness through inbreeding. The creation of wildlife corridors of dense vegetation is suggested to re-establish links between isolated populations. In some states the birds are hunted, so a control is suggested, followed by a complete ban on hunting if it results in increased numbers.

Family	Phasianidae
Size	17 in (43 cm)
Status	Vulnerable (decreasing)
Population	460,000
Range	Mid-western U.S.A. (endemic)
Habitat	Native prairie and oak woodland
Main threats	Conversion of native habitat to cropland, hunting

Gorgeted Wood-quail

Odontophorus strophium

This small wood-quail has distinctive black and white banding on the throat. The male has a short crest and bright chestnut breast and belly. Though more widespread than once thought, its range is nevertheless tiny—it is found only on certain mountain slopes in the East Andes of Colombia, mainly in large remnants of humid oak forest, between about 5,750 ft (1,750 m) and 6,700 ft (2,050 m). It forages for seeds, fruit, and invertebrate arthropods on the forest floor. Breeding occurs during the peak rainy times, from March to May and September to November. The bird is under threat from habitat destruction and exploitation that began in the 17th century. These forests have been in decline from logging and farming activities ever since. Conservation plans attempt to protect and improve the habitat, particularly at relevant altitudes for this species. Areas of intact forest remain above 6,400 ft (1,950 m) and it is also encouraging that some is regenerating after being abandoned.

Family	Odontophoridae
Size	9¾ in (25 cm)
Status	Endangered (decreasing)
Population	2,000–4,300
Range	Colombia (endemic)
Habitat	Humid, subtropical temperate forests dominated by oak and laurel trees
Main threats	Destruction of habitat by logging and agriculture

THREATENED PARTRIDGES, FRANCOLINS, QUAILS, AND ALLIES

FAMILY		COMMON NAME	SCIENTIFIC NAME	STATUS
PHASIANIDAE Partridges	■	Hainan Partridge	*Arborophila ardens*	VU
	■	White-necklaced Partridge	*Arborophila gingica*	VU
	■	Chestnut-breasted Partridge	*Arborophila mandellii*	VU
	■	White-faced Partridge	*Arborophila orientalis*	VU
	■	Sichuan Partridge	*Arborophila rufipectus*	EN
	■	Black Partridge	*Melanoperdix niger*	VU
PHASIANIDAE Forest-partridge	■	Udzungwa Forest-partridge	*Xenoperdix udzungwensis*	EN
PHASIANIDAE Francolins	■	Mount Cameroon Francolin	*Francolinus camerunensis*	EN
	■	Swamp Francolin	*Francolinus gularis*	VU
	■	Harwood's Francolin	*Francolinus harwoodi*	VU
	■	Nahan's Francolin	*Francolinus nahani*	EN
	■	Djibouti Francolin	*Francolinus ochropectus*	CR
	■	Swierstra's Francolin	*Francolinus swierstrai*	EN
PHASIANIDAE Quails	■	Himalayan Quail	*Ophrysia superciliosa*	CR
PHASIANIDAE Bush-quails	■	Manipur Bush-quail	*Perdicula manipurensis*	VU
PHASIANIDAE Prairie-chickens	■	Greater Prairie-chicken	*Tympanuchus cupido*	VU
	■	Lesser Prairie-chicken	*Tympanuchus pallidicinctus*	VU
PHASIANIDAE Sage-grouse	■	Gunnison Sage-grouse	*Centrocercus minimus*	EN
ODONTOPHORIDAE New World quails	■	Ocellated Quail	*Cyrtonyx ocellatus*	VU
	■	Bearded Wood-partridge	*Dendrortyx barbatus*	VU
	■	Black-fronted Wood-quail	*Odontophorus atrifrons*	VU
	■	Tacarcuna Wood-quail	*Odontophorus dialeucos*	VU
	■	Dark-backed Wood-quail	*Odontophorus melanonotus*	VU
	■	Gorgeted Wood-quail	*Odontophorus strophium*	EN

RED LIST ■ EX = Extinct ■ EW = Extinct in the Wild ■ CR = Critically Endangered ■ EN = Endangered ■ VU = Vulnerable

GUINEAFOWL, MEGAPODES, GUANS, AND CURASSOWS

Guineafowl (Numididae) are sociable chickenlike ground-feeding birds with long legs and a strong bill. Most have rather dark plumage, often gray or black with small white spots, and patches of bare skin around the head and neck. They move about in groups, sometimes hundreds strong, and make rather strident, cackling calls. They are found in Africa, mainly south of the Sahara desert, mostly in light woodland, savanna, and grassland. They nest on the ground, but normally take to the trees to roost in relative safety. They are sometimes domesticated and kept for their meat and eggs. One of the six species is threatened—the Vulnerable White-breasted Guineafowl found in the rain forests of tropical West Africa. It occurs in Sierra Leone (notably Gola Forest, where it is protected), Liberia, Côte d'Ivoire, and Ghana. The main threats to this species are habitat loss and hunting.

Megapodes (Megopodiidae) are large, turkeylike ground birds with powerful legs. They lay their eggs in mounds or burrows where they are incubated by the heat of the sun or from warmth resulting from the decomposition of vegetable matter, rather like in a garden compost heap. A female may lay as many as 30 eggs and the male attends the mound, adjusting the depth to keep the temperature suitable for egg development. Their young are highly precocial, and are able to run and even fly within hours of digging themselves out of the nest-mound. The family includes brush-turkeys and the famous Malleefowl. Of the 21 megapode species, five (including the Malleefowl) are Vulnerable and four Endangered.

Guans and curassows (Cracidae) are an American family of 50 species that also includes the chachalacas. They are unusual among gamebirds in that they live mainly in trees. They have short, rounded wings and a long, broad tail. Most have dark plumage and many sport crests. They construct nests of twigs and leaves in the low branches or on the ground. Eighteen members of this family are threatened, eight of them Endangered, six Vulnerable; three (White-winged Guan, Trinidad Piping-guan, and Blue-billed Curassow) Critically Endangered, one (Alagoas Curassow) Extinct in the Wild.

Malleefowl

Leipoa ocellata

This large, stocky bird is pale and mottled below with a central dark marking down its breast, mainly speckled brown above. Malleefowl pair for life and live in the "mallee" scrub of southern Australia. The preferred habitat consists of eucalyptus and acacia scrubland on sandy soil, ideally with a deep layer of leaf litter. In one of nature's most peculiar breeding techniques, the male bird (occasionally helped by his mate) creates a large mound of soil up to 5 ft (1.5 m) high, consisting of a core of moist leaf litter covered with sandy soil. Eggs are laid in the mound where they are incubated by the heat produced from the decomposing leaves, in this giant compost heap. The male carefully monitors the temperature so that it stays for most of the time at an optimum of 90–93°F (32–34°C), making adjustments by rearranging the amount of soil cover. When they hatch, the chicks dig their way out of the mound and are entirely independent, dispersing from the area and fending for themselves. Juvenile mortality is very high, made worse by introduced predators such as the Red Fox (*Vulpes vulpes*) and domestic cat. Malleefowl have declined quite steeply over the last few decades, from habitat loss and introduced predators. Nesting birds are easily disturbed by grazing animals.

Family Megapodiidae
Size 23½ in (60 cm)
Status Vulnerable (decreasing)
Population 100,000
Range Southern Australia (endemic)
Habitat Dry and sandy eucalyptus and acacia scrubland
Main threats Predation, disturbance by grazing animals, fire

White-winged Guan

Penelope albipennis

This stoutly built bird lives in dry forest in northwest Peru at an altitude of 980–4,250 ft (300–1,300 m), but its population and range are both very small, placing it firmly in the Critically Endangered category. The White-winged Guan is mainly black with white primary feathers and white flecks on its breast, long neck, and wing coverts. It has a black-tipped blue bill and purple skin around the eyes. The throat has a large bare purple or red-orange area and a double-lobed dewlap. At dawn, especially during the breeding season from January to August, it emits a deep croaky call that travels up to 1¾ miles (3 km). Birds lay clutches of two or three eggs. It prefers a permanent source of water nearby and consumes fruit, seeds, buds, leaves, and flowers. Hunting was probably the main cause of its decline and the species is now legally protected in a number of nature reserves. Captive-bred birds are being reintroduced to suitable sites, coupled with education and awareness campaigns.

Family Cracidae
Size 33½ in (85 cm)
Status Critically Endangered (decreasing)
Population 150–250
Range Peru (endemic)
Habitat Dry deciduous forest on slopes and ravines
Main threats Hunting, habitat destruction

Blue-billed Curassow

Crax alberti

This highly distinctive bird is large and black with a blue cere and hanging wattle. The male has a white vent and tail tip. Both sexes have a cream-colored bill with a blue base. The male has a curled black crest, the female's is black and white. The call is low and booming. They roost in forest trees close to good feeding grounds, using a roost for a number of days at a time. Their food includes fruit, shoots, invertebrates, and possibly carrion. Breeding is in the dry season. The favored habitat is humid rain forest, huge areas of which have been cleared, converted to plantations, grazing for livestock, and mining. The birds are hunted for meat and eggs. Reserves such as El Paujíl Bird Reserve and Paramillo National Park hold this rare species, but numbers are low and probably decreasing overall. El Paujíl Bird Reserve was established within its stronghold and surveys indicate that numbers here have increased.

Family Cracidae
Size 32¾–36½ in (83–93 cm)
Status Critically Endangered (decreasing)
Population 1,000–2,500
Range Colombia (endemic)
Habitat Humid forest to 3,950 ft (1,200 m)
Main threats Deforestation, hunting

THREATENED GUINEAFOWL, MEGAPODES, GUANS, AND CURASSOWS

FAMILY		COMMON NAME	SCIENTIFIC NAME	STATUS
NUMIDIDAE Guineafowl	■	White-breasted Guineafowl	*Agelastes meleagrides*	VU
MEGAPODIIDAE Megapodes	■	Bruijn's Brush-turkey	*Aepypodius bruijnii*	EN
	■	Moluccan Megapode	*Eulipoa wallacei*	VU
	■	Malleefowl	*Leipoa ocellata*	VU
	■	Maleo	*Macrocephalon maleo*	EN
	■	Biak Megapode	*Megapodius geelvinkianus*	VU
	■	Micronesian Megapode	*Megapodius laperouse*	EN
	■	Vanuatu Megapode	*Megapodius layardi*	VU
	■	Nicobar Megapode	*Megapodius nicobariensis*	VU
	■	Polynesian Megapode	*Megapodius pritchardii*	EN
CRACIDAE Guans and curassows	■	Blue-billed Curassow	*Crax alberti*	CR
	■	Red-billed Curassow	*Crax blumenbachii*	EN
	■	Wattled Curassow	*Crax globulosa*	EN
	■	Great Curassow	*Crax rubra*	VU
	■	Alagoas Curassow	*Mitu mitu*	EW
	■	Horned Guan	*Oreophasis derbianus*	EN
	■	Rufous-headed Chachalaca	*Ortalis erythroptera*	VU
	■	Helmeted Curassow	*Pauxi pauxi*	EN
	■	Horned Curassow	*Pauxi unicornis*	EN
	■	White-winged Guan	*Penelope albipennis*	CR
	■	Bearded Guan	*Penelope barbata*	VU
	■	White-browed Guan	*Penelope jacucaca*	VU
	■	Chestnut-bellied Guan	*Penelope ochrogaster*	VU
	■	Baudo Guan	*Penelope ortoni*	EN
	■	Cauca Guan	*Penelope perspicax*	EN
	■	Highland Guan	*Penelopina nigra*	VU
	■	Black-fronted Piping-guan	*Pipile jacutinga*	EN
	■	Trinidad Piping-guan	*Pipile pipile*	CR

RED LIST ■ EX = Extinct ■ EW = Extinct in the Wild ■ CR = Critically Endangered ■ EN = Endangered ■ VU = Vulnerable

Cranes

Cranes (family Gruidae) are tall graceful heronlike birds with long bill, neck, and legs. Most have white or gray plumage. Unlike herons, they extend their neck forward in flight. Strong fliers, most crane species make regular migrations to and from their breeding grounds, often in large, straggling flocks. As is the case with many other migrant birds, such movements expose cranes to various dangers as they migrate via a series of wetland staging posts. Some are shot, while others are victims of collisions with power lines or they ingest poison when feeding at badly polluted waters. In most countries wetland habitats are also under threat from drainage and development, and the number of suitable wintering wetland sites for cranes have diminished markedly in recent decades.

With their arresting plumage, extraordinary courtship dances, and loud bugling calls, these charismatic birds have entered firmly into human folklore in many parts of the world. The typical courtship dance begins with the pair standing side by side and calling, after which the birds bob and leap, spreading their tail feathers and wings, and sometimes hurling nearby objects into the air. Cranes are primarily birds of grasslands, marshes, and wetlands, and most species breed close to shallow water. The nest is typically a platform constructed from reeds and other vegetation, on the ground or in shallow water.

There are 15 living species of crane, of which 11 are threatened, mainly through loss or deterioration of their wetland habitats. Most (13) of the cranes are closely related, belonging to the same genus, *Grus*. Two species, Whooping Crane of North America and Red-crowned Crane from the Far East, are Endangered, while the magnificent Siberian Crane is Critically Endangered. The Siberian Crane is also the most reliant on wetland habitats, both in its breeding range and at its wintering sites and, with a population of a little over 3,000, it is a conservation priority. Six species of *Grus* are listed as Vulnerable. The remaining two species (both Vulnerable) are the crowned-cranes, African species classified in the genus *Balearica*. The crowned-cranes have dark plumage and a characteristic crest of golden, bristlelike feathers. The Black Crowned-crane is found mainly in sub-Saharan West Africa and the Sudan. The Gray Crowned-crane has a range in eastern and southern Africa, from Kenya to South Africa.

Whooping Crane
Grus americana

This large white crane has a red crown and black forehead. It has red facial skin, a cream bill, and black mustache. The primaries are also black and very obvious in flight. The only remaining native population is at Wood Buffalo National Park, on the border between Northwest Territories and Alberta, in north central Canada. These birds are migratory and winter in Texas, at or close to the Aransas National Wildlife Refuge. Two other reintroduced populations also exist in the eastern U.S.A., one resident in Florida and the other in Wisconsin, the latter migrating to Florida. Birds breed from April to May on prairie wetland with a range of pools, marsh, and willow, migrating to brackish coastal wetlands in the winter. The species has long been affected by human disturbance, land use, and hunting, and almost became extinct in the late 1930s. Thankfully, numbers are now slowly increasing. Conservation efforts include captive breeding programs, establishing further populations, and adapting power lines in an attempt to reduce fatal collisions.

Family	Gruidae
Size	4¼ ft (132 cm)
Status	Endangered (increasing)
Population	50–250
Range	Canada (wild), U.S.A. (introduced)
Habitat	Prairie wetlands
Main threats	Hunting, disturbance, power lines

Siberian Crane
Grus leucogeranus

This is a large, mainly white crane with black primaries and red legs. A red mask stretches between the bill and eyes. Though more numerous than the Whooping Crane, this species is classed as Critically Endangered largely because of threats to its wintering habitats following the enormous Three Gorges Dam project in China. There are two main breeding populations, both in Russia. Nearly all the birds breed in Siberia, migrating south and wintering to lakes in eastern China, notably Poyang Hu. They arrive at their breeding grounds in late May, lay eggs in June, and begin their return migration in September. They like wetlands for both breeding and wintering habitats, preferring large undisturbed sites. They feed on a variety of items including roots, rhizomes, seeds and shoots, invertebrates, fish, and even small rodents during the breeding season. The main threats are habitat loss and further degradation, including to the overwintering sites in China. The Three Gorges Dam may have a negative impact on the wetlands of the lower Yangtze, including Poyang Hu. Birds are still hunted on their migration though the birds are now legally protected. Captive breeding programs are also planned, using Whooping Cranes as a model.

Family	Gruidae
Size	4½ ft (140 cm)
Status	Critically Endangered (decreasing)
Population	3,750
Range	Russia, mainly Siberia (breeds) (endemic), China (winter)
Habitat	Wetlands
Main threats	Wetland loss and degradation, hunting

THREATENED CRANES				
FAMILY		**COMMON NAME**	**SCIENTIFIC NAME**	**STATUS**
GRUIDAE Cranes		Black Crowned-crane	*Balearica pavonina*	VU
		Gray Crowned-crane	*Balearica regulorum*	VU
		Whooping Crane	*Grus americana*	EN
		Sarus Crane	*Grus antigone*	VU
		Wattled Crane	*Grus carunculatus*	VU
		Red-crowned Crane	*Grus japonensis*	EN
		Siberian Crane	*Grus leucogeranus*	CR
		Hooded Crane	*Grus monacha*	VU
		Black-necked Crane	*Grus nigricollis*	VU
		Blue Crane	*Grus paradisea*	VU
		White-naped Crane	*Grus vipio*	VU

RED LIST EX = Extinct EW = Extinct in the Wild CR = Critically Endangered EN = Endangered VU = Vulnerable

Rails, Crakes, and Allies

The rails, crakes, and allies are a large family (Rallidae) that contains 135 species. In addition to rails, they include coots, moorhens, crakes, flufftails, and the unusual Weka and Takahe, the latter both threatened New Zealand endemics. Thirty-four members of this family are threatened, 12 of which are Endangered, four Critically Endangered, and one Extinct in the Wild. Several species that are restricted to oceanic islands tend to fly only weakly or have lost the ability to fly altogether, making them at risk from introduced predators. Most are small or medium-sized ground birds that skulk in dense vegetation and have mainly brown or grayish camouflaged plumage. The majority inhabit damp or wet habitats such as marshes, reedbeds, or damp forest and scrub, and most species nest on the ground. They communicate by means of squeals, whistles, or grunts and some species sound surprisingly mechanical. The family includes the world's smallest flightless bird, the Inaccessible Rail. The main threats to rails, crakes, and allies are habitat loss and predation.

Inaccessible Rail

Atlantisia rogersi

With the extinction of the closely related St Helena Rail, the Inaccessible Rail is the only representative of its genus, and also the world's smallest flightless bird, the size of a mouse. Rusty brown above and dark gray beneath, it has variable white barring on the flanks and belly. It has a red eye, black bill, and gray legs. It calls either with a loud, shrill alarm call or softer, general contact "tchick" noises. The birds breed from October to March, living in territorial family groups. They forage widely across the island, feeding on a broad range of items including insect larvae, worms, centipedes, moths, berries, and seeds. Pear-shaped nests are carefully constructed from vegetation and reached via a tunnel in the undergrowth. Found only on the isolated South Atlantic Inaccessible Island, it is still common, largely because the island remains free of Black rats (*Rattus rattus*), though these do live on Tristan da Cunha and could invade via fishing vessels. To reduce the chance of introductions there are strict controls on access to the island, which is a nature reserve.

Family	Rallidae
Size	6¾ in (17 cm)
Status	Vulnerable (stable)
Population	8,400
Range	Inaccessible Island (endemic)
Habitat	Grassland, thickets, heath
Main threats	Introduced predators (potential threat)

Zapata Rail

Cyanolimnas cerverai

The Zapata Rail has a yellow bill with a red base, and orange legs. It has a brown back, white undertail, and white throat, with blue underparts and gray hindflanks. It is medium-sized, short-winged, and almost flightless. This rail is named after the large Zapata Swamp in western Cuba, its main habitat. The swampland consists of flooded grassland and low trees. Nests are constructed on tussocks slightly above the water level. Breeding occurs during September and maybe also in December and January. The rails probably disperse during wet season floods, returning in the dry season. The swamp grasses were previously harvested for thatch, and this poses a potential problem should the activity return. Dry season burning also devastates areas and creates further habitat loss. Being virtually flightless, they are killed by introduced predators such as mongooses and rats. To tackle these problems, two protected areas have been set up and ecotourism is being explored. Corral de Santo Tomás Faunal Refuge and the Laguna del Tesoro are within the protected areas.

Family	Rallidae
Size	11½ in (29 cm)
Status	Critically Endangered (decreasing)
Population	250–1,000
Range	Southwest Cuba (endemic)
Habitat	Swamp, flooded grassland
Main threats	Predation; habitat loss, dry-season fires

THREATENED RAILS, CRAKES, AND ALLIES				
FAMILY		**COMMON NAME**	**SCIENTIFIC NAME**	**STATUS**
RALLIDAE Rails, crakes, and allies	■	Talaud Bush-hen	*Amaurornis magnirostris*	VU
	■	Sakalava Rail	*Amaurornis olivieri*	EN
	■	Brown Wood-rail	*Aramides wolfi*	VU
	■	Snoring Rail	*Aramidopsis plateni*	VU
	■	Inaccessible Rail	*Atlantisia rogersi*	VU
	■	Swinhoe's Rail	*Coturnicops exquisitus*	VU
	■	Zapata Rail	*Cyanolimnas cerverai*	CR
	■	Hawaiian Coot	*Fulica alai*	VU
	■	Gough Moorhen	*Gallinula nesiotis*	VU
	■	Samoan Moorhen	*Gallinula pacifica*	CR
	■	Makira Moorhen	*Gallinula silvestris*	CR
	■	Weka	*Gallirallus australis*	VU
	■	Calayan Rail	*Gallirallus calayanensis*	VU
	■	New Caledonian Rail	*Gallirallus lafresnayanus*	CR
	■	Okinawa Rail	*Gallirallus okinawae*	EN
	■	Guam Rail	*Gallirallus owstoni*	EW
	■	Lord Howe Woodhen	*Gallirallus sylvestris*	EN
	■	Blue-faced Rail	*Gymnocrex rosenbergii*	VU
	■	Talaud Rail	*Gymnocrex talaudensis*	EN
	■	Invisible Rail	*Habroptila wallacii*	VU
	■	Rusty-flanked Crake	*Laterallus levraudi*	EN
	■	Galapagos Rail	*Laterallus spilonotus*	VU
	■	Junín Rail	*Laterallus tuerosi*	EN
	■	Rufous-faced Crake	*Laterallus xenopterus*	VU
	■	Auckland Islands Rail	*Lewinia muelleri*	VU
	■	Takahe	*Porphyrio hochstetteri*	EN
	■	Henderson Crake	*Porzana atra*	VU
	■	Dot-winged Crake	*Porzana spiloptera*	VU
	■	Austral Rail	*Rallus antarcticus*	VU
	■	Madagascar Rail	*Rallus madagascariensis*	VU
	■	Bogota Rail	*Rallus semiplumbeus*	EN
	■	Plain-flanked Rail	*Rallus wetmorei*	EN
	■	White-winged Flufftail	*Sarothrura ayresi*	EN
	■	Slender-billed Flufftail	*Sarothrura watersi*	EN

RED LIST ■ EX = Extinct ■ EW = Extinct in the Wild ■ CR = Critically Endangered ■ EN = Endangered ■ VU = Vulnerable

Weka

Gallirallus australis

This large flightless rail is rather cootlike in overall shape. It is a strong swimmer and can also run fast. Wekas feed on fruit, seeds, invertebrates, eggs, and occasionally small mammals and small birds. There is great geographical variation in plumage, with pale, dark, chestnut, and black morphs in different regions. The birds reach maturity at one year, and breed all year round, making a shallow, cup-shaped nest from sedges or grass. Once much more common, Wekas are still found in scattered populations on both North Island and South Island, and also on many small offshore islands. They are however highly vulnerable to introduced mammalian predators, notably domestic cats, and have declined dramatically in some parts of their range. They are often killed on roads or from feeding on bait intended to control possums and rabbits. Conservation measures have involved relocating birds to a number of safe island locations, for example Chatham and Pitt Islands, where the population is estimated at 38,000–58,000.

Family	Rallidae
Size	18–23½ in (46–60 cm)
Status	Vulnerable (decreasing)
Population	120,000–187,000
Range	New Zealand (endemic)
Habitat	Forest, grassland, scrub, wetlands, semi-urban
Main threats	Introduced predators, habitat destruction, poisoning, road kill, disease

Takahe

Porphyrio hochstetteri

This magnificent large flightless bird has beautiful iridescent plumage. The head, neck, shoulders, and breast are peacock blue and the remaining body olive green. It has a huge red bill, adapted for shearing vegetation, topped by a bold red shield. The red legs and feet are sturdy, yet used in a dexterous way when feeding on the juicy bases of just a few species of grass and fern. Takahes can live for 14–20 years. Breeding usually starts at two years old with females laying clutches of two eggs.

It was believed to be extinct until 1948 when a small population was discovered in the Murchison Mountains in South Island's Fjordland. Since then the numbers have fluctuated but not increased, indicating this is the maximum for this small area. Competition for grazing from introduced Red Deer (*Cervus elaphus*) is a problem, as predation from alien carnivores such as Stoat (*Mustela erminea*) may also be. Deer have been controlled, and birds have also been translocated to offshore predator-free islands, backed up by a program of captive breeding.

Family	Rallidae
Size	24¾ in (63 cm)
Status	Endangered (increasing)
Population	150–220
Range	New Zealand (endemic)
Habitat	Alpine tussock grassland
Main threats	Grazing competition and predation, habitat loss

BUSTARDS, BUTTONQUAILS, PLAINS-WANDERER, KAGU, MESITES, AND FINFOOTS

The Bustard family (Otididae) consists of 25 species of large or medium-sized ground-living birds including the Great Bustard, the world's heaviest flying bird. They have long legs and necks, walk with a striding gait, and resemble geese in overall body shape. These birds are found mainly in open sites such as grassland, steppe, and savanna Six species are currently listed as threatened, two of which (Houbara Bustard and Great Bustard) are Vulnerable, three (Great Indian Bustard, Ludwig's Bustard, and Lesser Florican) are Endangered, and one (Bengal Florican) is Critically Endangered. A further six species are classed as Near Threatened.

Buttonquails (Turnicidae) comprise 16 species, three of which are threatened. They are small, quail-like, ground-dwelling birds with delicately camouflaged plumage. Distinct from ordinary quails, they have slender bills used to feed on insects and seeds. Unusually, just the male of a pair incubates the eggs.

The Endangered Plains-wanderer is the sole member of its family (Pedionomidae). Small, quail-like and endemic to certain grassland sites in Australia, its numbers fluctuate widely with rainfall patterns. It is a medium-sized bird with a body length of approximately 4¾ in (12 cm), resembling a buttonquail in many of its characteristics.

Also classed in a family of its own (Rhynochetidae), the strange chicken-sized Kagu is endemic to New Caledonia where it is Endangered, but still managing to hold its own in certain forest sites. This pale, nearly flightless bird superficially resembles a heron. However, it differs markedly from the majority of heron species in that its preferred habitat is within forests situated on mountainsides that have access to small streams.

There are only three species of mesite, all endemic to Madagascar and all Vulnerable, threatened mainly by habitat loss and hunting. These thrush-sized, dull brown birds feed on fruit and insects gleaned from the forest floor and they seldom fly. Their nests consist of a raised platform built between 3¼ and 6½ ft (1 and 2 m) above ground level.

Another three-species family—the finfoots (Heliornithidae)—are grebelike waterbirds with lobed toes and long, stiff tails. Their specialized feet make efficient paddles for swimming and diving while hunting for food—their diet consisting mainly of frogs and crustaceans. One species, the Masked Finfoot of Southeast Asia, is Endangered, mainly as a result of disturbance and alterations to rivers and other wetlands.

THREATENED BUSTARDS, BUTTONQUAILS, AND ALLIES				
FAMILY		**COMMON NAME**	**SCIENTIFIC NAME**	**STATUS**
OTIDIDAE Bustards	■	Great Indian Bustard	*Ardeotis nigriceps*	EN
	■	Houbara Bustard	*Chlamydotis undulata*	VU
	■	Bengal Florican	*Houbaropsis bengalensis*	CR
	■	Ludwig's Bustard	*Neotis ludwigii*	EN
	■	Lesser Florican	*Sypheotides indicus*	EN
	■	Great Bustard	*Otis tarda*	VU
TURNICIDAE Buttonquails	■	Sumba Buttonquail	*Turnix everetti*	VU
	■	Black-breasted Buttonquail	*Turnix melanogaster*	VU
	■	Buff-breasted Buttonquail	*Turnix olivii*	EN
PEDIONOMIDAE Plains-wanderer	■	Plains-wanderer	*Pedionomus torquatus*	EN
RHYNOCHETIDAE Kagu	■	Kagu	*Rhynochetos jubatus*	EN
MESITORNITHIDAE Mesites	■	Brown Mesite	*Mesitornis unicolor*	VU
	■	White-breasted Mesite	*Mesitornis variegatus*	VU
	■	Subdesert Mesite	*Monias benschi*	VU
HELIORNITHIDAE Finfoots	■	Masked Finfoot	*Heliopais personatus*	EN

RED LIST ■ EX = Extinct ■ EW = Extinct in the Wild ■ CR = Critically Endangered ■ EN = Endangered ■ VU = Vulnerable

Houbara Bustard
Chlamydotis undulata

The Houbara Bustard is a large upright bird living in semi-desert habitats in a range of North African countries, the Arabian Peninsula, and with an outlying population in the eastern Canary Islands. Adult birds have a distinctive black stripe down the side of the neck and the body is sandy with barred markings. During a distinctive breeding display, the male raises neck and crown feathers to create a white ruff and the head is laid down onto the back as the bird trots along. Houbara Bustards eat invertebrates, small vertebrates, and plant shoots, possibly migrating to find fresh vegetation, which may stimulate breeding. The nest is a scrape on the ground with a clutch of two to four eggs.

The continued tradition of hunting by Arab falconers, especially during the winter months—nowadays made worse by all-terrain vehicles, modern firearms, and other technology—is the main threat, compounded by degradation of habitats. Many birds are illegally trapped for supplying to falconers for training their falcons; many die in appalling transit conditions. Captive breeding programs maintain and supplement numbers, and future plans aim to provide protection in key areas and place controls on poaching and on the numbers hunted. Overgrazing is also a threat and reduction of grazing pressures is also beneficial.

Family	Otididae
Size	19 in (48 cm)
Status	Vulnerable (decreasing)
Population	49,000–62,000
Range	Canary Islands, North Africa, east from Israel, Palestine, Lebanon, Arabian Peninsula, Pakistan, Iran, Afghanistan
Habitat	Sandy and stony arid semi-desert, steppe
Main threats	Unsustainable hunting, habitat degradation

Kagu
Rhynochetos jubatus

The heronlike Kagu has pale gray feathers and an orange-red bill and legs. This unusual pale appearance within the shady forest habitat is probably why locals call it the "Ghost Bird." Almost flightless, they tend to walk or glide down slopes rather than fly. When open for display the wings show black-and-chestnut barred feathers. Kagus also display by raising their crest vertically like a fan. They live, nest, and forage on the forest floor making them vulnerable to predators, especially domestic dogs, which are the main threat for these isolated and fragmented populations. Adult birds live in pairs and build nests on the ground using fallen twigs and sticks. They have a number of contact calls, including a loud bark at daybreak along with a quieter range of rattling and hissing sounds. Kagus are ground feeders, locating their food by tapping the ground with their bill to find worms, snails, and lizards. There is evidence that the birds have long lives, living up to 15 years in the wild and as many as 30 years in captivity. Conservation plans aim to control introduced predator numbers, particularly domestic dogs, but also rats that are also known to take young birds. The native habitat has been degraded in the past by mining activities and is currently further damaged by grazing deer.

Family	Rhynochetidae
Size	21¾ in (55 cm)
Status	Endangered (stable)
Population	250–1,000
Range	New Caledonia (endemic)
Habitat	Forested mountainside, closed canopy scrub
Main threats	Introduced predators, habitat damage by deer, mining, deforestation

White-breasted Mesite
Mesitornis variegatus

This small rail-like bird is found at just five sites in north and west Madagascar (Menabe Forest, Ankarafantsika, Ankarana Special Reserve, Analamera Special Reserve, and Daraina Forest) and at Ambatovaky Special Reserve in the east. It has a low-slung gait, a small head and broad tail and is brown and gray above. The chest is white with dark crescent markings and a pale chestnut band. The majority of this small population inhabit deciduous forests of the north and south where they forage as small family groups of two to four individuals, mainly on small invertebrates and seeds. Food is found by flicking over leaves on the forest floor. The White-breasted Mesite is threatened by a number of factors including hunting and habitat destruction by slash-and-burn agriculture, as well as exploitation of the forests for logging and charcoal. This species appears not to repopulate recently burnt regenerating forest areas, preferring the fast-declining mature native habitat.

Family	Mesitornithidae
Size	12 in (31 cm)
Status	Vulnerable (declining)
Population	8,000
Range	Madagascar (endemic)
Habitat	Deciduous forest
Main threats	Slash-and-burn agriculture, logging, hunting

Plovers, Sandpipers, Stilts, Oystercatchers, Coursers, and Pratincoles

This section contains all the families that are commonly referred to as waders.

The plovers (Charadriidae) are one of the main families of waders, found mainly in wetland habitats across the globe. Of the 66 species of plover, seven are listed as threatened. The Madagascar Plover and the highly unusual Wrybill are Vulnerable; New Zealand Dotterel and Shore Plover are Endangered; and the remaining three, St Helena Plover, Sociable Lapwing, and Javan Lapwing, are Critically Endangered. The main threats are alteration of wetland habitats, and predation (especially in the case of island species).

Largest of the wader families, the sandpipers (Scolopacidae), have about 90 species, including curlews, snipes, and woodcocks. Thirteen species are listed as threatened, three of which (Spoon-billed Sandpiper, Eskimo Curlew, and Slender-billed Curlew) are Critically Endangered.

The stilt family (Recurvirostridae) includes the avocets. They are elegant waders with long, narrow bills, upturned in avocets. Of the nine species, only one, the Critically Endangered Black Stilt, is threatened.

Of the 11 species of oystercatcher (Haematopodidae), just one is threatened—the Chatham Oystercatcher (Endangered).

Coursers and pratincoles (Glareolidae) are unusual waders. There are 17 species of this bird, of which two are threatened: Jerdon's Courser (Critically Endangered) and Madagascar Pratincole (Vulnerable).

Spoon-billed Sandpiper
Eurynorhynchus pygmeus

This lovely little wader has a highly distinctive spoon-shaped bill, unique among waders. It has typical wader plumage, brownish on top and white below. Males alter in the breeding season to a red-brown head, neck, and streaked breast with the upper body feathers turning darker with a rufous fringe. Spoon-billed Sandpipers require a very specialized coastal breeding habitat—lagoon spits with vegetation featuring crowberry, lichens, sedges, and dwarf birch and willow. They feed mainly on the surrounding mudflats. They migrate through eastern Asia, wintering mainly on tidal mudflats in Bangladesh and Myanmar. Much of this land is being reclaimed and as a result, this species is under threat through loss of habitat. Industry and agriculture on reclaimed land also creates pollution, further degrading the remaining habitat. Locals traditionally trap larger waders for food, and some Spoon-billed Sandpipers are also caught. Proposals have been made to campaign to stop coastal trapping of this kind in Asian countries, alongside the implementation of wetland restoration projects. Future plans aim to gain legal protection for this species.

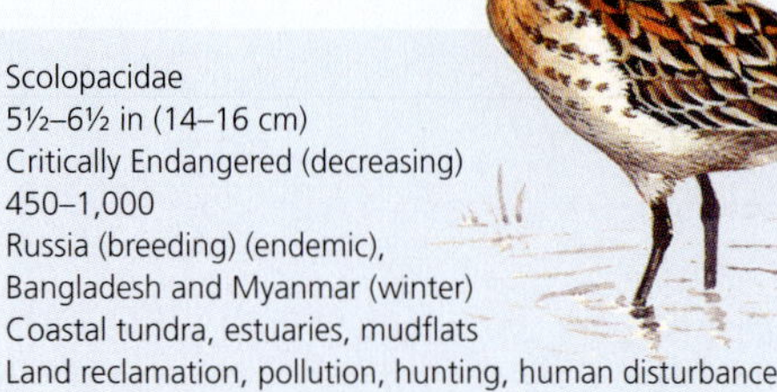

Family	Scolopacidae
Size	5½–6½ in (14–16 cm)
Status	Critically Endangered (decreasing)
Population	450–1,000
Range	Russia (breeding) (endemic), Bangladesh and Myanmar (winter)
Habitat	Coastal tundra, estuaries, mudflats
Main threats	Land reclamation, pollution, hunting, human disturbance

Wrybill
Anarhynchus frontalis

The Wrybill is the only bird known to have a bill that naturally curves to one side. This unusual plover uses its bill to probe for invertebrates under pebbles in shallow streams, close to which it also nests, laying its clutch in a scrape in the shingle. It is an attractive medium-sized bird, gray above and white below, with a black band on the throat and white forked eye stripes. The species is classed as Vulnerable, mainly due to threats to their nesting and feeding sites. Hydroelectric dams have altered the annual water flow feeding the channels, resulting in increased weed growth and reduction in suitable habitat.

Family Charadriidae
Size 7¾ in (20 cm)
Status Vulnerable (decreasing)
Population 4,500–5,000
Range New Zealand (endemic)
Habitat Shingle river banks, coastal areas
Main threats Introduced predators, habitat degradation, disturbance

Black Stilt
Himantopus novaezelandiae

New Zealand's Black Stilt have declined since the 1940s when there were some 500–1,000 birds. An elegant black wader with a very thin, long black bill and extremely long, thin red legs, it feeds on insects and small fish at swamps, braided riverbeds, and other wetlands. It breeds in the upper Waitaki Valley (South Island) where most birds are sedentary, though a small number also winter to coastal areas. The birds breed at three years, laying clutches of up to four eggs and prefer river shingle for nesting. These Critically Endangered birds are threatened by introduced predators, including stoats, ferrets, and cats.

Family Recurvirostridae
Size 7¾ in (20 cm)
Status Critically Endangered (increasing)
Population 40
Range New Zealand (endemic)
Habitat Braided river beds, swamp
Main threats Introduced predators, habitat loss, recreational use of rivers

THREATENED PLOVERS, SANDPIPERS, STILTS, OYSTERCATCHERS, AND ALLIES

FAMILY		COMMON NAME	SCIENTIFIC NAME	STATUS
CHARADRIIDAE Plovers	■	Wrybill	*Anarhynchus frontalis*	VU
	■	New Zealand Dotterel	*Charadrius obscurus*	EN
	■	St. Helena Plover	*Charadrius sanctaehelenae*	CR
	■	Madagascar Plover	*Charadrius thoracicus*	VU
	■	Shore Plover	*Thinornis novaeseelandiae*	EN
	■	Sociable Lapwing	*Vanellus gregarius*	CR
	■	Javan Lapwing	*Vanellus macropterus*	CR
SCOLOPACIDAE Sandpipers and allies	■	Great Knot	*Calidris tenuirostris*	VU
	■	Chatham Snipe	*Coenocorypha pusilla*	VU
	■	Spoon-billed Sandpiper	*Eurynorhynchus pygmeus*	CR
	■	Madagascar Snipe	*Gallinago macrodactyla*	VU
	■	Wood Snipe	*Gallinago nemoricola*	VU
	■	Eskimo Curlew	*Numenius borealis*	CR
	■	Far Eastern Curlew	*Numenius madagascariensis*	VU
	■	Bristle-thighed Curlew	*Numenius tahitiensis*	VU
	■	Slender-billed Curlew	*Numenius tenuirostris*	CR
	■	Tuamotu Sandpiper	*Prosobonia cancellata*	EN
	■	Ryukyu Woodcock	*Scolopax mira*	VU
	■	Moluccan Woodcock	*Scolopax rochussenii*	EN
	■	Spotted Greenshank	*Tringa guttifer*	EN
ROSTRATULIDAE Painted snipes	■	Australian Painted Snipe	*Rostratula australis*	EN
RECURVIROSTRIDAE Stilts	■	Black Stilt	*Himantopus novaezelandiae*	CR
HAEMATOPODIDAE Oystercatchers	■	Chatham Oystercatcher	*Haematopus chathamensis*	EN
GLAREOLIDAE Coursers and pratincoles	■	Madagascar Pratincole	*Glareola ocularis*	VU
	■	Jerdon's Courser	*Rhinoptilus bitorquatus*	CR

RED LIST ■ EX = Extinct ■ EW = Extinct in the Wild ■ CR = Critically Endangered ■ EN = Endangered ■ VU = Vulnerable

GULLS AND TERNS, AND AUKS

Gulls and terns (Laridae) are familiar birds of coasts and wetlands all over the world. Graceful in flight with long, narrow wings, most species have mainly white or pale plumage, though some are dark. They have webbed feet and swim well, and most species breed and feed at or close to wetlands or the sea. Terns are expert divers and typically plunge into the water to catch their prey, which usually consists of fish, small squid, or crustaceans. Compared with gulls, terns are slimmer and more angular in appearance, with shorter legs and usually a forked tail. Pollution and drainage, and development of wetlands and coastal waters are major threats to gulls and terns. These can lead to a variety of negative impacts on bird nesting sites, food sources, and habitat quality. There are 98 species, 11 of which are threatened. Endangered are Black-billed Gull, and Black-fronted and Peruvian Terns, while one species, Chinese Crested Tern, is Critically Endangered.

Auks (Alcidae) are dark, mostly small seabirds with compact bodies and short but powerful wings displaying a characteristic fast wing beat in flight. Most species breed colonially in the relative safety of remote coastal cliffs or islands. They feed on small fish and crustaceans, captured by diving from the surface of the sea. The wings are adapted for diving to feed, though many are also strong flyers. The webbed feet are set well back on an auk's body, creating a rather ducklike posture, and the birds also enjoy good buoyancy on the surface of the sea. All auks possess dense, waterproof plumage, with a distinctive smooth sheen. The family includes auks, puffins, guillemots, auklets, and murrelets. The extinct penguinlike flightless Greak Auk of the North Atlantic was last seen in 1852. This large bird had the misfortune of moving slowly on land, not possessing an instinctive fear of humans, and carrying plenty of meat that tasted good to explorers. Though hunting to extinction still comprises a major threat to some endangered birds, the range of problems generated by humans has widened considerably in the 20th and 21st centuries. Five auks are threatened, all species of murrelet. Most seriously at risk is Kittlitz's Murrelet of the Bering Sea region, where it is affected by oil spills, as a fisheries by-catch, and possibly by reduced abundance of prey, a possible consequence of climate change and glacier melt.

Relict Gull

Larus relictus

The Relict Gull has a small and fluctuating population that is affected greatly by land reclamation and changes in water levels. In mature breeding plumage the black hood extends further down the neck than in either Black-headed or Brown-headed Gulls. The broad, broken white eye-ring is also distinctive. The wings have white-tipped primaries, no white leading edge, and the birds are quite chunky in appearance. Relict Gulls breed on islets in saline lakes—habitats that offer some degree of isolation—in China, Kazakhstan, Russia, and Mongolia. Changes of vegetation, water levels, or accessibility all swiftly result in the habitat being abandoned. Some birds are known to migrate and winter to sites in China and South Korea, but little is really understood of these movements. Russian breeding sites are protected with no hunting allowed, but birds are under threat from extensive land reclamation plans in China and South Korea and tourist development in China. However, the main Chinese breeding site at Taolimiao-Alashan Nur is now partially a nature reserve, which should protect the land and reduce levels of disturbance.

Family Laridae
Size 17¼–17¾ in (44–45 cm)
Status Vulnerable (decreasing)
Population 2,500–10,000
Range China, Kazakhstan, Mongolia, Russia
Habitat Saline lakes, also estuaries and sandflats (winter)
Main threats Land reclamation, climate change, disturbance

Lava Gull

Larus fuliginosus

This unusual dark gull is found only in the Galápagos Islands of Ecuador, with the largest numbers on Santa Cruz, San Cristóbal, and Isabela. Lava Gulls are dark ash-gray in color with contrasting white eyelids and red inner mouthparts. The head is darkest gray, followed by the wings, with the belly and rump being slightly paler. Lava Gulls scavenge most of their food, but also feed on small fish and crustaceans, other seabird eggs, and also young Marine Iguanas (*Amblyrhynchus cristatus*). Nests are built alongside lagoons or calm pools where birds lay clutches of two eggs.

The small isolated populations are potentially threatened by the introduction of predators such as domestic dogs, domestic cats, and rats. The gulls tend to congregate at the main ports, which are good scavenging grounds. Galápagos National Park boundaries do not cover these ports, so conservation measures depend on careful population monitoring and maintaining controls over potential predators and disturbances.

Family Laridae
Size 20¾ in (53 cm)
Status Vulnerable (stable)
Population 600–800
Range Galápagos Islands, Ecuador (endemic)
Habitat Pools and coastal lagoons
Main threats Introduced predators

Marbled Murrelet

Brachyramphus marmoratus

With typical auk features, this medium-sized bird has a sturdy body and short neck. Dark brown above and marbled below, it is paler in winter, with contrasting dark breast stripes and a white eye-ring. It occurs along the Pacific northwest of the U.S.A. and Canada (California, Oregon, Washington, Alaska, Prince William Sound, Kenai Peninsula, Lower Cook Inlet, Barren Islands, Afognak, and Kodiak Islands) and British Columbia. It breeds in old-growth forest, sometimes at a distance from the coast. Though still numerous, it has suffered a large reduction in numbers over recent decades. A major threat is a notable reduction in the marine food reserves on which it depends. It feeds on fish including herring and sand lance, and also on invertebrates, at sea or in lagoons and coastal lakes. However, fish stocks have declined and smaller prey species such as krill have tended to become the main constituent of their diet. This seems to have lowered breeding productivity. The forest habitat required for nesting is threatened by logging, and birds are also caught in local fishing gill-nets. In addition, breeding birds suffer from nest predation by corvids. Conservation plans concentrate on improving breeding success through a proposal to increase fish stocks during the breeding season coupled with improved protection of nest sites. By relocating local campsites, predator numbers should also be reduced and the predation of nests diminished. Conservation also aims to reduce gill-net mortality and the threat of proposed logging in some areas.

Family Alcidae
Size 9¾ in (25 cm)
Status Endangered (decreasing)
Population 480,000–760,000
Range Northern U.S.A., Canada
Habitat Native old-growth forest (breeding), coastal waters (feeding)
Main threats Habitat loss, logging, reduced availability of prey, fishing nets

THREATENED GULLS AND TERNS, AND AUKS

FAMILY		COMMON NAME	SCIENTIFIC NAME	STATUS
LARIDAE Gulls and terns	■	Olrog's Gull	*Larus atlanticus*	VU
	■	Black-billed Gull	*Larus bulleri*	EN
	■	Lava Gull	*Larus fuliginosus*	VU
	■	Relict Gull	*Larus relictus*	VU
	■	Saunders's Gull	*Larus saundersi*	VU
	■	Red-legged Kittiwake	*Rissa brevirostris*	VU
	■	Indian Skimmer	*Rynchops albicollis*	VU
	■	Black-fronted Tern	*Sterna albostriata*	EN
	■	Chinese Crested Tern	*Sterna bernsteini*	CR
	■	Peruvian Tern	*Sterna lorata*	EN
	■	Fairy Tern	*Sterna nereis*	VU
ALCIDAE Auks	■	Kittlitz's Murrelet	*Brachyramphus brevirostris*	CR
	■	Marbled Murrelet	*Brachyramphus marmoratus*	EN
	■	Craveri's Murrelet	*Synthliboramphus craveri*	VU
	■	Xantus's Murrelet	*Synthliboramphus hypoleucus*	VU
	■	Japanese Murrelet	*Synthliboramphus wumizusume*	VU

RED LIST ■ EX = Extinct ■ EW = Extinct in the Wild ■ CR = Critically Endangered ■ EN = Endangered ■ VU = Vulnerable

DOVES AND PIGEONS

Doves and pigeons (Columbidae) are a large family of over 300 species found in most regions with the exception of the high latitudes of the north. Most live closely associated with trees and forests, making simple flimsy nests from twigs, and though the clutch is normally small, they often have several broods per year. As a general, though not rigid, rule the larger species are called pigeons and the smaller species doves. In size they range from the tiny ground-doves (genus *Columbina*), which are about 6¼ in (16 cm) in length, to the relatively huge crowned-pigeons (genus *Goura*), which grow to 30 in (75 cm) long. One of their most unusual features of this family (shared incidentally with flamingos) is their ability to produce a kind of "milk" somewhat comparable to mammalian milk, but secreted by the walls of the crop. The adult birds feed their young on this highly nutritious "crop-milk" for the first few days after hatching. Doves and pigeons feed on vegetable matter including leaves, shoots, seeds, and fruits, and some species are pests of crops. Fruit-pigeons that live in the tropics spend almost all their time in the trees, feeding mainly on fruit; they seldom go down to the ground to search for food.

Sixty species are currently listed as threatened, 15 of which are Endangered, nine are Critically Endangered and one (Socorro Dove) is Extinct in the Wild, with reintroduction planned.

Purple-winged Ground-dove

Claravis godefrida

The male of this species is slate-blue gray above and pale underneath, with a white face and belly. The tail is white with a gray center and the wings have two to three dark purple bands and white edges. The female has browner, duller plumage. This species was relatively common at the beginning of the 20th century when it occurred in Brazil, Argentina, and Paraguay, sometimes in flocks of up to 100 birds. There have been recent sightings in all three countries, but in such small numbers that it is now classed as Critically Endangered. The favored habitat is humid Atlantic forest with bamboo-rich undergrowth. The main threat is loss of this habitat, reducing the availability of seeds of bamboo and other grasses. Though Purple-winged Ground-doves have been observed feeding on sedge seeds, and fruit it is thought to rely, at least in Argentina, particularly on two bamboo species—Takuarusu (*Guadua chacoensis*) and Yatevo (*G. trinii*). Much of this native forest has been converted to plantations, leaving only small, fragmented remnants.

Family Columbidae
Size 7½–9 in (19–23 cm)
Status Critically Endangered (decreasing)
Population 50–250
Range Argentina, Brazil, Paraguay
Habitat Humid Atlantic forest
Main threats Habitat loss

THREATENED DOVES AND PIGEONS

FAMILY		COMMON NAME	SCIENTIFIC NAME	STATUS
COLUMBIDAE Doves and pigeons	■	Purple-winged Ground-dove	*Claravis godefrida*	CR
	■	Silvery Wood-pigeon	*Columba argentina*	CR
	■	Nilgiri Wood-pigeon	*Columba elphinstonii*	VU
	■	Pale-backed Pigeon	*Columba eversmanni*	VU
	■	White-tailed Laurel Pigeon	*Columba junoniae*	EN
	■	Yellow-legged Pigeon	*Columba pallidiceps*	VU
	■	Pale-capped Pigeon	*Columba punicea*	VU
	■	Maroon Pigeon	*Columba thomensis*	VU
	■	Sri Lanka Wood-pigeon	*Columba torringtoniae*	VU
	■	Blue-eyed Ground-dove	*Columbina cyanopis*	CR
	■	Tooth-billed Pigeon	*Didunculus strigirostris*	EN
	■	Polynesian Imperial-pigeon	*Ducula aurorae*	EN
	■	Vanuatu Imperial-pigeon	*Ducula bakeri*	VU

RED LIST ■ EX = Extinct ■ EW = Extinct in the Wild ■ CR = Critically Endangered ■ EN = Endangered ■ VU = Vulnerable

THREATENED DOVES AND PIGEONS *Continued*

FAMILY	COMMON NAME	SCIENTIFIC NAME	STATUS
COLUMBIDAE Doves and pigeons	Chestnut-bellied Imperial-pigeon	*Ducula brenchleyi*	VU
	Spotted Imperial-pigeon	*Ducula carola*	VU
	Timor Imperial-pigeon	*Ducula cineracea*	EN
	Marquesan Imperial-pigeon	*Ducula galeata*	EN
	Mindoro Imperial-pigeon	*Ducula mindorensis*	EN
	Gray Imperial-pigeon	*Ducula pickeringii*	VU
	Christmas Imperial-pigeon	*Ducula whartoni*	VU
	Mindanao Bleeding-heart	*Gallicolumba crinigera*	VU
	Polynesian Ground-dove	*Gallicolumba erythroptera*	CR
	Wetar Ground-dove	*Gallicolumba hoedtii*	EN
	Negros Bleeding-heart	*Gallicolumba keayi*	CR
	Caroline Islands Ground-dove	*Gallicolumba kubaryi*	VU
	Sulu Bleeding-heart	*Gallicolumba menagei*	CR
	Mindoro Bleeding-heart	*Gallicolumba platenae*	CR
	Marquesan Ground-dove	*Gallicolumba rubescens*	VU
	Santa Cruz Ground-dove	*Gallicolumba sanctaecrucis*	EN
	Shy Ground-dove	*Gallicolumba stairi*	VU

RED LIST EX = Extinct EW = Extinct in the Wild CR = Critically Endangered EN = Endangered VU = Vulnerable

Tooth-billed Pigeon
Didunculus strigirostris

This dark pigeon is greenish black with rust-colored upperparts. Its most unusual feature, reflected in its common name, is the robust hooked red-and-yellow bill with two upward projections on the lower mandible. These toothlike projections overlap the upper bill and enable it to slice through the fibrous skins of forest fruits, particularly *Dysoxylum*, to access the seeds and flesh. Small groups of Tooth-billed Pigeons are found at forest edges, clearings, and perched high in the forest canopy. The main threats to this species come from destruction of the limited remaining forest habitat. In 1990 a cyclone had a devastating effect on this already small isolated population and habitat by removing up to 75 percent of the forest canopy. This is compounded by ongoing deforestation by locals for agricultural purposes and invasion by non-native tree species. Some birds are also caught during seasonal hunts, though technically this species is legally protected.

Family Columbidae
Size 12 in (31 cm)
Status Endangered (decreasing)
Population 1,000–2,500
Range Samoa (endemic)
Habitat Primary forest
Main threats Deforestation, climate (cyclones), invasive alien trees, hunting

Victoria Crowned-pigeon
Goura victoria

This enormous pigeon has a remarkable tall crest of blue-gray lacy feathers tipped with white. The body is steely gray with darker maroon patches on the breast and belly. At rest, the wing shows a distinct pale blue-gray bar and the bird also has striking dark red-and-white scaly legs. The Victoria Crowned-pigeon is found in Indonesia (Papua and the islands of Biak and Yapen) and in Papua New Guinea. It is a ground-feeder, foraging on the rain forest floor for fallen fruit, seeds, and berries. It roosts in tree branches where it also builds a large platformlike nest of twigs and leaves. This species is threatened partly by its popularity as an easily caught local source of food. Further threats are continued logging of the forests, which reduces the habitat while also allowing hunters access to previously more impenetrable areas of forest. In Papua New Guinea it is already protected by law. Conservationists plan to enforce no hunting rules in uninhabited areas and make the bird a flagship species for local ecotourism.

Family Columbidae
Size 29 in (74 cm)
Status Vulnerable (decreasing)
Population 2,500–10,000
Range Indonesia (Papua, Biak, Yapen), Papua New Guinea
Habitat Lowland and swamp forest
Main threats Hunting, logging

Negros Bleeding-heart
Gallicolumba keayi

This colorful pigeon with iridescent green head, nape, and shoulders has a striking blood-red marking on its otherwise white breast. The remaining upperparts are colored deep chestnut and reddish purple and it has a pale band on the inner wing coverts. It is a ground-feeding bird found in native forest, nesting in native ferns that grow as epiphytes on the forest trees. This species was common until the 19th century, since when its numbers have been in decline. Previously found across the Philippines, it is now only known to exist on two islands—Panay and Negros. The primary forests of both these islands have been largely removed, though some of the remaining forest areas are now protected. The main threat comes from destruction of forest by clearance for agriculture, logging, and charcoal burning. The pigeon is also relatively easy to catch and has been hunted by locals for its meat and possibly for the cagebird trade. Conservation proposals include education campaigns on endemic species to reduce hunting, along with additional protection of remaining forest areas from illegal logging.

Family Columbidae
Size 11¾ in (30 cm)
Status Critically Endangered (decreasing)
Population 50–250
Range Philippines (endemic)
Habitat Dense, closed-canopy forest
Main threats Habitat degradation, hunting

Pink Pigeon
Nesoenas mayeri

This large pigeon is pale pinkish-gray with a darker back and rusty tail. By 1990 its population had fallen to just ten individuals, but extensive conservation efforts by the Durrell Wildlife Conservation Trust have brought it back from almost certain extinction to over 300. The Pink Pigeon survives in the Black River Gorges in southwest Mauritius, where the Black River National Park covers most of its range, as well as on the small Ile aux Aigrettes off the southeast coast. They inhabit the small remaining patches of native forest. The pigeons have also nested in introduced Japanese Red Cedar (*Cryptomeria*) though they prefer native trees. Even though the Pink Pigeon has had a close brush with extinction, it is still considered Endangered due to threats to its natural habitat by invasion of alien plant species as well as predation by alien carnivores such as the Crab-eating Macaque (*Macaca fascicularis*) and Indian Mongoose (*Herpestes auropunctatus*), rats and feral domestic cats. Cyclones have also destroyed nests. Alien pigeons have transmitted trichomoniasis, a potentially fatal disease, especially to young, vulnerable chicks.

Family Columbidae
Size 14–15 in (36–38 cm)
Status Endangered (fluctuating)
Population 360–400
Range Mauritius (endemic)
Habitat Native forest
Main threats Habitat loss, alien predators, disease, invasive plants, climate (cyclones)

THREATENED DOVES AND PIGEONS *Continued*

FAMILY		COMMON NAME	SCIENTIFIC NAME	STATUS
COLUMBIDAE Doves and pigeons	■	Gray-headed Quail-dove	*Geotrygon caniceps*	VU
	■	Tuxtla Quail-dove	*Geotrygon carrikeri*	EN
	■	Western Crowned-pigeon	*Goura cristata*	VU
	■	Southern Crowned-pigeon	*Goura scheepmakeri*	VU
	■	Victoria Crowned-pigeon	*Goura victoria*	VU
	■	New Britain Bronzewing	*Henicophaps foersteri*	VU
	■	Brown-backed Dove	*Leptotila battyi*	VU
	■	Tolima Dove	*Leptotila conoveri*	EN
	■	Ocher-bellied Dove	*Leptotila ochraceiventris*	VU
	■	Grenada Dove	*Leptotila wellsi*	CR
	■	Pink Pigeon	*Nesoenas mayeri*	EN
	■	Ring-tailed Pigeon	*Patagioenas caribaea*	VU
	■	Peruvian Pigeon	*Patagioenas oenops*	VU
	■	Mindanao Brown-dove	*Phapitreron brunneiceps*	VU
	■	Tawitawi Brown-dove	*Phapitreron cinereiceps*	EN
	■	Negros Fruit-dove	*Ptilinopus arcanus*	CR
	■	Makatea Fruit-dove	*Ptilinopus chalcurus*	VU

RED LIST ■ EX = Extinct ■ EW = Extinct in the Wild ■ CR = Critically Endangered ■ EN = Endangered ■ VU = Vulnerable

THREATENED DOVES AND PIGEONS *Continued*

FAMILY		COMMON NAME	SCIENTIFIC NAME	STATUS
COLUMBIDAE Doves and pigeons	■	Red-naped Fruit-dove	*Ptilinopus dohertyi*	VU
	■	Carunculated Fruit-dove	*Ptilinopus granulifrons*	VU
	■	Rapa Fruit-dove	*Ptilinopus huttoni*	VU
	■	Henderson Fruit-dove	*Ptilinopus insularis*	VU
	■	Flame-breasted Fruit-dove	*Ptilinopus marchei*	VU
	■	Cook Islands Fruit-dove	*Ptilinopus rarotongensis*	VU
	■	Mariana Fruit-dove	*Ptilinopus roseicapilla*	EN
	■	Blue-headed Quail-dove	*Starnoenas cyanocephala*	EN
	■	Large Green-pigeon	*Treron capellei*	VU
	■	Flores Green-pigeon	*Treron floris*	VU
	■	Pemba Green-pigeon	*Treron pembaensis*	VU
	■	Timor Green-pigeon	*Treron psittaceus*	EN
	■	Socorro Dove	*Zenaida graysoni*	EW

RED LIST ■ EX = Extinct ■ EW = Extinct in the Wild ■ CR = Critically Endangered ■ EN = Endangered ■ VU = Vulnerable

Blue-headed Quail-dove

Starnoenas cyanocephala

This cinnamon-brown dove has beautiful head markings with a white facial stripe, a black eye stripe, and a blue crown. It has a black gorget with a thin white border and mottled blue throat markings below. Blue-headed Quail-doves feed at ground level, foraging in leaf litter and on forest tracks for seeds, berries, and snails. It also nests low down in old tree stumps or between tree roots, breeding from April to June. Occasionally the bird is also found on lower, swampy areas. Once common, this endemic dove has suffered badly from habitat degradation, and one of the main populations is now found in the Zapata area. The habitat degradation comes not only from people, but also from extreme weather, including a significant hurricane in 1996 that felled many trees in this area. The bird is hunted for food and is still trapped using orange pips as bait, though this practice is currently illegal.

Family Columbidae
Size 11¾–13 in (30–33 cm)
Status Endangered (decreasing)
Population 1,000–2,500
Range Cuba (endemic)
Habitat Forest undergrowth including swamps
Main threats Habitat destruction, hurricanes, hunting

White-tailed Laurel Pigeon

Columba junoniae

This large pigeon has dark plumage, mainly brown and gray and a characteristic whitish terminal band to the tail. Endemic to the Canary Islands (Spain), it is found on La Palma, La Gomera, Tenerife, and El Hierro. The bulk of the population is on La Palma, especially in the north of the island, and it is also fairly common on La Gomera, again mainly in the north. It prefers steep terrain and inhabits a range of forest types, notably the rather unusual laurel forests for which the islands are famous. Much of the original forest has been lost, and felling and forest fragmentation remain a serious threat. Livestock grazing also degrades the habitat and predation of eggs and chicks by Black Rats (*Rattus rattus*) and feral domestic cats are a problem, especially on Tenerife. The species is particularly at risk from these predators as the birds nest on the ground, typically under rocky ledges or at the base of trees or beneath fallen tree trunks; both sites where they are easily detected by foraging predators. There are also losses to illegal hunting, and efforts are underway to control the hunting and to raise local awareness. The closely related Dark-tailed Laurel Pigeon (*Columba bollii*) is also found in similar habitats. It has a larger, stable population, though it is listed as Near Threatened.

Family Columbidae
Size 15 in (38 cm)
Status Endangered (decreasing)
Population 3,000–7,500
Range Canary Islands (endemic)
Habitat Forest and cultivated areas
Main threats Habitat loss, predation, hunting

NEW WORLD AMAZONS AND MACAWS

The parrot family (Psittacidae) is very large, with some 355 species, a staggering 96 of which are now considered to be threatened. Most parrots are native to the tropics and southern countries, with few occurring north of the Tropic of Cancer. For convenience we have divided this family into four groups, two from the New World (see pp.158–161) and two from the Old World (see pp.162–165).

Parrots and their close relations include some of the world's most beautiful and intelligent birds, and partly for this reason they have come under special threat by being trapped and traded for the cagebird market. In addition to the impact on the wild populations, this is a cruel business and many of the birds die in transit. This trade, coupled with habitat losses, has brought several species to the brink of extinction and has seriously endangered many more. Parrots are one of the most obviously recognizable groups of birds, characterized by their distinctive curved bill shape. The bill functions as an extremely versatile tool for tackling a variety of fruit and seeds. Like the woodpeckers, parrots have zygodactyl feet (with two toes facing forward and two back) that help them to climb in all directions. Parrots also use their feet very flexibly to grip food items and, rather unusually among birds, to pass them to their bill. Many species are collected and kept in captivity, either for their beauty, their amusing antics, or in some species, for their remarkable ability to mimic human speech and other sounds. Most live in tropical and subtropical woodland or forest—habitats that continue to be disturbed or felled—so it is not surprising that the group is under so much pressure. Some parrots are found only on islands where threats are from introduced pests such as domestic cats and rats.

Amazons and macaws—some of the most attractive members of the family—are found only in the New World: in Mexico, Central America, the Caribbean (where only amazons survive), and South America. Amazons (genus *Amazona*) are medium-sized, rather chunky parrots and most have colorful plumage, square tails, and bright speculum patches in the wings. Of the 30 species, 16 are threatened: five of these are Endangered and one, the endemic Puerto Rican Amazon, is Critically Endangered with a tiny wild population of fewer than 50. There are 17 species of macaw (genera *Ara*, *Anodorhynchus*, *Cyanopsitta*, *Orthopsittaca*, *Primolius*, and *Diopsittaca*), of which ten are threatened. Four are Endangered, and three (Glaucous, Spix's, and Blue-throated) are Critically Endangered. With their bright plumage and long, graduated tails, macaws are some of the largest and most spectacular of parrots. They have a large facial patch of bare, whitish skin. The beautiful Lear's Macaw (Endangered) of Brazil was brought back from the brink of extinction through habitat protection and prevention of illegal poaching.

St Vincent Amazon

Amazona guildingii

Endemic to St Vincent, where it is the only parrot, the St Vincent Amazon occurs in two color morphs. The commoner form is yellow-brown with a white head with yellow on the back of the crown, bronze upperparts and greenish vent, and orange and red wing coverts. The rarer green morph is duller, and lacks the orange. It inhabits moist, mainly low-altitude forest, between 410 ft (125 m) and 3,300 ft (1,000 m), where it feeds on fruits, flowers, and seeds. The peak breeding season is from February to May, and the birds nest mostly in holes in mature trees. Much of the original habitat has been lost, partly converted to plantations, notably bananas, reducing the nest sites. In addition, these parrots were also hunted for food. Amazon parrots are popular pets and trapping to supply the illegal trade in cagebirds seriously affected this species. It is now protected in the St Vincent Parrot Reserve and numbers are now increasing from a low of about 400 in the early 1980s.

Family	Psittacidae
Size	15¾ in (40 cm)
Status	Vulnerable (increasing)
Population	800
Range	St Vincent (endemic)
Habitat	Moist forest
Main threats	Habitat loss, hunting, trapping

Hyacinth Macaw

Anodorhynchus hyacinthinus

Parrots are very well-represented in South America and none is more impressive than the Hyacinth Macaw, the world's longest parrot. Well-named, its plumage is unusual in being a uniform hyacinth blue. Its large bill is dark and there is a patch of yellow skin on its face. Like most macaws it is a noisy and sociable bird, small groups foraging together in search of fruits and nuts, often communicating with loud, screeching calls. It lives mainly close to water in forest, savanna, and grassland, especially where there are plenty of palm trees to provide food. Nesting is either in holes in trees or on cliffs. Its size and beauty have made this species a prize target for hunters and trappers, mainly for the illegal cagebird trade. This and habitat loss have combined to reduce its numbers to fewer than 6,500 birds, most of which are found in Brazil, mainly in three areas: east Amazonia, a small area further south, and another close to the borders with Bolivia and Paraguay. The Pantanal holds the greatest numbers (about 5,000) and conservation efforts have resulted in an expansion of its range there.

Family	Psittacidae
Size	3–3¼ ft (95–100 cm)
Status	Endangered (decreasing)
Population	6,500
Range	Brazil, Bolivia, Paraguay
Habitat	Dry forest, savanna, flooded forest (*várzea*)
Main threats	Habitat loss, trapping

Great Green Macaw

Ara ambiguus

Four of the eight living species of macaw in this genus are threatened, two of which are Endangered (the Red-fronted Macaw and Great Green Macaw), and one Critically Endangered (Blue-throated Macaw). The Great Green Macaw is a large, impressive parrot with a huge, black bill topped by a red frontal band, and with black lines on the bare face patch. Its head, back, and upper wings are olive green, contrasting with the rest of the wings and tail, which are blue, the tail having a scarlet patch. Edges of lowland tropical rain forest are its favored habitat and it is associated with Mountain Almond (*Dipteryx panamensis*) trees, used for both food and nesting. In Ecuador it has been seen to feed partly on orchids. Illegal capture for food and feathers is a major threat and it is often shot, being regarded as a crop pest. Plantations, cattle ranching, and logging further degrade its habitat. Two strongholds for this beautiful macaw are Darién Biosphere Reserve (Panama) and Los Katíos National Park (Colombia).

Family Psittacidae
Size 33½–35½ in (85–90 cm)
Status Endangered (decreasing)
Population 1,000–2,500
Range Guatemala, Honduras, Nicaragua, Costa Rica, Panama, Colombia, Ecuador
Habitat Humid forest
Main threats Habitat loss, logging, hunting, capture

THREATENED AMAZONS AND MACAWS

FAMILY	COMMON NAME	SCIENTIFIC NAME	STATUS
PSITTACIDAE Amazons and macaws	Black-billed Amazon	*Amazona agilis*	VU
	Red-necked Amazon	*Amazona arausiaca*	VU
	Yellow-shouldered Amazon	*Amazona barbadensis*	VU
	Red-tailed Amazon	*Amazona brasiliensis*	VU
	Yellow-billed Amazon	*Amazona collaria*	VU
	Lilac-crowned Amazon	*Amazona finschi*	VU
	St Vincent Amazon	*Amazona guildingii*	VU
	Imperial Amazon	*Amazona imperialis*	EN
	Yellow-headed Amazon	*Amazona oratrix*	EN
	Red-spectacled Amazon	*Amazona pretrei*	VU
	Red-browed Amazon	*Amazona rhodocorytha*	EN
	Hispaniolan Amazon	*Amazona ventralis*	VU
	St Lucia Amazon	*Amazona versicolor*	VU
	Vinaceous Amazon	*Amazona vinacea*	EN
	Red-crowned Amazon	*Amazona viridigenalis*	EN
	Puerto Rican Amazon	*Amazona vittata*	CR
	Glaucous Macaw	*Anodorhynchus glaucus*	CR
	Hyacinth Macaw	*Anodorhynchus hyacinthinus*	EN
	Lear's Macaw	*Anodorhynchus leari*	EN
	Great Green Macaw	*Ara ambiguus*	EN
	Blue-throated Macaw	*Ara glaucogularis*	CR
	Military Macaw	*Ara militaris*	VU
	Red-fronted Macaw	*Ara rubrogenys*	EN
	Spix's Macaw	*Cyanopsitta spixii*	CR
	Blue-headed Macaw	*Primolius couloni*	VU

RED LIST EX = Extinct EW = Extinct in the Wild CR = Critically Endangered EN = Endangered VU = Vulnerable

NEW WORLD PARROTS, PARAKEETS, AND PARROTLETS

This section contains threatened parrots, parakeets, and parrotlets from Central and South America. The New World parrots include six threatened species, in three genera: *Hapalopsittaca* (two Vulnerable, one Critically Endangered); *Ognorhynchus* (one Endangered), and *Rhynchopsitta* (one Vulnerable, one Endangered). The genus *Hapalopsittaca* are medium-sized parrots that are characterized by bright green, red, and blue plumage, while the single species of *Ognorhynchus* and the two species of *Rhynchopsitta* resemble small macaws.

There are 15 threatened New World parakeets, in six genera. *Aratinga* (20 species: two are Vulnerable, two Endangered); *Bolborhynchus* (three species, of which one is Vulnerable); *Brotogeris* (eight species, of which one is Endangered); *Guaruba* (a single species, Endangered); *Leptosittaca* (a single species, Vulnerable); and *Pyrrhura* (20 species, of which three are Vulnerable, three Endangered, one Critically Endangered). Five New World parrotlets are threatened: *Forpus* (seven species of which one is Vulnerable); *Touit* (eight species, of which three are Vulnerable, one Endangered).

Gray-breasted Parakeet

Pyrrhura griseipectus

Rather long-tailed and slim, this parakeet is mainly green, with blue in the wings, and has a reddish-brown rump, belly, and tail. The shoulders bear bright red flashes. Its face is a dark plum color and the chest is silvery-gray with scalloped markings. A combination of a very small known range and population puts it firmly into the Critically Endangered category. Endemic to Brazil, it is known only from a single region of the northeast, namely the Serra do Baturité. In this region, "islands" of humid montane forests formed of trees about 65 ft (20 m) tall rise up from the semi-arid habitats of the lowlands. Sadly, these forests have been largely converted to plantations for coffee and other crops. The main threat though, both historically and current, is trapping of the birds for local and international trade. Luckily, the Gray-breasted Parakeet breeds well in captivity and reintroductions may be possible from cage-bred stocks.

Family Psittacidae
Size 9 in (23 cm)
Status Critically Endangered (decreasing)
Population 50–250
Range Brazil (endemic)
Habitat Montane humid forest
Main threats Trapping for cagebird trade, habitat loss, plantations

Yellow-eared Parrot

Ognorhynchus icterotis

Named for its large yellow ear-patches, this macawlike parrot is otherwise mainly green, except for the underside of the tail which is red-brown. It lives in humid montane forests of the Central Andes of Colombia, and possibly also in northwest Ecuador. It is closely dependent on the Wax Palm (*Ceroxylon quindiuense*) in which it feeds and nests. This, the tallest of all palms and the national tree of Colombia, can grow to 200 ft (60 m), and live to over 500 years: it also has a restricted distribution. Fronds from this tree are regularly gathered for Palm Sunday processions, a process that often involves felling the whole tree. Browsing by livestock also hinders regeneration of the palms. Decades of habitat degradation and hunting pressure have reduced Yellow-eared Parrot numbers to the point at which it is Endangered. Despite all these threats, conservation and education efforts have been successful in halting and even reversing its decline. Among other initiatives, breeding sites have been protected by fencing that allows regeneration of trees and nest-boxes have been provided for the parrots.

Family Psittacidae
Size 16½ in (42 cm)
Status Endangered
Population 600+
Range Colombia, Ecuador
Habitat Humid montane forest
Main threats Habitat loss, browsing by cattle, logging, hunting

THREATENED NEW WORLD PARROTS, PARAKEETS, AND PARROTLETS

FAMILY		COMMON NAME	SCIENTIFIC NAME	STATUS
PSITTACIDAE New World parrots	■	Rusty-faced Parrot	*Hapalopsittaca amazonina*	VU
	■	Indigo-winged Parrot	*Hapalopsittaca fuertesi*	CR
	■	Red-faced Parrot	*Hapalopsittaca pyrrhops*	VU
	■	Yellow-eared Parrot	*Ognorhynchus icterotis*	EN
	■	Thick-billed Parrot	*Rhynchopsitta pachyrhyncha*	EN
	■	Maroon-fronted Parrot	*Rhynchopsitta terrisi*	VU

RED LIST ■ EX = Extinct ■ EW = Extinct in the Wild ■ CR = Critically Endangered ■ EN = Endangered ■ VU = Vulnerable

Brown-backed Parrotlet
Touit melanonotus

Small and pretty, the Brown-backed Parrotlet is mainly green, with a dark brown back and brown ear coverts. Its tail is red with a black tip and central green feathers. In the wild it is confined to lower montane evergeen forest between about 1,600 ft (500 m) and 4,600 ft (1,400 m) in a handful of scattered sites in southwest Brazil (Bahia, Espírito Santo, Rio de Janeiro, and São Paulo). It seems to have been considered rare even in the 19th century, but is hard to spot and its presence all too easy to miss. Its diet includes seeds of forest trees, and species of mistletoe. The main threats to this bird are deforestation, urbanization, and road building. Fortunately it is found in a number of protected sites including Desengano and Pedra Branca State Parks and Tijuca National Park. However, its population still seems to be in decline.

Family Psittacidae
Size 6 in (15 cm)
Status Endangered (decreasing)
Population 250–1,000
Range Brazil (endemic)
Habitat Lower montane forest
Main threats Deforestation, urbanization, road building

Sun Parakeet
Aratinga solstitialis

This small parrot, also known as the Sun Conure, has striking, mainly golden-yellow plumage, with darker greenish-blue wings and tail, and contrasting patches of reddish-orange on the face, belly, and rump. It is partly this unusual color scheme that has made it so attractive as a cagebird. Sadly, this has led to a rapid reduction in the wild population over recent decades and the threat from trappers continues. Trapping is relatively easy too, as the birds readily come to feed on bait such as corn. It is restricted to dry semi-deciduous forests in central Guyana and in Brazil (Roraima), with an estimated 90 percent of the population in Brazil. It feeds mainly on fruit, buds, and flowers, and groups move about in flocks, making shrill, screeching calls. Trade in wild birds is an ongoing problem and has resulted in its virtual elimination from most of its range in Guyana. In captivity it is relatively common.

Family Psittacidae
Size 11¾ in (30 cm)
Status Endangered (decreasing)
Population 1,000–2,500
Range Guyana, Brazil
Habitat Dry, semi-deciduous forest
Main threats Trapping

THREATENED NEW WORLD PARROTS, PARAKEETS, AND PARROTLETS *Continued*

FAMILY		COMMON NAME	SCIENTIFIC NAME	STATUS
PSITTACIDAE New World parakeets	■	Socorro Parakeet	*Aratinga brevipes*	EN
	■	Hispaniolan Parakeet	*Aratinga chloroptera*	VU
	■	Cuban Parakeet	*Aratinga euops*	VU
	■	Sun Parakeet	*Aratinga solstitialis*	EN
	■	Rufous-fronted Parakeet	*Bolborhynchus ferrugineifrons*	VU
	■	Gray-cheeked Parakeet	*Brotogeris pyrrhoptera*	EN
	■	Golden Parakeet	*Guaruba guarouba*	EN
	■	Golden-plumed Parakeet	*Leptosittaca branickii*	VU
	■	White-necked Parakeet	*Pyrrhura albipectus*	VU
	■	Flame-winged Parakeet	*Pyrrhura calliptera*	VU
	■	Blue-throated Parakeet	*Pyrrhura cruentata*	VU
	■	Gray-breasted Parakeet	*Pyrrhura griseipectus*	CR
	■	El Oro Parakeet	*Pyrrhura orcesi*	EN
	■	Pfrimer's Parakeet	*Pyrrhura pfrimeri*	EN
	■	Santa Marta Parakeet	*Pyrrhura viridicata*	EN
PSITTACIDAE New World parrotlets	■	Yellow-faced Parrotlet	*Forpus xanthops*	VU
	■	Red-fronted Parrotlet	*Touit costaricensis*	VU
	■	Brown-backed Parrotlet	*Touit melanonotus*	EN
	■	Spot-winged Parrotlet	*Touit stictopterus*	VU
	■	Golden-tailed Parrotlet	*Touit surdus*	VU

RED LIST ■ EX = Extinct ■ EW = Extinct in the Wild ■ CR = Critically Endangered ■ EN = Endangered ■ VU = Vulnerable

Old World cockatoos, lorikeets, and parakeets

This group comprises the cockatoos, lorikeets, and parakeets of the Old World. Cockatoos are a distinctive group of some 20 species of parrot, sometimes classified in a separate family (Cacatuidae). They have an Australasian distribution being found mainly in Australia, New Guinea, and the Philippines. Most are medium-sized and rather chunky in build, with ornate, erectile crests, and some have colored facial skin. The common name "cockatoo" may have its origin in the Malay word "kakatuwah" meaning "vice" or "grip," a reference to their capable, chunky, sharp beaks, which can open and feed on a wide variety of foods, including seeds, tubers, corms, fruit, flowers, and insects. Australian cockatoos exploit stores of seeds found in many trees and shrubs such as the gums (eucalypts) and bottlebrushes in this way. Cockatoos often feed in flocks for safety, especially when feeding on the ground. Pairs are monogamous and nest mainly in old tree hollows.

Seven cockatoos are threatened, of which two (Baudin's Black-cockatoo and Short-billed Black-cockatoo) are Endangered, while Philippine and Yellow-crested Cockatoos are both Critically Endangered. Like many other parrots, cockatoos are much sought after as pets and the illegal trade is a major threat to some species.

Lorikeets are small to medium-sized parrots with rather pointed wings and tails. They have brush-tipped tongues that facilitate feeding on the nectar and pollen in flowers and on soft fruit, a feature shared with the closely related lories. Lorikeets fly well due to their pointed tails and tapered wings. Legs and claws are strong, enabling agile walking and gripping. The distinction between lories and lorikeets is rather subjective, but generally birds with shorter, blunter tails are called lories and those with longer ones are referred to as lorikeets.

Of the 35 species of lorikeet, eight are listed as threatened. Two are Endangered (Rimatara and Ultramarine Lorikeets) while three (Red-throated, New Caledonian, and Blue-fronted Lorikeets) are Critically Endangered.

The term "parakeet" is rather imprecise, being used somewhat loosely to refer to many small parrots. Of the Old World parakeets, nine are threatened, four of which are Endangered and one, Malherbe's Parakeet, is Critically Endangered. This New Zealand endemic has a very small population and suffered from alien predators, before being translocated to offshore predator-free islands. Some, such as the Rose-ringed Parakeet, native to Africa and south Asia, have established naturalized populations well outside their native range, from South America to Europe, made up of birds escaped or released from captivity. This species has established colonies in many European cities, and in parts of the U.S.A., Japan, and South Africa. In certain regions, such as in Kashmir, this parakeet raids fruit orchards and tends to be regarded as a pest.

Yellow-crested Cockatoo

Cacatua sulphurea

The Yellow-crested Cockatoo is one of the most flamboyant of parrots and, along with the commoner and related Sulfur-crested Cockatoo of Australia and New Guinea, is also one of the most sought after for the cagebird trade. Mainly white, this large parrot sports an impressive crest of bright yellow flowerlike feathers that arch dramatically forward. The undersides of its wings and tail are also yellow, as are the ear coverts. The skin around the eyes is bluish, the feet are gray, and the bill is stout and black. Its range is Indonesia and East Timor, where it inhabits a range of forest and scrub, nesting in cavities in trees. Its broad diet includes berries, fruits, nuts, and flowers. Like so many parrots it survives and breeds well in captivity and is severely threatened by capture for the illegal pet trade. Along with habitat loss and persecution as a pest (it occasionally feeds on crops), it is easy to see why it is decreasing.

Family	Psittacidae
Size	13–13¾ in (33–35 cm)
Status	Critically Endangered (decreasing)
Population	2,500–10,000
Range	Indonesia, East Timor
Habitat	Forest; scrub
Main threats	Trapping for cagebird trade, logging, pesticides

Blue-fronted Lorikeet

Charmosyna toxopei

A slender lorikeet with mostly green plumage, yellow-green on the breast, pale blue forecrown, and bright orange bill and legs, it shows a distinct yellow band across the underside of the secondaries. It is found only on the Indonesian island of Buru, and there apparently only in a very small area. The small range and very low numbers place this parrot firmly into the Critically Endangered category, and recent records are few and far between. Its original habitat was probably lower montane forest. While forests at higher levels remain, much of the lower altitude forests have been cleared, though it has also been seen in plantations as well as in secondary and logged forest. More information is required to confirm its current status and distribution.

Family	Psittacidae
Size	6¼ in (16 cm)
Status	Critically Endangered (decreasing)
Population	50–250
Range	Indonesia (endemic)
Habitat	Forest, plantation
Main threats	Habitat loss

Red-fronted Parakeet

Cyanoramphus novaezelandiae

This is a small, attractive parrot with a long, dark tail. The plumage is otherwise mainly bright green, with a red crown and cheek patch, and a flash of red behind the wings. This species was once found throughout much of the mainland of New Zealand, but it is now native only to a number of offshore islands, notably Chatham Island, Kermadec Islands, Three Kings, Kapiti Island, Stewart Island, Snares, and Antipodes Islands. It is not a fussy species with regard to habitat, being found in a variety of different environments including temperate rain forest, scrub, coastal forest, as well as in more open areas. The typical nest site is a hole in a large tree or cliff face, though it will use burrows in the ground. The food ranges from seeds, buds, and other plant material to insects and carrion. The major threat comes from habitat loss through forestry, especially the removal of mature trees with their associated nesting sites. Introduced mammals such as cats and rats also affect it, and introduced birds including starlings and mynahs compete for nest-sites.

Family Psittacidae
Size 10–11 in (25–28 cm)
Status Vulnerable
Population 22,000
Range New Zealand (several offshore islands)
Habitat Forest, scrub
Main threats Logging and burning, introduced predators

THREATENED COCKATOOS, LORIKEETS, AND PARAKEETS

FAMILY		COMMON NAME	SCIENTIFIC NAME	STATUS
PSITTACIDAE Old World cockatoos	■	White Cockatoo	*Cacatua alba*	VU
	■	Philippine Cockatoo	*Cacatua haematuropygia*	CR
	■	Salmon-crested Cockatoo	*Cacatua moluccensis*	VU
	■	Blue-eyed Cockatoo	*Cacatua ophthalmica*	VU
	■	Yellow-crested Cockatoo	*Cacatua sulphurea*	CR
	■	Baudin's Black-cockatoo	*Calyptorhynchus baudinii*	EN
	■	Short-billed Black-cockatoo	*Calyptorhynchus latirostris*	EN
PSITTACIDAE Old World lorikeets	■	Red-throated Lorikeet	*Charmosyna amabilis*	CR
	■	New Caledonian Lorikeet	*Charmosyna diadema*	CR
	■	Palm Lorikeet	*Charmosyna palmarum*	VU
	■	Blue-fronted Lorikeet	*Charmosyna toxopei*	CR
	■	Rimatara Lorikeet	*Vini kuhlii*	EN
	■	Blue Lorikeet	*Vini peruviana*	VU
	■	Henderson Lorikeet	*Vini stepheni*	VU
	■	Ultramarine Lorikeet	*Vini ultramarina*	EN
PSITTACIDAE Old World parakeets	■	Norfolk Island Parakeet	*Cyanoramphus cookii*	EN
	■	Chatham Parakeet	*Cyanoramphus forbesi*	EN
	■	Malherbe's Parakeet	*Cyanoramphus malherbi*	CR
	■	Red-fronted Parakeet	*Cyanoramphus novaezelandiae*	VU
	■	New Caledonian Parakeet	*Cyanoramphus saisetti*	VU
	■	Antipodes Parakeet	*Cyanoramphus unicolor*	VU
	■	Horned Parakeet	*Eunymphicus cornutus*	VU
	■	Uvea Parakeet	*Eunymphicus uvaeensis*	EN
	■	Mauritius Parakeet	*Psittacula eques*	EN

RED LIST ■ EX = Extinct ■ EW = Extinct in the Wild ■ CR = Critically Endangered ■ EN = Endangered ■ VU = Vulnerable

Old World parrots

The Old World parrots are many and varied, and include hanging-parrots, king-parrots, lories, lovebirds, racquet-tails, rosellas, shining-parrots, tiger-parrots, and the strange Kaka, Kea, and Kakapo, the latter three all endemic to New Zealand. Most species feed on a wide range of food including seeds, fruit, nectar, pollen, and buds. The specialized hooked bill of parrots is specially adapted mainly for crushing hard nuts and prizing open seed cases. In some, this strength is combined with careful use of a long tongue or agile foot to access or manipulate the food, providing a distinct advantage over many other birds. Parrots are also highly intelligent, and some are able to mimic calls of other species and other sounds, learn skills, and even use simple tools. In captivity, their health suffers greatly if understimulated. With their amusing abilities and their often bright plumage, many parrots are trapped in the wild and collected for sale to the cagebird trade, with large numbers dying in transit from the tropics. Their feathers combine pigmentation and light refraction to create a dazzling array of colors. However, not all parrots are brightly colored; those that live mainly on the ground, such as the Kakapo, have subtly mottled plumage creating camouflaged patterns rather than brilliant colors.

The Kaka, a close relative of the Kea, is a striking forest-parrot with dark plumage except for its white crown and crimson underwings, rump, and collar. Listed as an Endangered species, it inhabits forests made up of southern beech (*Nothofagus*). The species has declined in numbers mainly as a result of habitat loss, predation, and hunting.

A total of 21 species of these Old World parrots are threatened, and three of these are listed as Critically Endangered: Orange-bellied Parrot, Night Parrot, and Kakapo. The Orange-bellied Parrot is endemic to Tasmania where it breeds in forest bordering moorland, wintering to the southeast Australian mainland. This parrot is threatened by loss of wintering habitat and by competition for nesting holes from introduced common Starlings. The Night Parrot is a rare and enigmatic species of arid grasslands and scrub in the Australian interior. It is active mainly at night—a strategy that helps it retain valuable water in the hot, dry conditions. The Night Parrot has suffered from predation by feral domestic cats and water depletion by feral dromedary Camels (*Camelus dromedarius*)—the latter have now become a common sight in Australia's interior deserts and semi-desert regions.

Swift Parrot

Lathamus discolor

A slim bird with long, pointed wings and a tapering tail; it is rather like a large Budgerigar in overall shape. The plumage is mostly green, with a small blue patch on the crown, and a red forehead, throat, and under tail coverts. Swift Parrots are migratory, breeding mainly in southeastern Tasmania in the summer, and spreading north as far as the Australian mainland (mostly central Victoria) for the winter. In drought years the birds may travel long distances—as far as Queensland—in search of water and food. The numbers are very small and the species seems to be in decline. Swift Parrots tend to be closely associated with a small number of gum tree species, especially *Eucalyptus globulus* (Blue Gum), *E. ovata* (Swamp Gum), and *E. obliqua* (Stringybark). They feed on nectar and also on the sugary secretions left on the leaves by sap-sucking insects. Removal of eucalypt forest in Tasmania is partly responsible for the decline in numbers, and the specialist food requirement also renders this species particularly vulnerable. In some areas, introduced starlings may compete for available nesting holes. Recent increased frequency of droughts in its winter quarters has also impacted on its numbers.

Family	Psittacidae
Size	9¾ in (25 cm)
Status	Endangered (decreasing)
Population	1,000–2,500
Range	Australia (Tasmania) (endemic)
Habitat	Eucalypt woodland
Main threats	Habitat loss, drought

Kea

Nestor notabilis

The strange Kea, named for its shrill "kee-ah" call, is endemic to the South Island of New Zealand. It is one of the most unusual of all parrots. Large and somewhat dumpy, its plumage is mainly dark olive, with a green sheen, especially on the wings. The undersides of the wings and also the rump have bright scarlet patches, providing vivid contrast in flight. The upper mandible of its bill is extended into a sharp, decurved point, especially long in the male. Male Keas are polygamous, breeding with more than one female simultaneously. Unlike most parrots, it inhabits cool, alpine habitats where it specializes in a diet mainly of shoots and berries. Keas are opportunistic, and sometimes feed at rubbish dumps. They now number only about 5,000, but were once much more numerous. After some were found to be attacking sheep to eat their fat, many thousands were exterminated until the species came under formal protection in 1970, though some persecution persists. Keas are also affected by introduced mammals, notably possums, domestic cats, and stoats. They nest on or close to the ground, making them rather vulnerable to predators.

Family	Psittacidae
Size	19 in (48 cm)
Status	Vulnerable (decreasing)
Population	5,000
Range	New Zealand (endemic)
Habitat	Alpine pasture and forest
Main threats	Shooting, poisoning, introduced predators

Kakapo

Strigops habroptila

The world's heaviest and only flightless parrot, and also one of the most unusual and mysterious of all birds, the chunky Kakapo is nocturnal and the only parrot known to employ a lekking system of breeding. It is also very long-lived, potentially reaching 90 years or more. It has a heavy body, moss-green mottled plumage, and an owl-like face. Kakapos feed on stems, roots, fruit, leaves, nectar, and seeds and breed only every few years, determined by the periodic abundance of seeds and fruits. The males gather together at leks and call every night for about three months. Pairing is between January and early March. Endemic to New Zealand and once found on North, South, and Stewart Islands, it gradually retreated to remote areas in the Fjordland of South Island, with just a small remnant population on Stewart Island. By 1987 just three remained in Fjordland, all of them male, and by 1992 the Stewart Island birds were transferred to offshore islands, notably Codfish and Anchor Islands. It is now thought to be extinct in its natural range. Happily, the translocated populations stabilized, and by 2009 the total number of birds stood at about 120. On mainland sites many birds were killed by alien predators, especially domestic cats, and many were lost to disease.

Family Psittacidae
Size 22¾–25 in (58–64 cm)
Status Critically Endangered (recent slight increase)
Population 120
Range New Zealand (now only on certain offshore islands)
Habitat Forest
Main threats Disturbance, alien predators, disease

THREATENED OLD WORLD PARROTS

FAMILY	COMMON NAME	SCIENTIFIC NAME	STATUS
PSITTACIDAE Old World parrots	Black-cheeked Lovebird	*Agapornis nigrigenis*	VU
	Black-winged Lory	*Eos cyanogenia*	VU
	Red-and-blue Lory	*Eos histrio*	EN
	Swift Parrot	*Lathamus discolor*	EN
	Flores Hanging-parrot	*Loriculus flosculus*	EN
	Purple-naped Lory	*Lorius domicella*	VU
	Chattering Lory	*Lorius garrulus*	VU
	Orange-bellied Parrot	*Neophema chrysogaster*	CR
	Kaka	*Nestor meridionalis*	EN
	Kea	*Nestor notabilis*	VU
	Night Parrot	*Pezoporus occidentalis*	CR
	Superb Parrot	*Polytelis swainsonii*	VU
	Green Racquet-tail	*Prioniturus luconensis*	VU
	Blue-headed Racquet-tail	*Prioniturus platenae*	VU
	Blue-winged Racquet-tail	*Prioniturus verticalis*	EN
	Crimson Shining-parrot	*Prosopeia splendens*	VU
	Golden-shouldered Parrot	*Psephotus chrysopterygius*	EN
	Salvadori's Fig-parrot	*Psittaculirostris salvadorii*	VU
	Pesquet's Parrot	*Psittrichas fulgidus*	VU
	Kakapo	*Strigops habroptila*	CR
	Black-lored Parrot	*Tanygnathus gramineus*	VU

RED LIST ■ EX = Extinct ■ EW = Extinct in the Wild ■ CR = Critically Endangered ■ EN = Endangered ■ VU = Vulnerable

CUCKOOS AND TURACOS

Most people associate cuckoos (Cuculidae) with a parasitic breeding system in which their eggs are laid in the nest of another species, to be fostered by the host. In fact only about 45 species out of a total of some 140 use this reproductive strategy. The main cuckoo genera included are the "true cuckoos" (*Cuculus*), with 16 species; the "bronze-cuckoos" (*Chrysococcyx*) with 14 species; and the "lizard-cuckoos and relatives" (*Coccyzus*). The family also contains the coucals (*Centropus*), an Old World group with 28 species; couas (*Coua*), a group of nine species endemic to Madagascar; the New World "ground-cuckoos" (*Neomorphus*) with five species; and two species of the strange "roadrunners" (*Geococcyx*) of North and Central America. Roadrunners are curious creatures able to drop their body temperature at night when the surrounding air temperature falls. In the morning, to warm up again they expose dark areas of flesh to the Sun, almost in the manner of lizards. Nine species are listed as threatened, of which the Sumatran Ground-cuckoo and the Black-hooded Coucal are Critically Endangered.

Turacos are large and brightly colored, fruit-eating forest birds found only in one country: Africa. Their feathers lack a linking barb in their structure so appear somewhat shaggy, single, and hairlike rather than the continuous smooth surface as seen on the majority of birds. There are 23 species of turacos, two of which are threatened: Prince Ruspoli's Turaco (Vulnerable) and Bannerman's Turaco (Endangered).

Cocos Cuckoo
Coccyzus ferrugineus

Endemic to the Cocos Islands, some 300 miles (500 km) off the Pacific coast of Costa Rica, this brightly colored cuckoo has a small range and population. It is slatey gray-brown above, with an obvious dark mask. It has paler buff underparts, bright rufous wings, and a boldly patterned tail. Though there is no obvious decline, it is the island's rarest endemic land bird. Its favored habitat is second-growth forest vegetation that has regrown after the felling of primary forest. Typically this includes scrub such as hibiscus, and tangled vegetation along streams. It feeds mainly on large insects and small lizards. The island is suffering from degradation of its forests by White-tailed deer (*Odocoileus virginianus*), domestic pig and goats. Pigs dig in the forest floor, which inhibits regeneration. Introduced carnivores such as rats predate on nestlings, and domestic cats impact on all stages of its lifecycle. The island has been designated a National Park but further surveys, particularly on breeding habits, are needed before other conservation measures can be introduced.

Family Cuculidae
Size 13 in (33 cm)
Status Vulnerable (stable)
Population 250–1,000
Range Cocos Island (endemic)
Habitat Forest and scrub
Main Threats Introduced animals, habitat loss

Prince Ruspoli's Turaco
Tauraco ruspolii

A colorful forest bird, Prince Ruspoli's Turaco is still fairly common where it occurs, but its range is tiny (just a small region of southern Ethiopia) and it is suspected of being in decline. Favored habitats are mid-altitude woods and riverine forests, and also the juniper and olive forests north of Arero. It feeds mainly on the fruits of fig, juniper, and *Podocarpus* trees. Large and long-tailed, it is greenish-yellow above, with a dark back and tail. In flight it shows bright red patches on the wings. Its bill and eye-ring are also bright red and the crown bears an obvious pale crest. Since about 1995, habitat loss has accelerated, with increased frequency of fires and collection of fuel wood from the forests. A further threat comes from potential hybridization with the more common and much more widespread White-cheeked Turaco (*T. leucotis*) that has a range that overlaps. Illegal hunting is a further problem, as is the capture of these attractive birds for the cagebird trade.

Family Musophagidae
Size 15¾ in (40 cm)
Status Vulnerable (decreasing)
Population 2,500–10,000
Range Southern Ethiopia (endemic)
Habitat Mixed forests
Main Threats Habitat loss, fires, capture

THREATENED CUCKOOS AND TURACOS

FAMILY		COMMON NAME	SCIENTIFIC NAME	STATUS
CUCULIDAE Cuckoos	■	Sumatran Ground-cuckoo	*Carpococcyx viridis*	CR
	■	Green-billed Coucal	*Centropus chlororhynchus*	VU
	■	Sunda Coucal	*Centropus nigrorufus*	VU
	■	Short-toed Coucal	*Centropus rectunguis*	VU
	■	Black-hooded Coucal	*Centropus steerii*	CR

RED LIST ■ EX = Extinct ■ EW = Extinct in the Wild ■ CR = Critically Endangered ■ EN = Endangered ■ VU = Vulnerable

THREATENED CUCKOOS AND TURACOS *Continued*					
FAMILY		**COMMON NAME**	**SCIENTIFIC NAME**	**STATUS**
CUCULIDAE Cuckoos	■	Cocos Cuckoo	*Coccyzus ferrugineus*	VU
	■	Bay-breasted Cuckoo	*Coccyzus rufigularis*	EN
	■	Banded Ground-cuckoo	*Neomorphus radiolosus*	EN
	■	Red-faced Malkoha	*Phaenicophaeus pyrrhocephalus*	VU
MUSOPHAGIDAE Turacos	■	Bannerman's Turaco	*Tauraco bannermani*	EN
	■	Prince Ruspoli's Turaco	*Tauraco ruspolii*	VU

RED LIST ■ EX = Extinct ■ EW = Extinct in the Wild ■ CR = Critically Endangered ■ EN = Endangered ■ VU = Vulnerable

TYPICAL OWLS

Unmistakable, large-eyed, and mostly nocturnal predatory birds, owls vary in size from huge eagle-owls to tiny species such as the Elf Owl. They have a distinctive upright stance when perching and possess strong feet with sharp talons for gripping and feeding. Their eyes face forward and do not move much within their sockets so the whole head moves in order to view objects to the left or right. The neck of an owl is very flexible and the dense head feathers make the bulk of the head appear to swivel as it looks from side to side. Most owls hunt on virtually silent wings in the dusk or at night, listening for the rustling movements of their prey on the ground. The feathers, including the flight feathers, of most owl species are very soft-edged, an adaptation that reduces the sound produced by wing-flaps to an absolute mimimum, aiding surprise attack. The disk-shaped facial feathers help to focus sounds and allow owls to pinpoint their quarry with remarkable accuracy. Many species have earlike tufts of feathers, that are used as signals to each other and not for hearing. Their actual ears are located asymmetrically within the skull, an adaptation that enhances their directional awareness of sound. The prey taken varies from species to species: many owls feed primarily on small rodents, but the larger species such as eagle-owls can tackle medium-sized mammals, while fish-owls and fishing-owls feed mainly on fish and other aquatic prey. The smallest owls feed mainly on beetles, crickets, and other invertebrates. The nocturnal species spend the daytime roosting unnoticed. Many communicate by means of hooting calls that carry a long distance, others by harsh, screeching sounds, and some sound almost mechanical. While most have brown, speckled plumage giving good camouflage in woodland, those of open semi-desert country are usually paler, and the Snowy Owl of the northern tundra is mainly white.

This large family (Strigidae) has about 190 species. The major genera are: "scops-owls" (*Otus*); "wood-owls" (*Strix*); "screech-owls" (*Megascops*); "eagle-owls" (*Bubo*); and "pygmy-owls" (*Glaucidium*). In all, 27 species are threatened, of which six are Critically Endangered—all small species, including four species of scops-owl.

Usambara Eagle-owl

Bubo vosseleri

Large and long eared, this eagle-owl is a rich chestnut-brown with heavy barring, pale underparts spotted on the breast and finely barred on belly and flanks. The face is orange and the eyes orange-brown. The call is a series of low tones, rising and falling. The diet is mainly small mammals such as rodents. Endemic to the Eastern Arc Mountains of Tanzania, it is named for the Usambara Mountains, one of its strongholds. It is also found in Uluguru North Forest Reserve, Iwonde Forest, Kilombero Nature Reserve, and Udzungwa Scarp Forest Reserve. Overall it is thought to be in decline, though possibly under-recorded. All these forests, though protected, are under pressure through fuelwood collection. Other threats include forest clearance for agriculture, and grazing by livestock. There is evidence that the owls can tolerate some disturbance and it has been recorded in the forest edge alongside tea plantations and in forest with cardamom cultivation in the understorey.

Family	Strigidae
Size	19 in (48 cm)
Status	Vulnerable (decreasing)
Population	2,500–10,000
Range	Eastern Arc Mountains of Tanzania (endemic)
Habitat	Montane and submontane evergreen forest
Main Threats	Degradation of forests

Blakiston's Fish-owl

Ketupa blakistoni

Fascinating and rather mysterious, this massive owl haunts the montane forests of eastern Siberia, south to northeastern China (Heilongjiang, Jilin, and Inner Mongolia) and central and northeastern Hokkaido in Japan. It is a splendid bird, with long horizontal ear-tufts, orange-yellow eyes, and buff and dark brown wing bars. The upperparts and pale tail are also streaked with broad, dark brown markings. The call of this unusual owl is a short, deep two- or three-syllable hoot. Its population is hard to estimate, but is under a thousand. It requires dense forest with large old trees, and easy access to lakes, rivers, and streams. It feeds mainly on fish, but also takes small birds, mammals, amphibians, insects, and crustaceans. In most places it seems to be declining through disturbance, loss of ideal habitat, and pollution and damming of rivers. Collisions with power lines and drowning in fishing nets have also killed birds on Hokkaido. However, Hokkaido is one of its few strongholds, where most use artificial nest sites and receive some supplementary feeding. Another key site for the species is the Samarga Valley in eastern Siberia.

Family Strigidae
Size 23½–28 in (60–72 cm)
Status Endangered (decreasing)
Population 250–1,000
Range East Siberia, northeast China, north Japan
Habitat Dense forest, close to water
Main Threats Logging, damming of rivers, over-fishing

Sokoke Scops-owl

Otus ireneae

This remarkable owl has an extremely restricted range, being only found in a handful of forests near the coast of Kenya (notably Arabuko-Sokoke Forest), and in the East Usambara Mountains in Tanzania. A delightful, tiny owl, it measures only about 6 in (15 cm) and has heavily barred plumage. It occurs in two color forms: gray and rufous. It is very hard to spot, hiding deep in the forest by day. It has an indistinct compact profile, the only obvious feature being the very small projecting ear-tufts. It is usually identified by its highly characteristic call—a repeated soft "doo-doo-doo." It emerges at night to forage for beetles and other insects on the forest floor. The population of this bird has been estimated at about 2,500. Sadly, its numbers seem to be declining as the remaining forests continue to be logged, often illegally, and mining for minerals, such as titanium, also threatens the habitat. The main populations are in relatively undisturbed forest, with much lower densities in secondary stands. Forest clearance, logging, and disturbance are the main threats to this charming miniature owl.

Family Strigidae
Size 6 in (15 cm)
Status Endangered (decreasing)
Population 2,500
Range Kenya, Tanzania
Habitat Coastal forest
Main Threats Habitat loss and disturbance

THREATENED TYPICAL OWLS

FAMILY	COMMON NAME	SCIENTIFIC NAME	STATUS
STRIGIDAE Owls	Philippine Eagle-owl	*Bubo philippensis*	VU
	Usambara Eagle-owl	*Bubo vosseleri*	VU
	Albertine Owlet	*Glaucidium albertinum*	VU
	Pernambuco Pygmy-owl	*Glaucidium mooreorum*	CR
	Cloud-forest Pygmy-owl	*Glaucidium nubicola*	VU
	Forest Owlet	*Heteroglaux blewitti*	CR
	Blakiston's Fish-owl	*Ketupa blakistoni*	EN
	Giant Scops-owl	*Mimizuku gurneyi*	VU
	Fearful Owl	*Nesasio solomonensis*	VU
	Cinnabar Hawk-owl	*Ninox ios*	VU
	Christmas Island Hawk-owl	*Ninox natalis*	VU
	Russet Hawk-owl	*Ninox odiosa*	VU
	Flores Scops-owl	*Otus alfredi*	EN
	Javan Scops-owl	*Otus angelinae*	VU
	Biak Scops-owl	*Otus beccarii*	EN

RED LIST ■ EX = Extinct ■ EW = Extinct in the Wild ■ CR = Critically Endangered ■ EN = Endangered ■ VU = Vulnerable

THREATENED TYPICAL OWLS *Continued*				
FAMILY		COMMON NAME	SCIENTIFIC NAME	STATUS
STRIGIDAE Owls	■	Anjouan Scops-owl	*Otus capnodes*	CR
	■	São Tomé Scops-owl	*Otus hartlaubi*	VU
	■	Seychelles Scops-owl	*Otus insularis*	EN
	■	Sokoke Scops-owl	*Otus ireneae*	EN
	■	Moheli Scops-owl	*Otus moheliensis*	CR
	■	Grand Comoro Scops-owl	*Otus pauliani*	CR
	■	Pemba Scops-owl	*Otus pembaensis*	VU
	■	White-fronted Scops-owl	*Otus sagittatus*	VU
	■	Siau Scops-owl	*Otus siaoensis*	CR
	■	Serendib Scops-owl	*Otus thilohoffmanni*	EN
	■	Rufous Fishing-owl	*Scotopelia ussheri*	EN
	■	Long-whiskered Owlet	*Xenoglaux loweryi*	EN

RED LIST ■ EX = Extinct ■ EW = Extinct in the Wild ■ CR = Critically Endangered ■ EN = Endangered ■ VU = Vulnerable

BARN OWLS, NIGHTJARS, OWLET-NIGHTJARS

Barn owls (Tytonidae) differ in appearance from other owls mainly in their heart-shaped faces. There are 15 species in this family—13 species in the genus *Tyto*, and two species of "bay-owl" (*Phodilus*). The Barn Owl itself has the distinction of being one of the world's most widespread species, being found on every inhabited continent. It is closely associated with farming, and feeds mainly on small rodents such as rats and mice, common pests wherever grain is stored. However, this farmers' friend suffers through poisoning from agricultural chemicals, and is also often a victim of road traffic. Unlike most hawks, eagles, and falcons, owls swallow their prey whole and then regurgitate the indigestible material (mainly fur, feathers, and bones) as neat pellets. Analysis of owl pellets is thus a reliable guide to their diet.

Four species are listed as Vulnerable, and two are Endangered—the Taliabu Masked-owl, endemic to the Sula Islands of Indonesia, and the Congo Bay-owl of Democratic Republic of the Congo and nearby Rwanda.

Nightjars (Caprimulgidae) are rather mysterious birds. Secretive and nocturnal, they typically glide ghostlike over heath and scrub, catching moths and other flying insects in their wide, gaping mouths. The family contains about 90 species, the main groups being nightjars (*Caprimulgus*), nighthawks (e.g. *Chordeiles*), and eared-nightjars (*Eurostopodus*). Seven species are threatened, of which two, the Puerto Rican Nightjar and Jamaican Pauraque, are Critically Endangered.

The owlet-nightjars of Australia and New Guinea are well-named, as they look like a cross between a small owl and a nightjar. The nine species of owlet-nightjar all belong to the genus *Aegotheles*, one of which, New Caledonian Owlet-nightjar, is Critically Endangered.

Madagascar Red Owl
Tyto soumagnei

This owl is found only in Madagascar, whereas its close relative, the Barn Owl (*T. alba*), is one of the world's most widespread species. The Barn Owl also occurs in Madagascar and confusion has probably added to the difficulty of estimating the Red Owl's numbers. It resembles the Barn Owl, but its plumage is a rich orange-buff above, somewhat paler beneath. It is also slightly smaller. The facial disks are whitish. Its range follows the mountains down the eastern side of Madagascar where it mainly inhabits humid evergreen forest from sea level to about 6,600 ft (2,000 m). Strictly nocturnal, it is hard to census, but its numbers are in the thousands. It has recently been seen in the far southeast, in the lowlands of Tsitongambarika. It hunts from a perch in open forest or at the forest edge, and also over open rice-paddies and cultivated areas. For prey, it favors native mammals, notably Tsingy Tufted-tailed Rat (*Eliurus antsingy*), as well as insects, geckos, and frogs. Much of the lowland forest has been removed and logging and fire threaten much of the remaining rain forest.

Family	Tytonidae
Size	11¾ in (30 cm)
Status	Vulnerable (decreasing)
Population	2,500–10,000
Range	Madagascar (endemic)
Habitat	Evergreen rain forest
Main Threats	Deforestation, fire

White-winged Nightjar

Eleothreptus candicans

This is a small nightjar with sandy-gray plumage, brown face and throat, and gray crown, dark at the center. The male is mostly white below, with a white tail and wings, the latter tipped black. The female is browner with tawny wings and tail. Male White-winged Nightjars defend their territories and chase off intruding males, often alighting briefly on an anthill or termite mound and flying in an arc with butterfly wingbeats, making mechanical sounds. In common with other nightjars, it chases insects such as moths and beetles at dusk and at night using its large eyes. It forages in open areas, especially dry savanna with young vegetation, often in areas with abundant termite mounds and on sites regenerating following fire. Though minor fires open up more suitable habitat, more major fires destroy the breeding sites, threatening the species. The range is very small: it is known from just four locations including Emas National Park in Brazil and Mbaracayú Forest Nature Reserve in Paraguay.

Family Caprimulgidae
Size 7¾ in (20 cm)
Status Endangered (decreasing)
Population 1,000–2,500
Range Bolivia, Brazil, Paraguay
Habitat Wet grassland and scrub
Main Threats Grazing, agriculture, uncontrolled fires

THREATENED BARN OWLS, NIGHTJARS, AND OWLET-NIGHTJARS

FAMILY		COMMON NAME	SCIENTIFIC NAME	STATUS
TYTONIDAE Barn owls	■	Congo Bay-owl	*Phodilus prigoginei*	EN
	■	Bismarck Masked-owl	*Tyto aurantia*	VU
	■	Sulawesi Golden Owl	*Tyto inexspectata*	VU
	■	Manus Masked-owl	*Tyto manusi*	VU
	■	Taliabu Masked-owl	*Tyto nigrobrunnea*	EN
	■	Madagascar Red Owl	*Tyto soumagnei*	VU
CAPRIMULGIDAE Nightjars	■	Sunda Nightjar	*Caprimulgus concretus*	VU
	■	Puerto Rican Nightjar	*Caprimulgus noctitherus*	CR
	■	Itombwe Nightjar	*Caprimulgus prigoginei*	EN
	■	Nechisar Nightjar	*Caprimulgus solala*	VU
	■	White-winged Nightjar	*Eleothreptus candicans*	EN
	■	Sulawesi Eared-nightjar	*Eurostopodus diabolicus*	VU
	■	Jamaican Pauraque	*Siphonorhis americana*	CR
AEGOTHELIDAE Owlet-nightjars	■	New Caledonian Owlet-nightjar	*Aegotheles savesi*	CR

RED LIST ■ EX = Extinct ■ EW = Extinct in the Wild ■ CR = Critically Endangered ■ EN = Endangered ■ VU = Vulnerable

Swifts and Hummingbirds

Swifts (Apodidae) are among the most charismatic of all birds. Streamlined and rarely alighting except to nest, they are aerial acrobats par excellence, chasing and feeding on flying insects and capable of sudden bursts of speed, though they often glide relatively slowly. Observations of the Common Swift have shown that they often spend all night on the wing, presumably catching short bouts of "sleep" at high altitude. It is possible that (in this species at least) they may never land except when alighting at breeding sites, which implies that they may fly continuously from first fledging stage to first breeding age—a possible distance of 300,000 miles (500,000 km). In a lifetime, which is on average seven years, a Common Swift may fly around 1¼ million miles (2 million km). The oldest known bird of this species lived to the age of 21 years and it could have flown 3.7 million miles (6 million km), which is equivalent to flying to the Moon and back eight times. Most swift species are likely to fly up to 300 or more miles (500 km) a day in search of aerial insect prey.

The family contains just over 100 species, including true swifts (*Apus*), swiftlets (*Collocalia*), and others. Six species are threatened, one of which, the Guam Swiftlet, is Endangered. The four species of treeswifts of Southeast Asia belong to a separate family (Hemiprocnidae); none is threatened.

Equally fascinating are the hummingbirds (Trochilidae). Small and brightly colored, the members of this New World family of some 335 species fuel themselves with sugar-rich nectar, rather like flower-feeding insects. In flight, they use their flexible wings to hover and dart from flower to flower, or to indulge in elaborate aerial courtship displays. Many have colorful, often iridescent plumage, combined in some species with elaborate tail extensions. The common names reflect their remarkable attributes: brilliants, coquettes, emeralds, hermits, hillstars, incas, mangos, metaltails, mountain-gems, pufflegs, sabrewings, sapphires, starfrontlets, starthroats, sunangels, sunbeams, thornbills, trainbearers, woodnymphs, and woodstars. The smallest bird known, the Bee Hummingbird, endemic to Cuba and Near Threatened, is only 2 in (5 cm) long and weighs less than ½ oz (2 g), smaller than some Cuban insects.

Thirty species are threatened— mostly either Endangered or Critically Endangered, mainly as a result of habitat loss or degradation. Tiny tracking transmitters can be attached to hummingbirds to record their movements, and researchers have found that certain plants need hummingbirds to pollinate them. Where native forests no longer have enough *Heliconia* flowers, for example, the hummingbirds relocate to nearby banana plantations. This may indicate a reduction of biodiversity with potential loss of both bird and plant species.

Seychelles Swiftlet

Collocalia elaphra

This dainty swift is known from only three sites—all caves in the granitic Seychelles, on Mahé, Praslin, and La Digue. This genus, of which there are 30 species, includes birds whose nests are harvested in some parts of tropical Asia to make bird's-nest soup, as were the nests of the Seychelles Swiftlet in the past. The nests are built on the roofs of deep caves and just a single egg is laid. Small, and dark gray-brown, this endemic swift is classed as Vulnerable as its range is so small. The colonies fluctuate in numbers—roughly 50 nests in the La Digue cave, some 80 nests in the Praslin cave, and unknown numbers on Mahé. The birds hunt over the forests and wetlands, catching insects, notably flying ants. Past threats have included the use of insecticides, including highly toxic DDT. Quarrying has affected some cave systems, and introduced creatures such as domestic cats and Barn Owls are possible predators. Drainage of wetlands and the spread of introduced water plants may also in turn reduce the supply of insect prey.

Family	Apodidae
Size	4–4¾ in (10–12 cm)
Status	Vulnerable
Population	2,500–3,000
Range	Seychelles (endemic)
Habitat	Caves (breeding), forest, wetlands
Main threats	Alien predators, quarrying

Mangrove Hummingbird

Amazilia boucardi

This is a medium-sized hummingbird with, in the male, pale green crown and upperparts and bronze rump, bronze-green tail, and turquoise throat and chest. The female has mainly white underparts. Endemic to Costa Rica, it is patchily distributed in a handful of mangrove "forests" dotted along the Pacific coast. As in so many other places around the world, these special habitats are under severe threat, in this area specifically from developments including housing, road building, and construction of shrimp ponds. The mangroves are also affected by marine pollution in some areas, and also suffer from illegal logging, the wood from which is used to make charcoal. The Mangrove Hummingbird specializes on its preferred foodplant—the Pacific Mangrove (*Pelliciera rhizophorae*)—itself a rare species whose range has contracted. One proposed conservation measure that would highlight this species is a public awareness campaign to protect the mangrove "forests."

Family	Trochilidae
Size	3¾ in (9.5 cm)
Status	Endangered (decreasing)
Population	2,500–10,000
Range	Costa Rica (endemic)
Habitat	Mangroves
Main threats	Habitat loss

Scissor-tailed Hummingbird

Hylonympha macrocerca

As its name suggests, this hummingbird has an obvious forked tail—long in the male, somewhat shorter in the female. At an impressive 4 in (10 cm) in length, the male's tail is even longer than the body. The male is mainly dark green with a shiny violet cap and a bright green breast. The long tail is dark purple, the long, broad lateral feathers forming a scissor shape. The female is white below, spotted green, with a chestnut belly and undertail. The range is tiny—virtually restricted to the Paria Peninsula in Sucre, northeastern Venezuela, where it is still common locally. It feeds partly at bromeliad flowers and also on some insects they contain, mainly in humid montane forests, between 2,600 and 4,000 ft (800 and 1,200 m). It also catches insects in flight. As it prefers undisturbed forest, it has been badly affected by cultivation of coffee and cocoa as understorey crops, as well as forest removal, and uncontrolled burning.

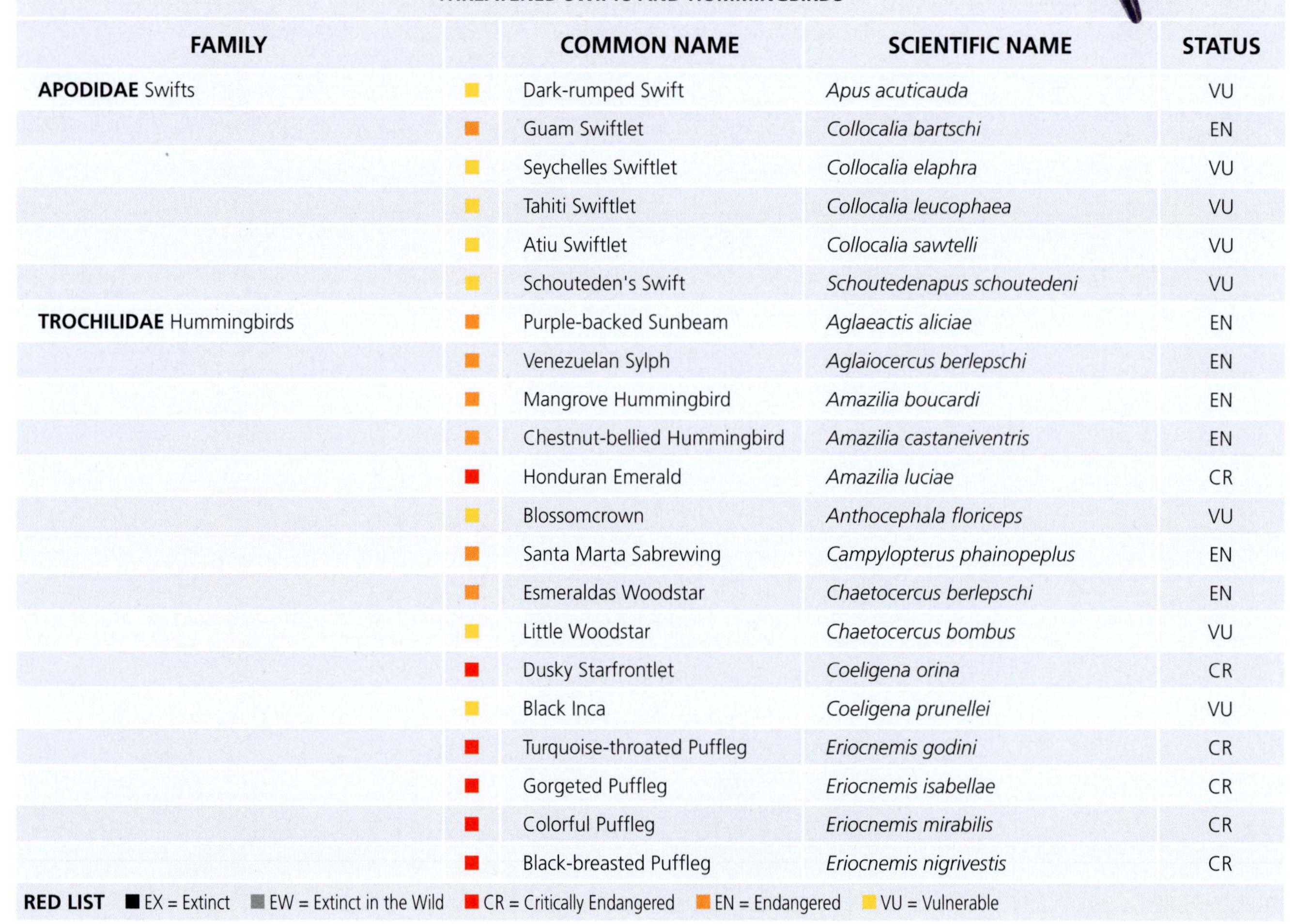

Family	Trochilidae
Size	7½ in (19 cm), 4 in (10 cm) of which is tail
Status	Endangered (decreasing)
Population	10,000–20,000
Range	Northeastern Venezuela (endemic)
Habitat	Montane forest
Main threats	Habitat loss, fire

THREATENED SWIFTS AND HUMMINGBIRDS

FAMILY		COMMON NAME	SCIENTIFIC NAME	STATUS
APODIDAE Swifts	■	Dark-rumped Swift	*Apus acuticauda*	VU
	■	Guam Swiftlet	*Collocalia bartschi*	EN
	■	Seychelles Swiftlet	*Collocalia elaphra*	VU
	■	Tahiti Swiftlet	*Collocalia leucophaea*	VU
	■	Atiu Swiftlet	*Collocalia sawtelli*	VU
	■	Schouteden's Swift	*Schoutedenapus schoutedeni*	VU
TROCHILIDAE Hummingbirds	■	Purple-backed Sunbeam	*Aglaeactis aliciae*	EN
	■	Venezuelan Sylph	*Aglaiocercus berlepschi*	EN
	■	Mangrove Hummingbird	*Amazilia boucardi*	EN
	■	Chestnut-bellied Hummingbird	*Amazilia castaneiventris*	EN
	■	Honduran Emerald	*Amazilia luciae*	CR
	■	Blossomcrown	*Anthocephala floriceps*	VU
	■	Santa Marta Sabrewing	*Campylopterus phainopeplus*	EN
	■	Esmeraldas Woodstar	*Chaetocercus berlepschi*	EN
	■	Little Woodstar	*Chaetocercus bombus*	VU
	■	Dusky Starfrontlet	*Coeligena orina*	CR
	■	Black Inca	*Coeligena prunellei*	VU
	■	Turquoise-throated Puffleg	*Eriocnemis godini*	CR
	■	Gorgeted Puffleg	*Eriocnemis isabellae*	CR
	■	Colorful Puffleg	*Eriocnemis mirabilis*	CR
	■	Black-breasted Puffleg	*Eriocnemis nigrivestis*	CR

RED LIST ■ EX = Extinct ■ EW = Extinct in the Wild ■ CR = Critically Endangered ■ EN = Endangered ■ VU = Vulnerable

Marvellous Spatuletail

Loddigesia mirabilis

The common name describes this hummingbird well. It is indeed a marvel, the male having a bright blue chin and crest, a green breast and belly, and bronze back. The tail is its most amazing feature—two long, pointed central feathers are framed by two even longer and extraordinary plumes, each with a narrow filament and ending in a spoon-shaped racket. The female also has the rackets, but has duller, paler plumage. It lives in scrub and forest edges between about 6,600 ft (2,000 m) and 10,000 ft (3,000 m), and the males are said to display at leks to attract the attention of the females. The range of this remarkable hummingbird is very small—just a handful of sites in northern Peru. Numbers are hard to estimate, but there are probably fewer than 1,000 birds. The main threat appears to be deforestation, and bizarrely, the miniscule dried hearts of the male birds are said by locals to act as an aphrodisiac, and some may be killed for this reason.

Family	Trochilidae
Size	4–6 in (10–15 cm)
Status	Endangered (decreasing)
Population	250–1,000
Range	Northern Peru (endemic)
Habitat	Forest, scrub
Main threats	Deforestation, hunting

THREATENED SWIFTS AND HUMMINGBIRDS *Continued*

FAMILY		COMMON NAME	SCIENTIFIC NAME	STATUS
TROCHILIDAE Hummingbirds	■	Chilean Woodstar	*Eulidia yarrellii*	EN
	■	Blue-capped Hummingbird	*Eupherusa cyanophrys*	EN
	■	White-tailed Hummingbird	*Eupherusa poliocerca*	VU
	■	Hook-billed Hermit	*Glaucis dohrnii*	EN
	■	Royal Sunangel	*Heliangelus regalis*	EN
	■	Scissor-tailed Hummingbird	*Hylonympha macrocerca*	EN
	■	Sapphire-bellied Hummingbird	*Lepidopyga lilliae*	CR
	■	Marvellous Spatuletail	*Loddigesia mirabilis*	EN
	■	Short-crested Coquette	*Lophornis brachylophus*	CR
	■	Violet-throated Metaltail	*Metallura baroni*	EN
	■	Perija Metaltail	*Metallura iracunda*	EN
	■	Glow-throated Hummingbird	*Selasphorus ardens*	VU
	■	Juan Fernandez Firecrown	*Sephanoides fernandensis*	CR
	■	Gray-bellied Comet	*Taphrolesbia griseiventris*	EN
	■	Mexican Woodnymph	*Thalurania ridgwayi*	VU

RED LIST ■ EX = Extinct ■ EW = Extinct in the Wild ■ CR = Critically Endangered ■ EN = Endangered ■ VU = Vulnerable

TROGONS, KINGFISHERS, HORNBILLS, GROUND-ROLLERS, AND MOTMOTS

The 40 species of trogon (Trogonidae) occur variously in Central and South America, tropical Africa and India, and in Southeast Asia. Medium-sized forest birds, they include a lot of fruit in their diet though some also feed on insects, catching them in midair. Most have brightly colored plumage and very broad, short bills. Trogons have short, weak legs and their feet have two toes facing forward and two facing back. The family also contains six species of quetzal, two of which, the Eared Quetzal and Resplendent Quetzal, are Near Threatened. The breeding male of the latter boasts tail feathers of up to 24 in (60 cm) in length and was sacred to the Mayas and Aztecs, representing the physical form of the god Quetzalcoatl. Only the Javan Trogon (Endangered) is threatened, but several are Near Threatened.

Kingfishers (Alcedinidae) contain 95 species. Small or medium-sized, most are associated with wetlands and feed mainly on fish caught by diving, though several inhabit forest or scrub and feed mainly on invertebrates or small terrestrial vertebrates such as lizards. They have large heads with large, dagger-shaped bills and many have bright colorful plumage. The family includes five species of kookaburra. Twelve species are listed as threatened, two of them, Tuamotu and Marquesan Kingfishers, are Critically Endangered island species that suffer from disturbance and predation by alien introductions.

Hornbills (Bucerotidae) are medium-sized to large mainly forest birds with long, heavy bills often topped by a horny, mainly hollow "casque" thought to amplify calls. In some species it is small or virtually imperceptible and serves only to strengthen the bill, while in others, different functions may be important—the Helmeted Hornbill has an unusual solid bill that is used as a battering ram in aerial battles between rival males. Long-tailed and broad-winged, hornbills mostly have striking black-and-white plumage. They are found in Africa and southern Asia, east to New Guinea. There are 55 species, nine of them threatened. The Rufous-headed Hornbill and Sulu Hornbill, both of the Philippines, are Critically Endangered, and the Sulu Hornbill is very close to extinction. The two species of ground-hornbill (Bucorvidae) are mainly ground-dwelling.

Ground-rollers (Brachypteraciidae) are five species, related to rollers (Coraciidae) and formerly included in the latter family. Medium-sized and rather crowlike, they are duller in plumage than the rollers and are endemic to Madagascar.

Motmots (Momotidae) comprise nine species, one of which, the Keel-billed Motmot is Vulnerable. They inhabit the forests of Central and South America and have bright plumage, long tails, and strong bills.

Silvery Kingfisher

Alcedo argentata

This is a small kingfisher endemic to the Philippines, where it is found on only a few islands of the central and southern Philippines, notably on Samar, Leyte, Bohol, Mindanao, Dinagat, Siargao, and Basilan. Once much more common, its habitat needs are rather specific—clear streams within lowland forest. However, being shy it is difficult to spot and survey, and so may well be under-recorded. Its plumage is striking: mainly blue-black and white, it shows a silvery rump and back as it flies. The small feet are bright red. The Philippines have suffered deforestation on a large scale and much suitable habitat has been destroyed to make way for crops such as bananas, rice-fields, and for oil palm. Many suitable streams now carry a high silt load through erosion, and this seems to exclude the Silvery Kingfisher, which needs clear water. It nests in holes in banks and feeds mainly on small fish and aquatic invertebrates.

Family	Alcedinidae
Size	5½ in (14 cm)
Status	Vulnerable (decreasing)
Population	2,500–10,000
Range	Philippines (endemic)
Habitat	Tropical forest streams
Main threats	Habitat loss, pollution

Rufous-headed Hornbill

Aceros waldeni

Critically Endangered, with a combination of low numbers and a tiny range, this is a medium-sized forest hornbill. The plumage is mainly black. The male has a rufous head, neck, and upper breast, and a red bill topped by a large wrinkled casque. The tail has a broad white central section. Once found on three of the central Philippine islands—Negros, Panay, and Guimares—it probably only survives on Panay, and possibly Negros, mainly in areas that are protected. Breeding is thought to be in March, with clutches of two to three eggs inside tree cavities. Rufous-headed Hornbills feed mainly on fruit, including figs. Although it was on the brink of extinction in 2001, careful conservation, including nest-guarding by local communities, brought numbers back up, at least on Panay. As well as loss of habitat, threats remain from hunting and also trapping for the pet trade. There is also a captive-breeding program, and efforts are underway to increase public awareness of the plight of this species.

Family	Bucerotidae
Size	23½–26½ in (60–65 cm)
Status	Critically Endangered (decreasing)
Population	1,000–2,500
Range	Philippines (endemic)
Habitat	Tropical forest
Main threats	Deforestation, hunting, capture

Scaly Ground-roller

Brachypteracias squamiger

All five species that make up the ground-roller family are endemic to Madagascar and three of these are threatened (all Vulnerable), including the Scaly Ground-roller. Rather thrushlike but with a stronger bill, it is well-named for the scaly appearance of its plumage, especially on its head. Intricately patterned, with copper-brown mantle, greenish wings, and marked with black crescents on its underparts, this unusual bird forages on the forest floor, mainly searching for invertebrates such as worms, snails, centipedes, spiders, ants, and beetles. Its habitat is lowland tropical rain forest in eastern Madagascar, ideally undisturbed forest with dark, lush undergrowth and rich leaf litter. It nests in a tunnel dug into a bank within the forest. Degradation of the forest is the major threat, mainly through slash-and-burn cultivation. In addition, much of the original forest has already been cleared, especially on the coastal plains. Predation by dogs from nearby villages poses an additional threat.

Family Brachypteraciidae
Size 10¾–12 in (27–31 cm)
Status Vulnerable (decreasing)
Population 2,500–10,000
Range Madagascar (endemic)
Habitat Lowland rain forest
Main threats Habitat loss and degradation

THREATENED TROGONS, KINGFISHERS, HORNBILLS, GROUND-ROLLERS, AND MOTMOTS

FAMILY		COMMON NAME	SCIENTIFIC NAME	STATUS
TROGONIDAE Trogons	■	Javan Trogon	*Apalharpactes reinwardtii*	EN
ALCEDINIDAE Kingfishers	■	Moustached Kingfisher	*Actenoides bougainvillei*	VU
	■	Blue-capped Kingfisher	*Actenoides hombroni*	VU
	■	Silvery Kingfisher	*Alcedo argentata*	VU
	■	Blue-banded Kingfisher	*Alcedo euryzona*	VU
	■	Bismarck Kingfisher	*Alcedo websteri*	VU
	■	Philippine Dwarf-kingfisher	*Ceyx melanurus*	VU
	■	Kofiau Paradise-kingfisher	*Tanysiptera ellioti*	EN
	■	Sombre Kingfisher	*Todiramphus funebris*	VU
	■	Tuamotu Kingfisher	*Todiramphus gambieri*	CR
	■	Marquesan Kingfisher	*Todiramphus godeffroyi*	CR
	■	Mangaia Kingfisher	*Todiramphus ruficollaris*	VU
	■	Rufous-lored Kingfisher	*Todiramphus winchelli*	VU
BUCEROTIDAE Hornbills	■	Sumba Hornbill	*Aceros everetti*	VU
	■	Narcondam Hornbill	*Aceros narcondami*	EN
	■	Rufous-necked Hornbill	*Aceros nipalensis*	VU
	■	Plain-pouched Hornbill	*Aceros subruficollis*	VU
	■	Rufous-headed Hornbill	*Aceros waldeni*	CR
	■	Palawan Hornbill	*Anthracoceros marchei*	VU
	■	Sulu Hornbill	*Anthracoceros montani*	CR
	■	Mindoro Hornbill	*Penelopides mindorensis*	EN
	■	Visayan Hornbill	*Penelopides panini*	EN
BUCORVIDAE Ground-hornbills	■	Southern Ground-hornbill	*Bucorvus cafer*	VU
BRACHYPTERACIIDAE Ground-rollers	■	Short-legged Ground-roller	*Brachypteracias leptosomus*	VU
	■	Scaly Ground-roller	*Brachypteracias squamiger*	VU
	■	Long-tailed Ground-roller	*Uratelornis chimaera*	VU
MOMOTIDAE Motmots	■	Keel-billed Motmot	*Electron carinatum*	VU

RED LIST ■ EX = Extinct ■ EW = Extinct in the Wild ■ CR = Critically Endangered ■ EN = Endangered ■ VU = Vulnerable

WOODPECKERS, TOUCANS AND BARBETS, AND JACAMARS

There are 218 species of woodpecker (Picidae), found throughout all regions, mainly in forests and wooded habitats. In size they range from the tiny piculets to substantial crow-sized woodpeckers. They have straight, chisel-like bills and a stiff tail, the latter used as a prop when clinging upright to a branch or trunk. Zygodactyl feet (with two toes facing forward, two to the rear) are another adaptation that helps them cling to vertical surfaces. Many have extremely long tongues (some barbed) that are used to extract prey from deep holes in wood or from ant-nests. The tongue when not in use is retracted into a special storage cavity in the skull. Many woodpeckers are boldly patterned, often black and white. Of the 11 threatened species, four are Critically Endangered: Okinawa, Kaempfer's, Imperial, and Ivory-billed. The latter two species are among the largest and most attractive. Neither has been seen for certain in recent years and one or both may now be extinct. Since they rely so heavily on trees, it is not surprising that woodpeckers are severely affected by widespread deforestation.

There are 122 species of toucans and barbets (Ramphastidae). Toucans are medium-sized birds of Central and South American forests. They have colorful plumage and some have large, heavy-looking bills. The bills actually have a lightweight honeycomb structure. These are used for feeding on a varied diet from fruit to the nestlings of other species. Barbets are medium-sized, plump, colorful birds found not only in tropical South America, but also in Africa, India, and Southeast Asia. They have a conical bill surrounded by bristles, and feed on fruit and insects. Of the five threatened species, two are Endangered: Yellow-browed Toucanet and White-mantled Barbet.

Jacamars (Galbulidae) are colorful with long, thin bills used for feeding on insects. Forest-dwelling birds from tropical America, they are almost the equivalent to the Old World bee-eaters. Similar to bee-eaters in appearance, they also excavate nesting burrows to breed, but their food consists of a wider range of insects. Two of the 18 species are listed as Vulnerable.

In recent years there has been a flurry of excitement among ornithologists in the wake of a number of tantalizing reports of sightings of the Ivory-billed Woodpecker on the U.S. mainland. Possible sightings were reported from Arkansas in 2004 and published in 2005, but intensive searches thereafter produced no confirmation. In June 2006, a $10,000 reward was offered for information leading to a definitive re-discovery, and in September 2006, a team of ornithologists claimed plausible sightings from northwestern Florida. However, subsequent searches in that area of Florida have also failed to produce definitive proof of the continued survival of this stunning bird. Then, in December 2008, the Cornell Laboratory of Ornithology announced a reward of $50,000 to "the person who can successfully lead a project biologist to a living Ivory-billed Woodpecker."

Despite some compelling indications of its possible survival, and a handful of hoaxes, in the absence of conclusive evidence, such as unambiguous video footage, photographs, or DNA samples, the re-discovery remains a tantalizing possibility. Nevertheless, the search has promoted conservation and led to the protection of its habitat. None of these sightings has been reliably confirmed, and until further definitive evidence is provided, the magnificent Ivory-billed Woodpecker may sadly be assumed extinct, though the case remains open and proven re-discovery a possibility.

Ivory-billed Woodpecker

Campephilus principalis

This elusive and charismatic large woodpecker teeters on the brink of extinction, indeed it may already have disappeared. It is very large and mainly black and white, with a bright red crest. Once found in the southeastern U.S.A. and also in Cuba, if it survives in either country the numbers are certainly tiny and its future is precarious. It was last confirmed in the U.S.A. in 1944, in northeastern Louisiana, and in Cuba in the late 1980s. Tantalizingly, however, it was reported from eastern Arkansas (Big Woods region) in 2004, and also from Florida (Choctawatchee River region) between 2005 and 2007, but neither of these records has been confirmed, though both are possible. In December 2009, it was allegedly found in the Sabine River region of Texas or Louisiana, but as this book goes to press these claims have not yet been verified. There is a $50,000 dollar prize for its re-discovery, which may also have the effect of encouraging hoaxes. In Cuba it may perhaps survive in the Sierra Maestra in the southeast, though the signs are not good.

Family	Picidae
Size	19-21 in (48–53 cm)
Status	Critically Endangered
Population	Fewer than 50
Range	U.S.A. and Cuba
Habitat	Forest
Main threats	Deforestation and forest degradation

Arabian Woodpecker

Dendrocopos dorae

Endemic to the Arabian Peninsula, this small woodpecker occurs locally in woodland, itself a rare habitat in this arid region. Thousands of years of removal and degradation of the woodland, especially by livestock grazing and cultivation, has left only fragmented pockets of suitable habitat. Luckily this species is not very fussy about the type of woodland and is found in groves of Date-palm (*Phoenix dactylifera*), fig (*Ficus*), and pandan (*Pandanus*), as well as coffee plantations and in orchards and mixed woodland on agricultural terraces. The main threats, both historical and at present, are logging and clearance of woodland, especially removal of the larger trees that are potential nesting sites. Grazing by livestock inhibits natural regeneration of woodland. This woodpecker occurs in protected areas, notably in Saudi Arabia in the Asir National Park and Raydah Reserve. There are also many private reserves that preserve some pockets of woodland, but these are not always well protected.

Family	Picidae
Size	7 in (18 cm)
Status	Vulnerable (decreasing)
Population	2,500–10,000
Range	Saudi Arabia and Yemen
Habitat	Woodland
Main threats	Habitat degradation and woodland clearance

White-mantled Barbet

Capito hypoleucus

Mainly blue-black and pale with a heavy bill, a scarlet crown, and a white mantle, this forest bird has a very small and extremely fragmented range. Endemic to Colombia, it occurs in montane humid forests at the north tip of the Central Andes and in some nearby ranges as well as along the eastern slope of the Central Andes and the western slope of the East Andes. In the small number of places where it does occur it may be common or even abundant, but the population overall is thought to be declining. Though it prefers primary forest, it is also found in areas of secondary forest and plantations. The main diet of this bird consists of fruit, seeds, and insects. The northern Andes have been steadily deforested since the 19th century, with many of the slopes cleared for crops and also for mining and settlement. One of its strongholds is in the Serranía de los Yariguíes National Park.

Family	Ramphastidae
Size	7½ in (19 cm)
Status	Endangered (decreasing)
Population	2,500–10,000
Range	Colombia (endemic)
Habitat	Montane humid forest
Main threats	Deforestation, mining

THREATENED WOODPECKERS, TOUCANS AND BARBETS, AND JACAMARS

FAMILY	COMMON NAME	SCIENTIFIC NAME	STATUS
PICIDAE Woodpeckers	Imperial Woodpecker	*Campephilus imperialis*	CR
	Ivory-billed Woodpecker	*Campephilus principalis*	CR
	Kaempfer's Woodpecker	*Celeus obrieni*	CR
	Fernandina's Flicker	*Colaptes fernandinae*	VU
	Arabian Woodpecker	*Dendrocopos dorae*	VU
	Okinawa Woodpecker	*Dendrocopos noguchii*	CR
	Sulu Woodpecker	*Dendrocopos ramsayi*	VU
	Helmeted Woodpecker	*Dryocopus galeatus*	VU
	Great Slaty Woodpecker	*Mulleripicus pulverulentus*	VU
	Red-cockaded Woodpecker	*Picoides borealis*	VU
	Speckle-chested Piculet	*Picumnus steindachneri*	VU
RAMPHASTIDAE Toucans and barbets	Yellow-browed Toucanet	*Aulacorhynchus huallagae*	EN
	White-mantled Barbet	*Capito hypoleucus*	EN
	Five-colored Barbet	*Capito quinticolor*	VU
	Scarlet-banded Barbet	*Capito wallacei*	VU
	Zambian Barbet	*Lybius chaplini*	VU
GALBULIDAE Jacamars	Coppery-chested Jacamar	*Galbula pastazae*	VU
	Three-toed Jacamar	*Jacamaralcyon tridactyla*	VU

RED LIST ■ EX = Extinct ■ EW = Extinct in the Wild ■ CR = Critically Endangered ■ EN = Endangered ■ VU = Vulnerable

Broadbills, Scrub-birds, and Ovenbirds

The families treated in the remainder of this section all belong to a single order—the Passeriformes—often referred to as the passerines, or somewhat misleadingly as perching birds. Those families from swallows and martins onward (see p. 188) form the suborder Oscines, often referred to as the songbirds.

Of the 15 species of broadbill (Eurylaimidae), three are threatened and classed as Vulnerable. Broadbills are found in Africa and also in the Himalayan region and in Southeast Asia. They are sturdy birds with a broad, flattened bill and several are brightly colored. They feed opportunistically on a range of invertebrates and small vertebrates though the three species in the *Calyptomena* genus feed mainly on fruit.

Only two species make up the scrub-bird family (Atrichornithidae) and both are Vulnerable. Endemic to Australia, they are small, brown and long-tailed, and generally hard to see as they are ground dwelling, with shy and secretive behavior. It is thought that they may possibly be related to the more ancient lyrebirds and bowerbirds.

Ovenbirds (Furnariidae) are a large and varied family from South and Central America, with 238 species, of which 26 are threatened. Critically Endangered are Masafuera Rayadito, Royal Cinclodes, White-bellied Cinclodes, and Alagoas Foliage-gleaner. They are medium-sized, mostly rather drab-colored birds and many of them feed mainly on the ground. The true ovenbirds or horneros (*Furnarius*) build extremely strong, domed mud-nests rather like miniature ovens—hence the name. These are cleverly constructed with a side entrance, on the ground, made from a mixture of woven vegetation and mud. The family also includes miners, earthcreepers, cinclodes, tit-spinetails, thistletails, canasteros, spinetails, thornbirds, cacholotes, foliage-gleaners, treehunters, and leaftossers—groups named mainly after their appearance or habits. The various types of ovenbirds have evolved to fill a wide range of ecological niches. Earthcreepers behave in a similar way to larks, while cinclodes are dipperlike and associated with water. Cacholotes construct huge nests, up to 5 ft (1.5 m) across, out of thorny twigs.

Rufous Scrub-bird

Atrichornis rufescens

The Rufous Scrub-bird lives in the Australian eucalypt or rain forest of Queensland and New South Wales. The ideal habitat is one of dense ground cover with a thick layer of leaf litter and a high humidity. The closely-related Noisy Scrub-bird (*A. clamosus*) is found only in the far southwest of Australia, so the two species are separated by a long distance. This suggests that the scrub-birds once had a much wider distribution and became separated as the climate became more arid. Mainly small and brown, it has rufous-and-black barring on the leading edge of its wings, back and tail. When alert, the tail is held in an upright, wrenlike posture. It has a pale patch under the chin, and the eye and bill are dark. The breast is paler and mottled brown and buff. Its favored habitat contains moist leaf-litter and roughly 3 ft- (1 m-) tall ground cover. It tends to forage in natural clearings with open canopy, such as where occasional trees have fallen or been felled.

The true population is unknown, and estimates vary between 730 and 12,000 pairs, though a decrease has been observed since the 19th century when large areas of land were cleared. Southern populations have been noticed moving to similar habitat at a higher altitude, and this is thought to be a possible response to climate change. The existing habitats are threatened by logging and burning, but proposals are underway to implement fire management plans and to ban further logging.

Family	Atrichornithidae
Size	6¼-7 in (16–18 cm)
Status	Vulnerable (decreasing)
Population	10,000–20,000
Range	Australia (endemic)
Habitat	Rain forest and wet eucalypt forest
Main threats	Logging, fire

Bolivian Spinetail

Cranioleuca henricae

The Bolivian Spinetail, which was only discovered as recently as 1993, has a rufous crown, and a whitish stripe over the eye. The wings and spiny tail are rufous, while the mantle, nape, and rump are olive-brown. It has a gray face, chin, and underparts apart from the vent and flanks, which are olive, and olive-yellow legs and feet. The prime habitats are in the dry rain-shadow valleys of the eastern slopes of the Andes, between around 6,000 ft (1,800 m) and 10,000 ft (3,000 m) in West Bolivia. Here it is found in the understorey of deciduous forests, where it is thought to nest in the "gray beard" epiphytic bromeliads. These epiphytes are found only in the humid valleys, typically close to streams, where they grow on the branches of large trees, with the older trees tending to support the richest bromeliad growth. Even where the forest floor has been grazed, there is evidence that the birds can survive as long as the old forest trees with their "gray beard" bromeliads remain. This habitat is threatened by eucalypt plantations that have dramatically affected the area's hydrology, reducing the overall humidity and creating problems with landslides and soil erosion. Other areas are overgrazed by goats, and further degraded by people collecting firewood, and by fires and logging. These activities all reduce the regenerative power of the forests. Road developments will also make the forest more accessible for locals collecting wood for charcoal.

Family	Furnariidae
Size	5¾ in (14.5 cm)
Status	Endangered (decreasing)
Population	1,000–2,500
Range	Bolivia (endemic)
Habitat	Forest
Main threats	Deforestation, grazing

African Green Broadbill

Pseudocalyptomena graueri

This small, rotund, mainly leaf-green bird has a blue breast, throat, and rump. It has a buff-colored forehead with fine black streaks and a narrow black eyestripe. This broadbill prefers habitat high up in primary forest between 25 ft (7 m) and 65 ft (20 m) from the ground. Here it feeds on a diet of snails, small insects, such as beetles and larvae, seeds, buds, and flowers. The birds usually search for food alone or in small flocks. When displaying, it trembles and puffs out its feathers. African Green Broadbills are usually solitary or paired, though they sometimes join small, mixed flocks with other species.

The species was first discovered in 1908, but was then not seen again for some 20 years. It is known only from three locations within two African countries: Bwindi Impenetrable Forest (Uganda) and Mount Kahuzi and the Itombwe Mountains (both in the Democratic Republic of Congo). Each site has problems, being threatened by mining, farming, and hunting. Political problems and conflict in the areas have also added to pressures with large relocated populations of refugees from Burundi and Rwanda setting up camps close to the mountain habitat in the Democratic Republic of Congo. Famine from recent crop failures in Uganda also places additional stress on the forest as locals clear new areas in order to grow food.

Family Eurylaimidae
Size 4 in (10 cm)
Status Vulnerable (decreasing)
Population 2,500–10,000
Range Democratic Republic of Congo, Uganda
Habitat Rain forest
Main threats Deforestation and degradation

Masafuera Rayadito

Aphrastura masafuerae

Typical of the family, the Masafuera Rayadito has a rather slender body and bill. Its plumage is a mixture of gray, brown, and buff with more distinctive black wings with a cinnamon bar. The tail is also dark with a red-brown patch and spiky terminal feathers. The species is endemic to a small island, Alejandro Selkirk (Más Afuera) in the Juan Fernández Islands, Chile. It lives in habitats dominated by the large tree fern (*Dicksonia externa*) that grow to 15 ft (5 m) tall, mainly at altitudes of 2,600–4,300 ft (800–1,300 m). It nests in rock crevices alongside stream courses that run through the highland forests, and feeds on insects plucked mainly from the tree ferns.

Threats come from potential loss of this isolated tree fern habitat. The ferns and surrounding forest are affected by fires, timber cutting, and trampling by uncontrolled goats. Introduced predators include house mice and rats, that predate nesting birds, while domestic cats impact on young and adults. Another predator is the Red-backed Hawk (*Buteo polyosoma*), whose numbers have increased, partly as a result of the availability of the introduced rodents.

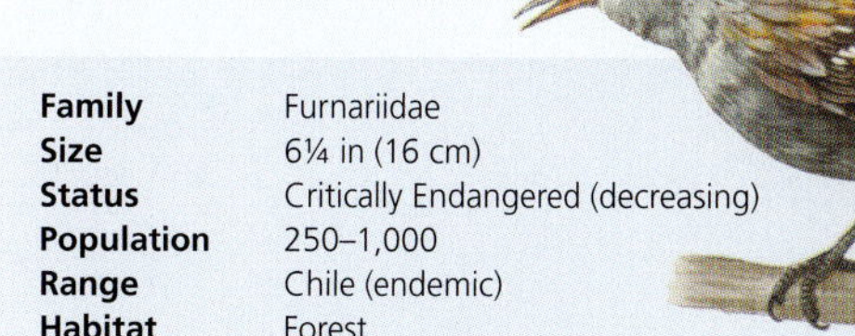

Family Furnariidae
Size 6¼ in (16 cm)
Status Critically Endangered (decreasing)
Population 250–1,000
Range Chile (endemic)
Habitat Forest
Main threats Alien predators, fire

THREATENED BROADBILLS, SCRUB-BIRDS, AND OVENBIRDS

FAMILY		COMMON NAME	SCIENTIFIC NAME	STATUS
EURYLAIMIDAE Broadbills	■	Mindanao Broadbill	*Eurylaimis steerii*	VU
	■	African Green Broadbill	*Pseudocalyptomena graueri*	VU
ATRICHORNITHIDAE Scrub-birds	■	Noisy Scrub-bird	*Atrichornis clamosus*	VU
	■	Rufous Scrub-bird	*Atrichornis rufescens*	VU
FURNARIIDAE Ovenbirds	■	Pink-legged Graveteiro	*Acrobatornis fonsecai*	VU
	■	Masafuera Rayadito	*Aphrastura masafuerae*	CR
	■	Cipo Canastero	*Asthenes luizae*	VU
	■	Royal Cinclodes	*Cinclodes aricomae*	CR
	■	White-bellied Cinclodes	*Cinclodes palliatus*	CR
	■	Campo Miner	*Geositta poeciloptera*	VU
	■	Henna-hooded Foliage-gleaner	*Hylocryptus erythrocephalus*	VU
	■	White-browed Tit-spinetail	*Leptasthenura xenothorax*	EN
	■	Chestnut-backed Thornbird	*Phacellodomus dorsalis*	VU
	■	Alagoas Foliage-gleaner	*Philydor novaesi*	CR
	■	White-throated Barbtail	*Premnoplex tatei*	VU
	■	Bolivian Spinetail	*Cranioleaca henricae*	EN

RED LIST ■ EX = Extinct ■ EW = Extinct in the Wild ■ CR = Critically Endangered ■ EN = Endangered ■ VU = Vulnerable

THREATENED BROADBILLS, SCRUB-BIRDS, AND OVENBIRDS *Continued*				
FAMILY		**COMMON NAME**	**SCIENTIFIC NAME**	**STATUS**
FURNARIIDAE Ovenbirds	■	Perija Thistletail	*Schizoeaca perijana*	EN
	■	Great Spinetail	*Siptornopsis hypochondriaca*	VU
	■	Apurimac Spinetail	*Synallaxis courseni*	VU
	■	Rusty-headed Spinetail	*Synallaxis fuscorufa*	VU
	■	Pinto's Spinetail	*Synallaxis infuscata*	EN
	■	Hoary-throated Spinetail	*Synallaxis kollari*	EN
	■	Maranon Spinetail	*Synallaxis maranonica*	VU
	■	Blackish-headed Spinetail	*Synallaxis tithys*	EN
	■	Bahia Spinetail	*Synallaxis whitneyi*	VU
	■	Russet-bellied Spinetail	*Synallaxis zimmeri*	EN
	■	Rufous-necked Foliage-gleaner	*Syndactyla ruficollis*	VU
	■	Russet-mantled Softtail	*Thripophaga berlepschi*	VU
	■	Orinoco Softtail	*Thripophaga cherriei*	VU
	■	Striated Softtail	*Thripophaga macroura*	VU

RED LIST ■ EX = Extinct ■ EW = Extinct in the Wild ■ CR = Critically Endangered ■ EN = Endangered ■ VU = Vulnerable

ANTBIRDS

Antbirds (Thamnophilidae) are a diverse family of 215 species found in Central and South America, mainly in forest habitats. They are related to antpittas and ant thrushes (*Formicariidae*). Most are rather small, with rounded wings and strong legs. A key feature is the anatomy of their front toes, which are joined at the base. Many species spend much of their time hunting for mainly invertebrate prey on the forest floor. The habit of some members of this large family to follow swarms of ants is the probable origin of their common name. Some antbirds obtain about half of their food in this way, gaining rich pickings from other invertebrates disturbed by army ants as they march through the undergrowth. Some species occasionally also take small vertebrates, some join mixed-species flocks to feed, and others feed higher in the canopy. Antbirds usually lay two eggs that both parents rear equally until fledged, then each will concentrate on feeding just one individual.

The family contains antbirds, antshrikes, antvireos, antwrens, bushbirds, fire-eyes, and bare-eyes. In some species the sexes have very different plumage. Habitat loss is the main problem affecting antbirds, and 24 species are now listed as threatened, with nine of these Endangered and three (Restinga, Rio de Janeiro, and Alagoas Antwrens) Critically Endangered.

Black-hooded Antwren

Formicivora erythronotos

The male of this small antwren is mainly slate back with white flanks hidden under black wings with three white bars. The back is chestnut colored. Females are similar, but with olive feathers replacing the black, and a creamier underside. Believed extinct until 1987, this Brazilian endemic was rediscovered along a small coastal strip in south Rio de Janeiro state. It lives mainly in swampy secondary forest with a rich undergrowth comprised largely of pioneer plant species including grasses, mangroves, *Lantana camara*, and species of *Cecropia*, *Morus*, *Rubus*, and *Vernonia*. It has also been found on abandoned banana plantations. However, it appears very sensitive to human activities, including using land for pasture, plantations, and tourist developments. These pressures are all rendering the habitat unsuitable and are responsible for its decline. Future land management plans propose protecting the dwindling resources and habitat for this endangered endemic.

Family	Thamnophilidae
Size	4½ in (11.5 cm)
Status	Endangered (decreasing)
Population	1,000–2,500
Range	Brazil (endemic)
Habitat	Swampy coastal forest and scrub
Main threats	Habitat loss

Fringe-backed Fire-eye
Pyriglena atra

The male of this species is black, with a white-fringed dorsal patch and fiery-red eyes. Females have a black tail, but are chestnut above, buff underneath, and have gray scapular feathers. They live in small areas of the narrow strip of Atlantic forest in eastern Brazil. They seem to prefer areas with low-growing trees and like to perch on horizontal branches near the ground within forest clearings. Fringe-backed Fire-eyes often follows swarms of army ants, which they pursue as they pass through the low habitat, feeding mainly on other invertebrates disturbed by these aggressive ants. They sometimes move about in small groups. The main breeding season is thought to be from October to March. The main threats seem to be from the loss of habitat and the associated vulnerability of remaining small isolated populations. However, their use of secondary forest has meant this bird has a few more habitat options than some species, though these too are declining in number and quality.

Family	Thamnophilidae
Size	7 in (17.5 cm)
Status	Endangered (decreasing)
Population	1,000–2,500
Range	Brazil (endemic)
Habitat	Lowland forest
Main threats	Habitat loss

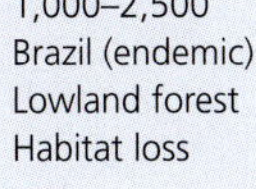

White-bearded Antshrike
Biatas nigropectus

The male has a black cap and lower throat and breast, and white ear coverts joining with a buff collar. It is otherwise mainly brown with rufous wings and olive-brown below. The female has a rufous cap. This species eats a wide range of food including ants, seeds, insect larvae, and spiders.

It has a rather scattered and restricted range in northeastern Argentina and southeastern Brazil. Brazilian birds are found in the remaining areas of native forest, at an altitude of 4,000 ft (1,200 m), preferring open areas with edges of tall thicket. The species is closely associated with certain species of bamboo—notably *Guadua trinii* in Argentina. The bamboo has a 30-year life cycle after which it experiences rather synchronous seeding and die-off. Combined with areas of bamboo being small and isolated, this poses a significant threat to the Argentinian populations. In Brazil, large areas of forest are being replaced by agriculture, mining, and coffee, banana, and rubber plantations. Forest land is also being taken for urbanization and road-building projects.

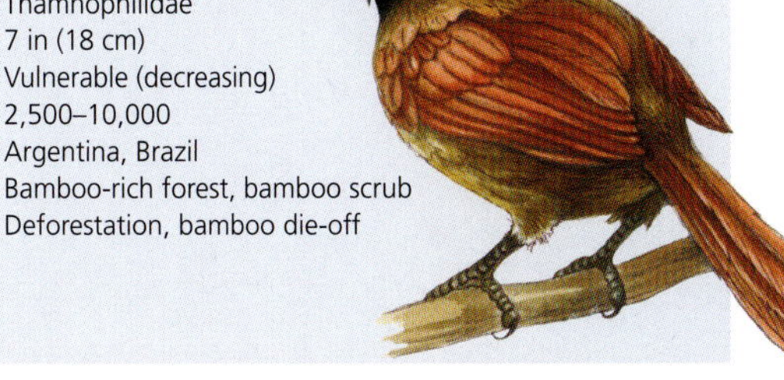

Family	Thamnophilidae
Size	7 in (18 cm)
Status	Vulnerable (decreasing)
Population	2,500–10,000
Range	Argentina, Brazil
Habitat	Bamboo-rich forest, bamboo scrub
Main threats	Deforestation, bamboo die-off

THREATENED ANTBIRDS

FAMILY	COMMON NAME	SCIENTIFIC NAME	STATUS
THAMNOPHILIDAE Antbirds	White-bearded Antshrike	*Biatas nigropectus*	VU
	Bananal Antbird	*Cercomacra ferdinandi*	VU
	Recurve-billed Bushbird	*Clytoctantes alixii*	EN
	Rondonia Bushbird	*Clytoctantes atrogularis*	VU
	Bicolored Antvireo	*Dysithamnus occidentalis*	VU
	Plumbeous Antvireo	*Dysithamnus plumbeus*	VU
	Black-hooded Antwren	*Formicivora erythronotos*	EN
	Restinga Antwren	*Formicivora littoralis*	CR
	Ash-throated Antwren	*Herpsilochmus parkeri*	EN
	Pectoral Antwren	*Herpsilochmus pectoralis*	VU
	Bahia Antwren	*Herpsilochmus pileatus*	VU
	Gray-headed Antbird	*Myrmeciza griseiceps*	VU
	Scalloped Antbird	*Myrmeciza ruficauda*	EN
	Rio de Janeiro Antwren	*Myrmotherula fluminensis*	CR
	Salvadori's Antwren	*Myrmotherula minor*	VU
	Alagoas Antwren	*Myrmotherula snowi*	CR
	Band-tailed Antwren	*Myrmotherula urosticta*	VU

RED LIST EX = Extinct EW = Extinct in the Wild CR = Critically Endangered EN = Endangered VU = Vulnerable

THREATENED ANTBIRDS *Continued*				
FAMILY		**COMMON NAME**	**SCIENTIFIC NAME**	**STATUS**
THAMNOPHILIDAE Antbirds	■	Allpahuayo Antbird	*Percnostola arenarum*	VU
	■	Fringe-backed Fire-eye	*Pyriglena atra*	EN
	■	Slender Antbird	*Rhopornis ardesiacus*	EN
	■	Parana Antwren	*Stymphalornis acutirostris*	EN
	■	Yellow-rumped Antwren	*Terenura sharpei*	EN
	■	Orange-bellied Antwren	*Terenura sicki*	EN
	■	Spiny-faced Antshrike	*Xenornis setifrons*	VU

RED LIST ■ EX = Extinct ■ EW = Extinct in the Wild ■ CR = Critically Endangered ■ EN = Endangered ■ VU = Vulnerable

ANTPITTAS, PITTAS, WOODCREEPERS, NEW ZEALAND WRENS, AND ASITIES

Antpittas (Formicariidae) are closely related to the antbirds and are also found in Central and South America. There are 63 species, and the family also contains the ant thrushes. They have an upright posture, a short tail, long powerful legs, and are mainly rather drab in plumage. They typically hunt amongst the leaf litter of tropical forests for invertebrate prey, many species specializing in ants, as their name implies. Like the antbirds, antpittas are threatened mainly by destruction or alteration of their forest habitats. Three of the 12 threatened species are Endangered, while one, the Tachira Antpitta, is Critically Endangered.

Pittas (Pittidae) are a family of 31 species found in a range of mainly wooded habitats from Africa, and South and East Asia to New Guinea and Australia. Rather thrushlike in build, but with long legs and a short tail, most pittas have bright, colorful plumage. Nine pittas are threatened, of which one, Gurney's Pitta, is Endangered.

The 52 species of woodcreepers (Dendrocolaptidae) of Central and South America contain one threatened species—Moustached Woodcreeper of northeast Brazil (Vulnerable).

There are just two living species of endemic New Zealand wrens (Acanthisittidae): Rifleman and the South Island Wren (Vulnerable), the latter suffering from predation by alien mammals, notably the House Mouse and Stoat. The Rifleman is one of New Zealand's smallest birds and feeds treecreeper-like from the bark of trees. Both birds are poor fliers and it is thought they may possibly have evolved from a more ancient lineage of flightless birds.

Asities (Philepittidae), a family endemic to Madagascar, contains four species, one of which, Yellow-bellied Asity, is threatened. Asities feed on a variety of foods, from insects to fruit, and are adapted to exploit a range of forest habitats, at both high and low altitude.

Giant Antpitta
Grallaria gigantea

This very large antpitta is olive-brown above and chestnut underneath. The back of the crown and nape are gray and the sides of the head are chestnut. The throat and breast feathers have black edging so these parts appear subtly barred from a distance. The bill is dark and heavy, used for feeding mainly on invertebrates such as beetle larvae, slugs, and giant earthworms (*Rhynodrilus*). The Giant Antpitta inhabits mainly subtropical and temperate forest at an altitude of 7,200 ft (2,200 m) to 8,500 ft (2,600 m), and is known only from southwestern Colombia and northern Ecuador. In both countries its range has been reduced and numbers have declined considerably, mainly as a result of habitat loss and it is now recorded from only a handful of locations. It prefers swampy areas in the humid cloud-forest understorey but also forages in pastures and in secondary forest. Threats to its habitat have existed since the early 20th century with lower forest being converted to crop growing. More recently there has been extensive deforestation due to logging, agriculture, and plantations. Some fragments of remaining habitat are now protected, and research is underway to find new areas for potential relocation.

Family	Formicariidae
Size	10½ in (26.5 cm)
Status	Vulnerable (decreasing)
Population	1,000–2,500
Range	Colombia, Ecuador
Habitat	Humid montane forest
Main threats	Deforestation and degradation

Gurney's Pitta

Pitta gurneyi

The male of this attractively marked pitta has a black central breast and belly, and a black head with a blue crown. The flanks are yellow with black barring. The upper wings and back are a rich brown, while the tail is deep blue with black undercoverts. It is found in lowland semi-evergreen forests, with Salacca Palms (*Salacca zalacca*) in the undergrowth, in which it nests. It breeds during the wet season and lives in permanently damp gulleys with access to forest edges. In 1997 it was feared this species had been reduced to just a handful of pairs in Thailand, but in 2003 it was discovered at a number of sites in southern Myanmar. Estimates based on suitable habitat put its numbers at between 5,000 and 8,000 pairs. The habitat has suffered due to massive lowland forest clearances in order to grow crops of rubber, oil palm, coffee, and fruit. In Thailand the birds are also hunted for the cagebird trade. Efforts are concentrating on conserving lowland forest areas in Myanmar, as problems are anticipated with land earmarked for future oil-palm plantation.

Family Pittidae
Size 7¼–8 in (18.5–20.5 cm)
Status Endangered (decreasing)
Population 10,000–16,000
Range Thailand, Myanmar
Habitat Lowland forest
Main threats Deforestation, trapping

Yellow-bellied Asity

Neodrepanis hypoxantha

This tiny jewel-like beauty has, as its name says, a yellow belly that it displays aggressively to any intruder, even to humans. The upper plumage of the male is black with iridescent blue edges, and it has a vibrant turquoise-and-ultramarine wattle from the bill base to behind the eye. The long decurved bill is suited to sipping plant nectar in the manner of a sunbird, though it also feeds on arthropods. It lives only in the high altitude forests of eastern Madagascar, and is found up to the treeline. An active feeder, it forages in both the woodland canopy and in the mossy damp habitats below. The higher woodland habitats of Madagascar are less threatened than the more accessible lowlands due to their reduced value for agriculture or timber. However, previous deforestation activities have left very fragmented habitats that are prone to fire damage from slash-and-burn farming methods. Future plans aim to improve the remaining habitats and alter current agricultural burning practices.

Family Philepittidae
Size 3½–4 in (9–10 cm)
Status Vulnerable (decreasing)
Population 10,000–20,000
Range Madagascar (endemic)
Habitat Montane wet woodland
Main threats Habitat fragmentation, fire

THREATENED ANTPITTAS, PITTAS, WOODCREEPERS, NEW ZEALAND WRENS, AND ASITIES

FAMILY	COMMON NAME	SCIENTIFIC NAME	STATUS
FORMICARIIDAE Antpittas	Moustached Antpitta	*Grallaria alleni*	VU
	Santa Marta Antpitta	*Grallaria bangsi*	VU
	Tachira Antpitta	*Grallaria chthonia*	CR
	Great Antpitta	*Grallaria excelsa*	VU
	Giant Antpitta	*Grallaria gigantea*	VU
	Cundinamarca Antpitta	*Grallaria kaestneri*	VU
	Brown-banded Antpitta	*Grallaria milleri*	EN
	Jocotoco Antpitta	*Grallaria ridgelyi*	EN
	Bicolored Antpitta	*Grallaria rufocinerea*	VU
	Hooded Antpitta	*Grallaricula cucullata*	VU
	Ocher-fronted Antpitta	*Grallaricula ochraceifrons*	EN
	Masked Antpitta	*Hylopezus auricularis*	VU
PITTIDAE Pittas	Black-faced Pitta	*Pitta anerythra*	VU
	Blue-headed Pitta	*Pitta baudii*	VU
	Gurney's Pitta	*Pitta gurneyi*	EN
	Whiskered Pitta	*Pitta kochi*	VU
	Fairy Pitta	*Pitta nympha*	VU
	Schneider's Pitta	*Pitta schneideri*	VU

RED LIST ■ EX = Extinct ■ EW = Extinct in the Wild ■ CR = Critically Endangered ■ EN = Endangered ■ VU = Vulnerable

THREATENED ANTPITTAS, PITTAS, WOODCREEPERS, NEW ZEALAND WRENS, AND ASITIES *Continued*				
FAMILY		**COMMON NAME**	**SCIENTIFIC NAME**	**STATUS**
PITTIDAE Pittas	■	Azure-breasted Pitta	*Pitta steerii*	VU
	■	Superb Pitta	*Pitta superba*	VU
	■	Graceful Pitta	*Pitta venusta*	VU
DENDROCOLAPTIDAE Woodcreepers	■	Moustached Woodcreeper	*Xiphocolaptes falcirostris*	VU
ACANTHISITTIDAE New Zealand wrens	■	South Island Wren	*Xenicus gilviventris*	VU
PHILEPITTIDAE Asities	■	Yellow-bellied Asity	*Neodrepanis hypoxantha*	VU

RED LIST ■ EX = Extinct ■ EW = Extinct in the Wild ■ CR = Critically Endangered ■ EN = Endangered ■ VU = Vulnerable

TYRANT-FLYCATCHERS

The tyrant-flycatchers (Tyrannidae) are a huge New World family with some 412 species, the bulk of them found in Central and South America. One of the largest and most diverse of all bird families, they exhibit wide-ranging physical characteristics, range, and distribution. Mainly small to medium-sized birds of forest, woodland, and savanna, they contain many groups, including attilas, chat-tyrants, elaenias, flatbills, flycatchers, ground-tyrants, kingbirds, pewees, pygmy-tyrants, shrike-tyrants, spadebills, tit-tyrants, tody-flycatchers, tody-tyrants, and tyrannulets. The smaller species can weigh as little as ¼ oz (7 g), have bodies less than 2¾ in (7 cm) long, while larger species can be up to 1 ft (29 cm) long and weigh around 3¼ oz (90 g). Most are dull or olive-colored with subtle camouflage patterned variations, though several have bright plumage. Many species have crests that they use raised in displays, the most elaborate being that of the Pacific Royal Flycatcher. Tyrant flycatchers include insects in their diet and some catch their prey in flight, though some also feed on fruit and the larger species eat small vertebrates such as lizards and frogs. Thirty species are currently listed as threatened, though none is Critically Endangered. Ten are listed as Endangered, including four tyrannulets and the impressive rather shrikelike Giant Kingbird, the latter known only from a handful of sites on Cuba where it is decreasing. The main threats to these birds come from deforestation and habitat degradation.

Strange-tailed Tyrant

Alectrurus risora

Well-named, the male of this unusual bird has stunning black tail streamers, increasing the overall length by a third to 11¼ in (30 cm). The male is mainly black and white, and has a featherless patch on the neck that takes on a brilliant reddish-pink color during the breeding season. The female is browner, with streaked plumage, buff underneath, and has shorter brown tail feathers. This gregarious species has a complex lek-like mating system and males may be monogamous or polygamous. The typical habitats are marshy wetlands with tall grasses up to 5 ft (1.5 m) in height where they usually live in groups of 20–50 individuals. They feed on a range of invertebrates and have been observed following other predators such as one of the army ants (*Labidus praedator*) and the Six-banded Armadillo (*Euphractus sexcinctus*) to find food. Threats stem from loss of habitat by spreading agriculture (soybean, corn, and wheat). Eucalypt plantations have altered land use and caused hydrology problems. The grass-burning techniques used to improve grazing also impact on this species.

Family Tyrannidae
Size 8 in (20 cm) plus 4 in (10 cm) tail in male
Status Vulnerable (decreasing)
Population 10,000–20,000
Range Argentina, Paraguay

Kaempfer's Tody-tyrant

Hemitriccus kaempferi

This small, dainty flycatcher is brownish to olive green above and yellow-green below. The throat is pale yellow, and the secondaries a greenish-yellow. The wings and tail are dark, and it has a creamy-yellow wing bar. Endemic to Brazil, it is known from a number of fragmented locations in the southeast of the country. It lives in pairs in humid lowland evergreen forest, showing a preference for areas where rivers run though the dense vegetation. It feeds mainly on small insects, either caught in flight, or by hovering to pick insects from the foliage. The Atlantic forests of Brazil have been extensively cleared and many of the remnants are under threat from conversion to plantations of bananas, rice, and timber trees. Many coastal sites are also being developed for housing and roads. Rising sea levels are also a possible future threat to the small fragmented habitats by the coast.

Family Tyrannidae
Size 4 in (10 cm)
Status Endangered (decreasing)
Population 9,000–18,500
Range Brazil (endemic)
Habitat Lowland Atlantic forest
Main threats Deforestation

Pacific Royal Flycatcher

Onychorhynchus occidentalis

This flycatcher is dull brown in color above with a rufous rump and tail. Underneath it is dull orange apart from a white throat. However, its most impressive feature is its fanlike crest—mainly red in the male and yellow in the female. Usually, the crest rests furled flat across the head, but when raised and unfurled during courtship or territorial displays or in alarm (as when handled for ringing) it is spectacular. It inhabits lowland humid deciduous forest in western Ecuador (Esmeraldas to El Oro) and adjacent northwest Peru (Tumbes), but is scarce. Birds have been observed breeding in January and April, building nests that hang from branches and vines over shaded watercourses. They sometimes forage in mixed-species flocks. Preferring to forage below an altitude of 4,000 ft (1,200 m), this species is particularly threatened by deforestation in these lowland areas. Grazing by cattle and goats damages the forest beyond its ability to regenerate, and logging and illegal settlements add further pressure. Traditional methods used to burn grazing lands also create problems of degradation from unintentional and controlled fire damage.

Family Tyrannidae
Size 6¼–6½ in (16–16.5 cm)
Status Vulnerable (decreasing)
Population 2,500–10,000
Range Ecuador, Peru
Habitat Lowland forest, scrub
Main threats Deforestation, livestock grazing, fire

THREATENED TYRANT-FLYCATCHERS

FAMILY		COMMON NAME	SCIENTIFIC NAME	STATUS
TYRANNIDAE Tyrant-flycatchers	■	White-tailed Shrike-tyrant	*Agriornis albicauda*	VU
	■	Strange-tailed Tyrant	*Alectrurus risora*	VU
	■	Cock-tailed Tyrant	*Alectrurus tricolor*	VU
	■	Ash-breasted Tit-tyrant	*Anairetes alpinus*	EN
	■	Tawny-chested Flycatcher	*Aphanotriccus capitalis*	VU
	■	Ocheraceous Attila	*Attila torridus*	VU
	■	Rufous Twistwing	*Cnipodectes superrufus*	VU
	■	Sharp-tailed Tyrant	*Culicivora caudacuta*	VU
	■	Noronha Elaenia	*Elaenia ridleyana*	VU
	■	Fork-tailed Pygmy-tyrant	*Hemitriccus furcatus*	VU
	■	Kaempfer's Tody-tyrant	*Hemitriccus kaempferi*	EN
	■	Buff-breasted Tody-tyrant	*Hemitriccus mirandae*	VU
	■	Gray-breasted Flycatcher	*Lathrotriccus griseipectus*	VU
	■	Rufous Flycatcher	*Myiarchus semirufus*	EN
	■	Santa Marta Bush-tyrant	*Myiotheretes pernix*	EN
	■	Cocos Flycatcher	*Nesotriccus ridgwayi*	VU
	■	Pacific Royal Flycatcher	*Onychorhynchus occidentalis*	VU
	■	Atlantic Royal Flycatcher	*Onychorhynchus swainsoni*	VU

RED LIST ■ EX = Extinct ■ EW = Extinct in the Wild ■ CR = Critically Endangered ■ EN = Endangered ■ VU = Vulnerable

THREATENED TYRANT-FLYCATCHERS *Continued*				
FAMILY		COMMON NAME	SCIENTIFIC NAME	STATUS
TYRANNIDAE Tyrant-flycatchers	■	Urich's Tyrannulet	*Phyllomyias urichi*	EN
	■	Yungas Tyrannulet	*Phyllomyias weedeni*	VU
	■	Bahia Tyrannulet	*Phylloscartes beckeri*	EN
	■	Alagoas Tyrannulet	*Phylloscartes ceciliae*	EN
	■	Restinga Tyrannulet	*Phylloscartes kronei*	VU
	■	Antioquia Bristle-tyrant	*Phylloscartes lanyoni*	EN
	■	Minas Gerais Tyrannulet	*Phylloscartes roquettei*	EN
	■	Russet-winged Spadebill	*Platyrinchus leucoryphus*	VU
	■	Lulu's Tody-flycatcher	*Poecilotriccus luluae*	VU
	■	Giant Kingbird	*Tyrannus cubensis*	EN
	■	Black-and-white Monjita	*Xolmis dominicanus*	VU
	■	Mishana Tyrannulet	*Zimmerius villarejoi*	VU

RED LIST ■ EX = Extinct ■ EW = Extinct in the Wild ■ CR = Critically Endangered ■ EN = Endangered ■ VU = Vulnerable

MANAKINS, COTINGAS, AND TAPACULOS

These three families are all forest birds living in regions of Central and South America.

There are 54 species of manakins (Pipridae) found from southern Mexico to northern Argentina mostly in lowland rain forests. In many species the males are brightly colored and they typically raise the feathers of their crests or throat patches in display. They are small or medium-sized and have rather a large head and small bill. Always on the move, these active birds search through the foliage for fruits and insects. Four species are threatened, of which one, the Araripe Manakin, is Critically Endangered.

Cotingas (Cotingidae), comprising 96 species, have a similar distribution. Most inhabit tropical forests, ranging from lowland rain forests to montane cloud forests. In size they range from the very small Kinglet Calyptura (Critically Endangered) to the large umbrellabirds. Most cotingas have a broad bill, hooked at the tip, and rounded wings, and in many the males have brightly colored plumage. In addition to cotingas, the family also contains cock-of-the-rocks, becards, bellbirds, fruitcrows, fruiteaters, and pihas. Eighteen species are threatened.

Tapaculos (Rhinocryptidae) are mainly dull-colored ground-birds of dense forests. Most species are found in the temperate forests of the southern slopes of the Andes. There are 57 species, six of which are threatened, two of these Critically Endangered: Bahia Tapaculo and Stresemann's Bristlefront.

THREATENED MANAKINS, COTINGAS, AND TAPACULOS				
FAMILY		COMMON NAME	SCIENTIFIC NAME	STATUS
PIPRIDAE Manakins	■	Araripe Manakin	*Antilophia bokermanni*	CR
	■	Golden-crowned Manakin	*Lepidothrix vilasboasi*	VU
	■	Wied's Tyrant-manakin	*Neopelma aurifrons*	VU
	■	Black-capped Piprites	*Piprites pileata*	VU
COTINGIDAE Cotingas	■	Kinglet Calyptura	*Calyptura cristata*	CR
	■	Yellow-billed Cotinga	*Carpodectes antoniae*	EN
	■	Black-headed Berryeater	*Carpornis melanocephala*	VU
	■	Bare-necked Umbrellabird	*Cephalopterus glabricollis*	VU
	■	Long-wattled Umbrellabird	*Cephalopterus penduliger*	VU

RED LIST ■ EX = Extinct ■ EW = Extinct in the Wild ■ CR = Critically Endangered ■ EN = Endangered ■ VU = Vulnerable

Araripe Manakin
Antilophia bokermanni

One of the most recently identified and rarest of birds, the Araripe Manakin was scientifically described in 1998, two years after its discovery. Its exceptionally restricted area of specialized habitat is a mere 11 sq miles (28 sq km) along the Araripe Plateau in northeast Brazil. It nests along streams and springs among branches overhanging water and feeds largely on small fruits. Like many manakins, the male and female are strikingly different. The former is white except for black wings and tail, with a bright crimson-carmine streak running from the forehead feather tuft over the crown, nape, and shoulders to the upper back; the female is shades of olive-green, paler above, with a smaller frontal tuft. This species' already tiny range has been further reduced by general deforestation for timber and plantations such as banana and maize, and especially tourist development with apartments and swimming pools reducing water availability. Conservation measures are supported by, among others, oil giant British Petroleum and wildlife presenter Sir David Attenborough.

Family	Pipridae
Size	6 in (15.5 cm)
Status	Critically Endangered (decreasing)
Population	250–1,000
Range	Brazil (endemic)
Habitat	Moist forest
Main threats	Deforestation, development

Three-wattled Bellbird
Procnias tricarunculatus

This usually secretive, chiefly frugivorous bird is betrayed by the male's astonishingly penetrating breeding-season call. As the common name implies, this is bell-like, but more a hollow "boing" than a resonant "ding," often accompanied by shrill whistles. Its other unique trait, also reflected in the name, is the male's three long, thin, black, flaccid wattles that dangle from the bill base. The male has a white head, neck, upper back, and chest, with a dark eye surround, and is otherwise chestnut or rufous, while the wattleless female is an attractive mixed olive with yellow feather edges above and yellow with olive flecks below. This species feeds mainly on fruits such as berries and the birds move about following fruit availability. The birds breed in moist forests between 2,500 ft (750 m) and 6,600 ft (2,000 m), mostly between March and September. Outside the breeding season they undertake complex migrations, ranging up to 10,000 ft (3,000 m), then descending to the foothills toward both Pacific and Caribbean coasts. Clearance of these lowlands for ranches, crops such as bananas, and tourism present the major threats. It occurs in a number of reserves, notably Sierra de Agalta National Park (Honduras), La Amistad International Park (Costa Rica and Panama), and Monteverde Biological Reserve (Costa Rica).

Family	Cotingidae
Size	10¼–12¼ in (26–31cm)
Status	Vulnerable (decreasing)
Population	10,000–20,000
Range	Costa Rica, Panama, Nicaragua, Honduras
Habitat	Moist forest
Main threats	Habitat loss, logging

THREATENED MANAKINS, COTINGAS, AND TAPACULOS *Continued*

FAMILY	COMMON NAME	SCIENTIFIC NAME	STATUS
COTINGIDAE Cotingas	Banded Cotinga	*Cotinga maculata*	EN
	Turquoise Cotinga	*Cotinga ridgwayi*	VU
	Chestnut-bellied Cotinga	*Doliornis remseni*	VU
	Bay-vented Cotinga	*Doliornis sclateri*	VU
	Scimitar-winged Piha	*Lipaugus uropygialis*	VU
	Chestnut-capped Piha	*Lipaugus weberi*	EN
	Slaty Becard	*Pachyramphus spodiurus*	EN
	Peruvian Plantcutter	*Phytotoma raimondii*	EN
	Bare-throated Bellbird	*Procnias nudicollis*	VU
	Three-wattled Bellbird	*Procnias tricarunculatus*	VU
	Gray-winged Cotinga	*Tijuca condita*	VU
	White-winged Cotinga	*Xipholena atropurpurea*	EN
	White-cheeked Cotinga	*Zaratornis stresemanni*	VU

RED LIST ■ EX = Extinct ■ EW = Extinct in the Wild ■ CR = Critically Endangered ■ EN = Endangered ■ VU = Vulnerable

Bahia Tapaculo

Eleoscytalopus psychopompus

Originally considered not just extremely rare, but possibly extinct from the 1980s, this exceedingly shy species has recently been found at several restricted and fragmented locations in coastal Bahia State, eastern Brazil —including Una Biological Reserve with another Critically Endangered species, the Yellow-breasted Capuchin monkey (*Cebus xanthosternos*). Possibly this timid tapaculo exists at other sites, but general loss of lowland rain forests to timber, agriculture, livestock grazing, and water and mineral extraction all pose severe risks. Small, dumpy, and darting, the species prefers dense tangles of shrubs and vines over or close to water, where it hunts a variety of small invertebrate foods, especially insects, and takes some plant matter, including nectar. Its coloration is dark slate-gray above, with a white chin extending to the chest and belly, a white spot between each eye and the dark bill (loral spot), and a distinctive reddish streak along each side of the lower chest and belly, between the white below and the slate above.

Family	Rhinocryptidae
Size	4½ in (11.5cm)
Status	Critically Endangered (decreasing)
Population	50–250
Range	Brazil (endemic)
Habitat	Wet lowland forest
Main threats	Deforestation

THREATENED MANAKINS, COTINGAS, AND TAPACULOS *Continued*

FAMILY		COMMON NAME	SCIENTIFIC NAME	STATUS
RHINOCRYPTIDAE Tapaculos	■	Bahia Tapaculo	*Eleoscytalopus psychopompus*	CR
	■	Stresemann's Bristlefront	*Merulaxis stresemanni*	CR
	■	Marsh Tapaculo	*Scytalopus iraiensis*	EN
	■	Tacarcuna Tapaculo	*Scytalopus panamensis*	VU
	■	Ecuadorian Tapaculo	*Scytalopus robbinsi*	EN
	■	Upper Magdalena Tapaculo	*Scytalopus rodriguezi*	EN

RED LIST ■ EX = Extinct ■ EW = Extinct in the Wild ■ CR = Critically Endangered ■ EN = Endangered ■ VU = Vulnerable

SWALLOWS AND MARTINS, LARKS, AND PIPITS

These three families, and all those following, are members of the suborder Oscines (songbirds).

Swallows and martins (Hirundinidae) are 83 species of agile, streamlined birds that hunt flying insects, catching their prey in flight, mostly over fields and scrub or over water. They have a global distribution and are found mainly in open habitats. The majority breed in temperate regions and migrate to warmer areas to spend the winter. They mostly have dark upperparts and paler underparts and notched or forked tails, sometimes with streamers, and their wings are long and pointed. They either construct adhesive nests from mud, or nest in holes or burrows, some species in colonies. In addition to swallows and martins, the family also contains species known as crag-martins, river-martins, saw-wings, and striped-swallows. Seven species are listed as threatened, of which the Bahama Swallow is Endangered and the White-eyed River-martin is Critically Endangered.

Larks (Alaudidae), which are made up of 93 species, are small, mostly brown, ground-dwelling birds that are found mainly in open habitats such as grassland, scrub, and semi-deserts. Their nests are usually hidden among grasses and other ground vegetation, and the adults can also be hard to spot until they fly. They are known for their musical songs and calls and often deliver these from display flights over their territories. Seven are threatened, three of which, Archer's, Raso, and Sidamo Larks, are Critically Endangered.

Pipits (Motacillidae) comprise 61 species of slender birds of open habitats. They have a long tail and long claws, indeed some species are known as longclaws. The family also contains the wagtails. Pipits are mostly brown and streaked, while wagtails tend to be more boldly marked and have longer tails. Five species are threatened: four pipits and one longclaw.

Blue Swallow

Hirundo atrocaerulea

At a glance this small swallow looks black, but a close view reveals it as deep metallic blue. In flight the outer tail feathers are markedly long, even for a swallow. It is a migrant within Africa, its widely spaced breeding areas in the southeast ranging from South Africa and Swaziland northward to southeastern Democratic Republic of Congo and southwestern Tanzania. Most populations journey north for the non-breeding months to northeastern Democratic Republic of Congo, Uganda, and west Kenya. The breeding habitat is relatively specific: upland damp savanna with creeks, pools, and swamps, where insects develop as aquatic larvae before becoming winged adults—and the swallow's food. The nest is a typical swallow mud cup, but it is sited in a most unusual place for a swallow—underground, attached to the roof or wall of an earth crack, a pothole, or the burrow of an Aardvark (*Orycteropus afer*). Similarly its nonbreeding savanna areas are also patchworked with wetlands. Drainage and conversion of these grasslands to stock grazing and agriculture—especially sugar cane and maize—along with human settlements and alien species are threats to this already relatively uncommon species.

Family Hirundinidae
Size 7–9¾ in (18–25 cm)
Status Vulnerable (decreasing)
Population 2,000
Range Southern and Central Africa
Habitat Grassland
Main threats Afforestation, agriculture, fire

Rudd's Lark

Heteromirafra ruddi

Mainly buff, apart from the brown-flecked chest and dark brown wings, this smallish lark also has darker brown cheeks, eyes, and crown, with its buff crest best seen when erect. Its proportions, compared to other larks, show relatively long legs, large head, and short tail. The species is endemic to eastern South Africa, with the main population centers southwest of Lesotho and west of Swaziland. Typically a grassland bird, it prefers damp and moderately-grazed savanna with lack of rocky patches, bushes, or trees. It can tolerate well-managed mixed livestock grazing regimes, but suffers after intensive grazing and frequent burning, whether deliberate or natural wildfires. It requires patches of slightly longer grasses for nest concealment and a ready supply of small invertebrates such as grasshoppers and other insects, spiders, and worms, especially for the chicks. Rudd's Lark is one of many species at risk from natural grassland clearance for crops, housing, and associated building, mining, and encroachment of the grassland by forestry species such as wattles and invasives such as bracken.

Family Alaudidae
Size 5½ in (14 cm)
Status Vulnerable (decreasing)
Population 2,500–5,000
Range South Africa (endemic)
Habitat Grassland
Main threats Habitat loss, afforestation, mining, fire

THREATENED SWALLOWS, MARTINS, LARKS, AND PIPITS

FAMILY		COMMON NAME	SCIENTIFIC NAME	STATUS
HIRUNDINIDAE Swallows and martins	■	White-eyed River-martin	*Eurochelidon sirintarae*	CR
	■	Blue Swallow	*Hirundo atrocaerulea*	VU
	■	White-tailed Swallow	*Hirundo megaensis*	VU
	■	Galapagos Martin	*Progne modesta*	VU
	■	Peruvian Martin	*Progne murphyi*	VU
	■	Bahama Swallow	*Tachycineta cyaneoviridis*	EN
	■	Golden Swallow	*Tachycineta euchrysea*	VU
ALAUDIDAE Larks	■	Raso Lark	*Alauda razae*	CR
	■	Red Lark	*Certhilauda burra*	VU
	■	Archer's Lark	*Heteromirafra archeri*	CR
	■	Rudd's Lark	*Heteromirafra ruddi*	VU
	■	Sidamo Lark	*Heteromirafra sidamoensis*	CR
	■	Ash's Lark	*Mirafra ashi*	EN
	■	Botha's Lark	*Spizocorys fringillaris*	EN

RED LIST ■ EX = Extinct ■ EW = Extinct in the Wild ■ CR = Critically Endangered ■ EN = Endangered ■ VU = Vulnerable

Sprague's Pipit

Anthus spragueii

A migratory species, this pipit breeds in summer in central states neighboring the Canada-U.S.A. border, then travels south to winter in the southernmost U.S.A., especially Texas, and southward to include the northern two-thirds of Mexico. Epidemic conversion of North America's natural, well-drained, mixed-length prairie grasslands to farm crops such as wheat and rye, and grazing of cattle and other stock, devastated this pipit's habitat, mainly during the late 19th and early-mid 20th centuries. Its ground nest, with stems folded over to form a tentlike structure, is especially vulnerable to modern agricultural practice. Another threat is that invasive shrubs and trees colonize its preferred open habitat. Sprague's Pipit has a buff background well streaked with dark brown on the chest and face, and the opposite coloration on the crown and wings, a pale face with brownish cheeks, and a pale eyebrow stripe. The mainstay of its diet is insects and other small invertebrates, especially when breeding, supplemented by seeds, buds, and other plant matter in winter.

Family	Motacillidae
Size	6¼ in (16 cm)
Status	Vulnerable (decreasing)
Population	870,000
Range	Canada, U.S.A.
Habitat	Grassland
Main threats	Agriculture, grazing, fire

THREATENED SWALLOWS, MARTINS, LARKS, AND PIPITS *Continued*

FAMILY		COMMON NAME	SCIENTIFIC NAME	STATUS
MOTACILLIDAE Pipits	■	Yellow-breasted Pipit	*Anthus chloris*	VU
	■	Ocher-breasted Pipit	*Anthus nattereri*	VU
	■	Sokoke Pipit	*Anthus sokokensis*	EN
	■	Sprague's Pipit	*Anthus spragueii*	VU
	■	Sharpe's Longclaw	*Macronyx sharpei*	EN

RED LIST ■ EX = Extinct ■ EW = Extinct in the Wild ■ CR = Critically Endangered ■ EN = Endangered ■ VU = Vulnerable

BULBULS, BUSH-SHRIKES AND HELMET-SHRIKES, CUCKOOSHRIKES, VANGAS, SHRIKES, AND LEAFBIRDS

Bulbuls (Pycnonotidae) are an Old World family of mostly tropical songbirds with 124 species found from Africa to the Far East. The family also contains bristlebills, finchbills, greenbuls, and nicators. Most species are small to medium-sized with fairly drab plumage and slightly curved bills. Their songs are mostly rather pleasant and musical and they tend to keep up a constant stream of contact calls as they search through foliage, often in small groups. Adaptable birds, they take a wide range of food such as invertebrates, fruit, nectar, and scraps. Ten species are listed as threatened, with one, the Liberian Greenbul, Critically Endangered.

Bush-shrikes and helmet-shrikes (Malaconotidae) are 52 species of small or medium-sized birds with a strong, shrikelike, usually hooked or notched bill. They are found mostly in Africa, with one also in southern Arabia. Most are boldly marked, and some are colorful. Helmet-shrikes have black-and-white plumage, while the bush-shrikes are varied, some being very colorful with red, yellow, and green markings. The family also contains boubous, gonoleks, puffbacks, and tchagras. Seven are threatened.

Cuckooshrikes (Campephagidae) are found in a wide area from eastern Africa to South and East Asia, Australasia, and the western Pacific islands. There are 84 species and the family also includes cicadabirds, flycatcher-shrikes, minivets, trillers, and woodshrikes. They have broad bills and a rather long tail and some species are brightly colored. Five are threatened, including the Critically Endangered Reunion Cuckooshrike.

Vangas (Vangidae) are a varied family of small to medium-sized shrikelike woodland birds, almost all restricted to Madagascar (one also in the Comoro Islands). There are 22 species, of which five are threatened. This family also contains the newtonias.

Shrikes (Laniidae) have slender bodies, rounded wings, a long, mobile tail, and hooked bill. Resembling miniature birds of prey, they are quite fierce predators, and their prey ranges from large insects such as beetles to small birds and reptiles. There are 30 species, and some are known as fiscals. Only one species—the São Tomé Fiscal (Critically Endangered)—is threatened.

Leafbirds (Chloropseidae) are small or medium-sized tree-living birds with a decurved bill and brush-tipped tongue; these birds are predominantly green in plumage. There are 11 species found in India and Southeast Asia. One is threatened—the Philippine Leafbird (Vulnerable).

Mauritius Black Bulbul

Hypsipetes olivaceus

The Mauritius Black Bulbul is indeed black—but only on the cap. The remainder of its plumage is dark gray or occasionally gray-brown, enlivened by the yellow-to-pink bill and legs, which may be brightened with an orange tint. A relatively omnivorous bird, it relies mainly on foods such as berries, buds, and seeds, but it also eats a range of small invertebrates, especially insects, and the occasional small lizard or frog. Threats in past decades were clearance of this bulbul's main habitat of forests, including native evergreen broadleaf forest and some introduced deciduous stands, as well as mixed shrubland, and also predation by introduced rats, Common Mynahs (*Acridotheres tristis*), Crab-eating Macaques (*Macaca fascicularis*) and other species. Its numbers were probably fewer than 200 pairs in the 1970s. Since then, with many conservation efforts to protect Mauritius' precious remaining forest areas, the population has stabilized or very gradually increased.

Family	Pycnonotidae
Size	8¾–9 in (22–23 cm)
Status	Vulnerable (stable)
Population	560
Range	Mauritius (endemic)
Habitat	Forest
Main threats	Habitat degradation, alien predators

Uluguru Bush-shrike

Malaconotus alius

Confined to the Uluguru Mountains of east Tanzania, the main known populations of this species in the Uluguru North Forest Reserve were supplemented by its 2007 discovery in the neighboring Uluguru South Forest Reserve. This bush-shrike has bright yellow underparts, merging to a pale olive rump and tail, olive upperparts, a distinctive black cap extending to the upper neck, and the typical shrike's heavy, hook-tipped bill. Surveys are in progress to determine the precise extent of the species' range. Its prime habitat is upland and mountain damp-to-wet forests between 3,300 ft (1,000 m) and 5,000 ft (1,500 m) altitude, with occasional sightings even higher. It hunts mainly in the canopy for large insects such as beetles, crickets, and cockroaches, and possibly the threatened Uluguru Tree Frog (*Leptopelis uluguruensis*). The bush-shrike's main threat is the slow creep of forest clearance up the slopes, for farming and settlements, as well as illegal timber cutting within the reserve for firewood and construction, and water abstraction for nearby human population centers, especially Dar-es-Salaam.

Family	Malaconotidae
Size	8¾–9½ in (22–24 cm)
Status	Critically Endangered (decreasing)
Population	2,400
Range	Tanzania (endemic)
Habitat	Submontane and montane forest
Main threats	Habitat degradation

Western Wattled Cuckooshrike

Campephaga lobata

Originally a species of the lowland rainforest along the southernmost parts of West Africa, this cuckooshrike has to some extent adapted to degraded habitats such as secondary growth logged forests and plantations. However, the scale of forest destruction in its area is vast, especially for timber and conversion to farming, and presents the chief and increasing threat. It is a brightly colored bird: the male has a green back and wings that lighten toward the tips, an extensive black cap, nape, and chin strap, dark tail, and red-to-orange cheeks and underparts. The female is only slightly duller and both sexes have a conspicuous orange-to-yellow gape. They forage through the canopy for a mixed diet of seeds, buds, and small animal fare. Among its few remaining sanctuaries are Taï National Park, famous for its long-term chimpanzee studies, and Gola Forest Reserve, both in Sierra Leone.

Family Campephagidae
Size 8¼ in (21 cm)
Status Vulnerable (decreasing)
Population 20,000–50,000
Range West Africa
Habitat Lowland rain forest
Main threats Logging, disturbance

THREATENED BULBULS, BUSH-SHRIKES, HELMET-SHRIKES, CUCKOOSHRIKES, VANGAS, SHRIKES, AND LEAFBIRDS

FAMILY		COMMON NAME	SCIENTIFIC NAME	STATUS
PYCNONOTIDAE Bulbuls	■	Green-tailed Bristlebill	*Bleda eximius*	VU
	■	Prigogine's Greenbul	*Chlorocichla prigoginei*	EN
	■	Yellow-bearded Greenbul	*Criniger olivaceus*	VU
	■	Mauritius Black Bulbul	*Hypsipetes olivaceus*	VU
	■	Streak-breasted Bulbul	*Ixos siquijorensis*	EN
	■	Liberian Greenbul	*Phyllastrephus leucolepis*	CR
	■	Taiwan Bulbul	*Pycnonotus taivanus*	VU
	■	Yellow-throated Bulbul	*Pycnonotus xantholaemus*	VU
	■	Straw-headed Bulbul	*Pycnonotus zeylanicus*	VU
	■	Hook-billed Bulbul	*Setornis criniger*	VU
MALACONOTIDAE Bush-shrikes and helmet-shrikes	■	Gabela Bush-shrike	*Laniarius amboimensis*	EN
	■	Orange-breasted Bush-shrike	*Laniarius brauni*	EN
	■	Uluguru Bush-shrike	*Malaconotus alius*	CR
	■	Green-breasted Bush-shrike	*Malaconotus gladiator*	VU
	■	Mount Kupe Bush-shrike	*Malaconotus kupeensis*	EN
	■	Yellow-crested Helmet-shrike	*Prionops alberti*	VU
	■	Gabela Helmet-shrike	*Prionops gabela*	EN
CAMPEPHAGIDAE Cuckooshrikes	■	Western Wattled Cuckooshrike	*Campephaga lobata*	VU
	■	Black-bibbed Cicadabird	*Coracina mindanensis*	VU
	■	Reunion Cuckooshrike	*Coracina newtoni*	CR
	■	White-winged Cuckooshrike	*Coracina ostenta*	VU
	■	Mauritius Cuckooshrike	*Coracina typica*	VU
VANGIDAE Vangas	■	Red-shouldered Vanga	*Calicalicus rufocarpalis*	VU
	■	Helmet Vanga	*Euryceros prevostii*	VU
	■	Red-tailed Newtonia	*Newtonia fanovanae*	VU
	■	Bernier's Vanga	*Oriolia bernieri*	VU
	■	Van Dam's Vanga	*Xenopirostris damii*	EN
LANIIDAE Shrikes	■	São Tomé Fiscal	*Lanius newtoni*	CR
CHLOROPSEIDAE Leafbirds	■	Philippine Leafbird	*Chloropsis flavipennis*	VU

RED LIST ■ EX = Extinct ■ EW = Extinct in the Wild ■ CR = Critically Endangered ■ EN = Endangered ■ VU = Vulnerable

Wrens, Nuthatches, Dippers, and Tits

Wrens (Troglodytidae) are small, active birds, mostly with brown streaked plumage, slightly decurved bill, and loud, penetrating songs. Most species have fairly short tails that they often hold cocked upright. Active insectivores, wrens typically search in thick vegetation for food such as insects and spiders and can be difficult to spot, especially as most have brown or gray mottled and striped camouflaged plumage. Most wrens construct an oval-shaped nest from grasses and moss, with an entrance hole at one side. There are 79 species, of which the majority are restricted to the Americas, though one, the Winter Wren, is widespread across much of the northern hemisphere from North America to Europe and Asia. Eight species are threatened, two of which are Endangered and two Critically Endangered.

The 25 species of nuthatch (Sittidae) are agile woodland birds with a large head and powerful bill. Like woodpeckers, they mainly feed by searching for invertebrates on and around tree trunks and branches, climbing and clinging on using their powerful claws. Many species are gray or blue-gray above and pale buff beneath, with an obvious black or dark eyestripe. They typically forage on the branches of trees, though some inhabit rocks, and have the rather unusual ability to clamber downward as well as upward. Five species are threatened, including the large Giant Nuthatch (Vulnerable) of southwestern China, Myanmar, and Thailand.

Dippers (Cinclidae) are five species of dumpy birds that specialize in diving, swimming, and even walking underwater, feeding on aquatic invertebrates. Their feathers are well supplied with oil, making their plumage particularly waterproof and they use their short but powerful wings to swim under the water, which can occur even against the current. The aquatic larvae of insects such as mayflies, stoneflies, and caddisflies form a large part of their diet. They are found in and around fast-flowing rivers and streams. One, the Rufous-throated Dipper (Vulnerable) of Bolivia and Argentina, is threatened.

Tits (Paridae), also known as "titmice" or "chickadees," consist of 53 species of small, rather dainty songbirds with a global distribution. They are agile birds, often seen feeding on small insects and spiders, or seeds, and many regularly visit feeding stations in gardens. Most are dull colored with black-and-white head markings, though some are colorful. Only one species is threatened, the White-naped Tit (Vulnerable).

Zapata Wren

Ferminia cerverai

Like many wrens, the Zapata Wren has a song that seems far too loud for such a small bird—even though it is one of the larger wren species. The song is warbling and high-pitched, and its calls include a harsh "chip-chip-chip". This wren is named for, and endemic to, the Zapata Swamp, a large tract that includes the Zapata Biosphere Reserve (Ramsar Site) in the southwest of the island of Cuba. Here it inhabits several kinds of freshwater swamp, especially those flooded annually, and those with plentiful Jamaica Swamp Sawgrass (*Cladium jamaicensis*). It is chiefly insectivorous, taking a wide range of small insects as well as spiders, worms, snails, little lizards and frogs, and some plant matter, particularly berries and soft seeds. The Zapata Wren has gray underparts and is otherwise mid-brown, striped with very dark brown or black. Compared to more familiar wrens, it has long legs, tail, and bill.

Family	Troglodytidae
Size	6¼ in (16 cm)
Status	Endangered (decreasing)
Population	1,000–2,500
Range	Cuba (endemic)
Habitat	Freshwater marshes
Main threats	Fire, drainage

Algerian Nuthatch

Sitta ledanti

First discovered in the early 1970s, and officially described and named in 1976, this species is thought to be not only Algeria's sole nuthatch, but also the country's only endemic bird. It inhabits just a handful of areas in the mountain forests of the northeast, from 1,600 ft (500 m) to higher than 6,000 ft (1,800 m). These locations include Taza National Park and Mount Babor, and are all within a 35-mile (60-km) radius. The Algerian Nuthatch feeds on small creatures such as insects and spiders through the summer, switching largely to seeds, berries, and nuts in winter, catching some of these small creatures through the cold season. Like other nuthatches, it can clamber in any direction up and down tree trunks. It has buff-yellow underparts, a pale throat, and gray-blue upperparts, with a white eyebrow stripe separating the faint black crown from the typical nuthatch black eye-stripe. In the female, these black areas are more grayish. The species is at risk from fires, from degradation of the forest by livestock, and human disturbance.

Family	Sittidae
Size	5 in (12.5 cm)
Status	Endangered (decreasing)
Population	250–1,000
Range	Algeria (endemic)
Habitat	Forest
Main threats	Fire, disturbance

White-naped Tit

Parus nuchalis

Also known as White-winged Tit, the White-naped Tit is endemic in India. This bird has a complex black-and-white coloration with black mantle, black-and-white primary to tertial wing feathers, and white sides tinged with yellow or orange; the black wing coverts are distinctive. Its diet includes insects, especially caterpillars, beetles, weevils, and grubs, and also fruits and nectar. The species occupies two main regions, one in the south, centered on Karnataka, Andhra Pradesh, and Tamil Nadu, and the other in the northwest, from Rajasthan to Gujarat. However, within these regions, the populations are small, scattered, and fragmented, and threatened by loss of their specialized acacia-dominated dry tropical thorn-scrub forest. This is cleared for firewood and timber products that range from matchsticks to furniture to charcoal. The areas are also being cleared for other purposes, notably mineral mining and quarrying. Another problem is the invasion of the thorn-scrub forest by an alien small tree, the Honey Mesquite (*Prosopis glandulosa*), which has the dubious distinction of being one of the IUCN's World's 100 Worst Weeds.

Family Paridae
Size 4¾ in (12 cm)
Status Vulnerable (decreasing)
Population 2,500–10,000
Range India (endemic)
Habitat Dry scrub-forest
Main threats Habitat loss and fragmentation, mining

THREATENED WRENS, NUTHATCHES, DIPPERS, AND TITS

FAMILY		COMMON NAME	SCIENTIFIC NAME	STATUS
TROGLODYTIDAE Wrens	■	Apolinar's Wren	*Cistothorus apolinari*	EN
	■	Zapata Wren	*Ferminia cerverai*	EN
	■	Munchique Wood-wren	*Henicorhina negreti*	CR
	■	Nava's Wren	*Hylorchilus navai*	VU
	■	Niceforo's Wren	*Thryothorus nicefori*	CR
	■	Cobb's Wren	*Troglodytes cobbi*	VU
	■	Santa Marta Wren	*Troglodytes monticola*	VU
	■	Clarion Wren	*Troglodytes tanneri*	VU
SITTIDAE Nuthatches	■	Beautiful Nuthatch	*Sitta formosa*	VU
	■	Algerian Nuthatch	*Sitta ledanti*	EN
	■	Giant Nuthatch	*Sitta magna*	VU
	■	White-browed Nuthatch	*Sitta victoriae*	EN
	■	Corsican Nuthatch	*Sitta whiteheadi*	VU
CINCLIDAE Dippers	■	Rufous-throated Dipper	*Cinclus schulzi*	VU
PARIDAE Tits	■	White-naped Tit	*Parus nuchalis*	VU

RED LIST ■ EX = Extinct ■ EW = Extinct in the Wild ■ CR = Critically Endangered ■ EN = Endangered ■ VU = Vulnerable

THRUSHES, MOCKINGBIRDS AND THRASHERS

Thrushes (Turdidae) are found all over the world; there are about 170 species, of which 19 are threatened. Medium-sized songbirds, they are similar in overall shape, having a relatively long tail and a fairly stocky build. Most species have mainly brown plumage, often with spotted or streaked undersides, though the family also contains species that are very colorful. Thrushes typically feed on the ground, often pulling out worms and other invertebrates; many also include berries and other fruits in their diet. The Song Thrush, a common Eurasian species, has perfected the art of smashing snails against a hard surface and extracting the soft body inside. Its technique is to pick up the snail in its bill and break the shell with a repeated flick against a stone or similar surface. Many thrushes have loud, musical songs and rather chattering alarm calls. In Europe, the pleasant songs of the Eurasian Blackbird and Song Thrush are familiar sounds of spring, as is the fluting repertoire of the Wood Thrush in eastern North America. The American Robin is also a species of thrush, and was confusingly named by early European colonists for its bright orange breast. The varied family also contains alethes, ant-thrushes, bluebirds, cochoas, ground-thrushes, nightingale-thrushes, shortwings, solitaires, and whistling-thrushes. Two Hawaiian thrushes, Olomao and Puaiohi, are Critically Endangered, as is the Taita Thrush, restricted to a few tiny patches of montane cloud-forest in southern Kenya.

Mockingbirds and thrashers (Mimidae) contain 35 species that reside in the Americas and Caribbean. Most are medium-sized, with a long bill and long tail, with a build not unlike that of the thrushes. However, their bodies are thinner and more elongated and their tail is proportionately longer. Many have mainly brown or gray plumage, which is often heavily streaked. Being active birds, they spend much of their time searching among ground vegetation and leaf litter for a wide range of prey, varying from insects and other invertebrates, to small lizards, in addition to seeds and fruit. They are well known for their musical and elaborate songs and the ability of many to mimic the calls and phrases of other birds. The Northern Mockingbird of North America has the distinction of being one of the world's great mimics, and builds phrases of other bird species into its own repertoire, even including sounds from machinery and other animal sounds such as the barking of dogs. Individual birds add these "alien" sounds to their repertoire with remarkable speed, and embellish their songs accordingly. In common with many impressive songsters, the Northern Mockingbird has fairly drab, mainly pale gray plumage. The family also contains catbirds and tremblers. Seven species are threatened, three of which are Critically Endangered.

Javan Cochoa

Cochoa azurea

One of the most distinctive cochoas, the male is deep, glowing azure-blue above and black elsewhere, including the bill, eyes, and legs, while the female is similar, but with a less intense blue, and dark brown rather than black. The four cochoa species have been traditionally grouped, as here, with the thrushes, Turdidae, but some experts propose they should be transferred to the Muscicapidae (Old World flycatchers). As its name implies, the Javan Cochoa is endemic to the Southeast Asian island, being found mainly in the far west. It is an unobtrusive, generally subdued bird of mountain forests at altitudes of about 3,300 ft (1,000 m) to 10,000 ft (3,000 m). Habitat loss in the lower altitude zone is a great threat as the forests are invaded for timber, tourism, and geothermal energy projects, causing isolation of higher forest into habitat "islands." The consequent problems for the birds are many, including limited breeding opportunities and genetic impoverishment.

Family	Turdidae
Size	9 in (23 cm)
Status	Vulnerable (decreasing)
Population	2,500–10,000
Range	Indonesia (Java) (endemic)
Habitat	Montane rain forest
Main threats	Deforestation, development

Izu Thrush

Turdus celaenops

This thrush is named after one of its two major areas, the Izu Islands (a National Park) stretching south from Tokyo in Japan—specifically the isles between Izu Oshima and Aogashima. It also inhabits some of the Nansei Shoto (Ryukyu-Okinawa) Islands, a chain to the country's far southwest. There are very rare sightings of the species on Japan's main islands, usually on Honshu and in winter. The male Izu Thrush has a deep orange-red lower breast and flanks with a white-centered belly, black head, neck, and upper breast, and is brown elsewhere. The female's head is brown and the black throat white-streaked. The main habitat is deciduous woodland, but this bird also frequents shrublands, and quiet parks and gardens. Despite varied habitat tolerance and a relatively omnivorous diet—focusing on invertebrates in summer and seeds and fruits in winter—the Izu Thrush faces enormous pressure from nest predation. Its principal adversaries are introduced species including the Siberian Weasel (*Mustela sibirica*) and domestic cat, and also the Large-billed Crow (*Corvus macrorhynchos*).

Family	Turdidae
Size	9 in (23 cm)
Status	Vulnerable (decreasing)
Population	2,500–10,000
Range	Japan (endemic)
Habitat	Forest and woodland, gardens
Main threats	Alien predators, habitat loss

Floreana Mockingbird

Mimus trifasciatus

Named after the relatively large island of Floreana in the Galápagos Islands, East Pacific, this species became extinct there by the 1880s. It clings to survival on two islets, Gardener-by-Floreana and Champion, both east of Floreana and both less than 0.4 sq. mi (1 sq. km) in area. Such a precarious range and population mean that adverse weather conditions such as hurricanes, drought, and the La Niña/El Niño oscillation, have drastic effects on its continued existence. This mockingbird has whitish underparts, dark gray white-tipped wing and tail feathers, and a bold black eye stripe below a narrower white eyebrow stripe. It prefers dry scrub, especially with cacti and euphorbias. It is primarily an insectivore, but also takes larger prey such as lizards and crabs, as well as nectar and pollen, and even the regurgitated food of local seabirds such as boobies. A victim of nest predation on Floreana, especially by rats, the species is part of a ten-year action plan to reintroduce it to this larger island.

Family	Mimidae
Size	10 in (25 cm)
Status	Critically Endangered (decreasing)
Population	45–60
Range	Ecuador (Galápagos) (endemic)
Habitat	Cactus scrub
Main threats	Alien predators (Black Rat), disease

THREATENED THRUSHES, MOCKINGBIRDS AND THRASHERS

FAMILY	COMMON NAME	SCIENTIFIC NAME	STATUS
TURDIDAE Thrushes	Thyolo Alethe	*Alethe choloensis*	EN
	White-bellied Shortwing	*Brachypteryx major*	VU
	Bicknell's Thrush	*Catharus bicknelli*	VU
	Forest Thrush	*Cichlherminia lherminieri*	VU
	Javan Cochoa	*Cochoa azurea*	VU
	Sumatran Cochoa	*Cochoa beccarii*	VU
	Olomao	*Myadestes lanaiensis*	CR
	Omao	*Myadestes obscurus*	VU
	Puaiohi	*Myadestes palmeri*	CR
	Sri Lanka Whistling-thrush	*Myophonus blighi*	EN
	Izu Thrush	*Turdus celaenops*	VU
	Gray-sided Thrush	*Turdus feae*	VU
	Taita Thrush	*Turdus helleri*	CR
	Somali Thrush	*Turdus ludoviciae*	VU
	Yemen Thrush	*Turdus menachensis*	VU
	La Selle Thrush	*Turdus swalesi*	EN
	Ashy Thrush	*Zoothera cinerea*	VU
	Spotted Ground-thrush	*Zoothera guttata*	EN
	Guadalcanal Thrush	*Zoothera turipavae*	VU
MIMIDAE Mockingbirds and thrashers	Socorro Mockingbird	*Mimus graysoni*	CR
	Espanola Mockingbird	*Mimus macdonaldi*	VU
	San Cristobal Mockingbird	*Mimus melanotis*	EN
	Floreana Mockingbird	*Mimus trifasciatus*	CR
	White-breasted Thrasher	*Ramphocinclus brachyurus*	EN
	Bendire's Thrasher	*Toxostoma bendirei*	VU
	Cozumel Thrasher	*Toxostoma guttatum*	CR

RED LIST EX = Extinct · EW = Extinct in the Wild · CR = Critically Endangered · EN = Endangered · VU = Vulnerable

Babblers and Parrotbills, and Rockfowl

Babblers and parrotbills (Timaliidae) are a large and very varied family of about 320 species, from Africa and Asia, though there is also a single North American species, the Wrentit. Most species have short, rounded wings, a long tail, and stout bill, quite short in some species. In size they range from very small to crow-sized. They are found in habitats ranging from forest and scrub to grassland, reeds, and semi-deserts. Babblers are mostly social birds and often travel about in flocks, making regular contact calls. Such flocks may consist of as many as 30 individuals and cohesion of the group is maintained partly through these calls. Many species are co-operative breeders, using nonbreeding, probably closely -related individuals as "helpers at the nest" to assist in incubating eggs and feeding the nestlings. The family also contains barwings, fulvettas, laughingthrushes, liocichlas, minlas, scimitar-babblers, shrike-babblers, sibias, tit-babblers, wren-babblers, and yuhinas. The scimitar-babblers (main genus *Pomatorhinus*) are named for their long, rather stout, decurved bills. They reach their highest diversity in the Himalayan region where they are found mainly in dense vegetation in hill country. Wren-babblers (main genus *Spelaeornis*) are tiny short-tailed babblers that live mainly in moist forests, where they actively hunt for invertebrates among the leaf litter. Parrotbills (main genus *Paradoxornis*) are about 20 species of small, active birds that are named for their bill that is shaped a little like that of a parrot. Highly social, they typically move about in close-knit flocks, feeding in the undergrowth or in reedbeds. The Bearded Parrotbill (formerly known as the Bearded Tit or Bearded Reedling) is a familiar reedbed bird of Europe and Asia. The laughingthrushes of tropical and subtropical Asia comprise a large genus (main species *Garrulax*). Large and very vocal, and named for their raucous, laughing calls, they are often captured in China and other parts of Asia and kept in cages—a sad life, especially for such social birds. A total of 31 species from this family are listed as threatened, one of which, the attractive Blue-crowned Laughingthrush of China, is Critically Endangered.

By contrast, there are just two species of rockfowl (Picathartidae) that are found only in tropical West Africa, and both are listed as Vulnerable. The White-necked Picathartes is found in Guinea, Sierra Leone, Liberia, Côte d'Ivoire, and Ghana. Also known as bald crows or picathartes, these strange-looking birds have bare, unfeathered heads. They nest colonially on cliffs or in caves, and forage mainly in the leaf-litter of rain forests. These species are threatened mainly by logging and disturbance.

Collared Laughingthrush

Garrulax yersini

This member of the Old World babbler family is the subject of taxonomic scrutiny, with suggestions to move it from the genus *Garrulax* to a newly -created genus, as *Trochalopteron yersini*. Also known as Yersin's Laughingthrush, it is an almost garishly colored bird with a black hood, silver cheeks, dull orange body, dark to black primary coverts, pale green outer wings, and darker olive-green rump and tail. It inhabits forests, both native and disturbed, as well as scrub, at altitudes between 5,000 ft (1,500 m) and 8,200 ft (2,500 m) in the De Lat uplands of southeastern Vietnam, northwest of Ho Chi Minh City. This region is subject to government encouraged human settlement where new homes, farms, logging, firewood collection, and mineral exploitation are having enormously degrading effects on the local habitats. General clearance has the effect of removing the undergrowth where birds search for many kinds of food, from insects to berries, and further fragments its already highly scattered populations.

Family	Timaliidae
Size	10¼–11 in (26–28 cm)
Status	Endangered (decreasing)
Population	2,500–10,000
Range	Vietnam (endemic)
Habitat	Montane forest
Main threats	Logging, forest degradation

Rusty-throated Parrotbill

Paradoxornis przewalskii

This Old World babbler is known from very restricted mountain areas on the border between southern Gansu and northern Sichuan Provinces in central China. Detailed information is lacking because sightings have always been sparse. It has been recorded from several high-altitude habitats, generally between 6,600 ft (2,000 m) and 10,000 ft (3,000 m), including bamboo thickets, and conifer forest, both evergreen and deciduous larch. It has a grayish cap and cheeks, a black streak from the loral region past the eye to the nape, an orange throat and upper chest, and olive and brown wings and tail. It is thought to feed mainly on insects. Like other wildlife of the area, including the famed Giant Panda (*Ailuropoda melanoleuca*), its habits and distribution may be influenced by episodic flowering and die-back of bamboo stands. With such widespread deforestation, this parrotbill may be one of the species to benefit from safe areas for Giant Pandas and Sichuan Golden Snub-Nosed Monkeys (*Rhinopithecus roxellana*), such as the Jiuzhaigou (Jiuzhai Valley) Nature Reserve.

Family	Timaliidae
Size	5–5¾ in (13–14.5 cm)
Status	Vulnerable (decreasing)
Population	2,500–10,000
Range	China (endemic)
Habitat	Woodland, bamboo thickets
Main threats	Habitat fragmentation

Omei Shan Liocichla
Liocichla omeiensis

Also called the Emei Shan Liocichla or Gray-faced Liocichla, this species is restricted to a few mountain localities in the south and middle of China's southwestern province of Sichuan. These areas include localities in the Emei/Omei Shan, Xiaoxiang Ling, and Daliang Shan ranges. The bird frequents ground vegetation and undergrowth in broadleaved forests, bamboo stands, and mixed scrub, with a tendency to move to higher altitudes, up to 8,200 ft (2,500 m) in the summer months, then descends to sheltered lowlands below 3,300 ft (1,000 m) for winter. Its basic coloration is grayish-olive with a red tinge around the lower crown and nape, and bright red wing patches; the wings and tail are more yellow-trimmed in the female. Massive deforestation for farmland, building, and timber plantations has removed much of this Old World babbler's best habitat. Its main hopes of survival rest on sanctuary areas such as the Emei Shan Protected Scenic Site, and the Mamize, Heizhugou, Mabian Dafengding, and Laojun Shan Nature Reserves. Despite legal measures aimed to protect the Omei Shan Liocichloa against capture, it is still occasionally trapped for the cagebird trade.

Family	Timaliidae
Size	8 in (20.5 cm)
Status	Vulnerable (decreasing)
Population	2,500–10,000
Range	China (endemic)
Habitat	Forest and scrub
Main threats	Habitat loss and fragmentation, disturbance

Gray-necked Picathartes
Picathartes oreas

The Gray-necked Picathartes or Gray-necked Rockfowl of Africa is strikingly marked in two-tone blue-gray on the neck, back, wings, and tail, with orange-tinged white throat and underparts, and a very distinctive head that is lilac to violet on the forehead and front crown, changing abruptly to a carmine-red headcap, with a black side triangle centered on the eye. It has a large powerful bill and long strong legs and feeds mainly on the ground, taking sizeable prey such as snails, lizards, frogs, and bird nestlings, as well as insects and worms. It has characteristic hopping movements and makes short flights between trees, but uses its wings mainly to balance on low branches. Its range extends from southeastern Nigeria through Cameroon south to Gabon and Republic of Congo, where it lives in closed-canopy primary rain forest. Despite its varied diet and habitat, it is limited by its special nest site requirements: a ledge, cave opening, or cliff with a rain-protecting overhang above and a predator-deterring smooth, steep drop below. The immense habitat degradation in its region is a major worry, as well as predation by primates.

Family	Picathartidae
Size	13¾ in (35 cm)
Status	Vulnerable (decreasing)
Population	2,500–10,000
Range	West Africa
Habitat	Rain forest
Main threats	Deforestation, disturbance

THREATENED BABBLERS, PARROTBILLS, AND ROCKFOWL

FAMILY		COMMON NAME	SCIENTIFIC NAME	STATUS
TIMALIIDAE Babblers and parrotbills	■	Black-crowned Barwing	*Actinodura sodangorum*	VU
	■	Gold-fronted Fulvetta	*Alcippe variegaticeps*	VU
	■	Jerdon's Babbler	*Chrysomma altirostre*	VU
	■	Gray-crowned Crocias	*Crocias langbianis*	EN
	■	Flame-templed Babbler	*Dasycrotapha speciosa*	EN
	■	Sumatran Laughingthrush	*Garrulax bicolor*	VU
	■	White-speckled Laughingthrush	*Garrulax bieti*	VU
	■	Rufous-breasted Laughingthrush	*Garrulax cachinnans*	EN
	■	Ashy-headed Laughingthrush	*Garrulax cinereifrons*	VU
	■	Blue-crowned Laughingthrush	*Garrulax courtoisi*	CR
	■	Chestnut-eared Laughingthrush	*Garrulax konkakinhensis*	VU
	■	Golden-winged Laughingthrush	*Garrulax ngoclinhensis*	VU
	■	Snowy-cheeked Laughingthrush	*Garrulax sukatschewi*	VU
	■	Collared Laughingthrush	*Garrulax yersini*	EN
	■	White-throated Mountain-babbler	*Kupeornis gilberti*	EN
	■	Bugun Liocichla	*Liocichla bugunorum*	VU

RED LIST ■ EX = Extinct ■ EW = Extinct in the Wild ■ CR = Critically Endangered ■ EN = Endangered ■ VU = Vulnerable

THREATENED BABBLERS, PARROTBILLS, AND ROCKFOWL *Continued*				
FAMILY		**COMMON NAME**	**SCIENTIFIC NAME**	**STATUS**
TIMALIIDAE Babblers and parrotbills	■	Omei Shan Liocichla	*Liocichla omeiensis*	VU
	■	Black-breasted Parrotbill	*Paradoxornis flavirostris*	VU
	■	Rusty-throated Parrotbill	*Paradoxornis przewalskii*	VU
	■	Gray-hooded Parrotbill	*Paradoxornis zappeyi*	VU
	■	Marsh Babbler	*Pellorneum palustre*	VU
	■	Falcated Wren-babbler	*Ptilocichla falcata*	VU
	■	Bornean Wren-babbler	*Ptilocichla leucogrammica*	VU
	■	Rusty-faced Babbler	*Robsonius rabori*	VU
	■	Gray-banded Babbler	*Robsonius sorsogonensis*	VU
	■	Rusty-throated Wren-babbler	*Spelaeornis badeigularis*	VU
	■	Tawny-breasted Wren-babbler	*Spelaeornis longicaudatus*	VU
	■	Negros Striped-babbler	*Stachyris nigrorum*	EN
	■	Snowy-throated Babbler	*Stachyris oglei*	VU
	■	Hinde's Pied-babbler	*Turdoides hindei*	VU
	■	Slender-billed Babbler	*Turdoides longirostris*	VU
PICATHARTIDAE Rockfowl	■	White-necked Picathartes	*Picathartes gymnocephalus*	VU
	■	Gray-necked Picathartes	*Picathartes oreas*	VU
RED LIST ■ EX = Extinct ■ EW = Extinct in the Wild ■ CR = Critically Endangered ■ EN = Endangered ■ VU = Vulnerable				

OLD WORLD WARBLERS, CISTICOLAS AND ALLIES, AND GNATCATCHERS

Old World warblers (Sylviidae) are a large family of 290 species. Most are small and slender-bodied with rather dull, camouflaged plumage, either brown or greenish. They have rather narrow bills, and most feed on insects and their larvae though some also take seeds, nectar, and fruit. Many have loud, musical songs. They are found in a range of habitats, typically woodland and scrub, though many, notably the reed-warblers, are typical of reedbeds and other wetland sites. Most warblers build cup-shaped nests in the safety of dense vegetation and breed in monogamous pairs. This family also includes bush-warblers, brush-warblers, crombecs, eremomelas, fernbirds, grassbirds, grasshopper-warblers, hyliotas, leaf-warblers, longbills, scrub-warblers, stubtails, tailorbirds, tesias, and thicketbirds. Tailorbirds typically hold their tails upright in a wrenlike posture. Known for their distinctive nests, they stitch a cradle of leaves together using spider silk or plant fibers to support their grass nest. The total of threatened species stands at 36, 15 of which are reed-warblers (*Acrocephalus*). Three are Critically Endangered: the Millerbird, Nightingale Reed-warbler, and Long-billed Tailorbird.

Cisticolas and allies (Cisticolidae) contain 114 species of small insectivorous birds found from Europe and Africa east to Asia and Australia. Most live in grassland, scrub, or marshy areas, though some have adapted to habitats such as road verges and weedy areas of cultivated ground. Many species, especially in the genus *Cisticola*, have characteristic songs that are repetitive and rather mechanical, and these are reflected in their common names—Bubbling, Chattering, Chirping, Churring, Croaking, Piping, Rattling, Siffling, Singing, Tinkling, Trilling, Wailing, Whistling, Winding, and Zitting Cisticolas. The generic name *Cisticola* refers to their nest-building skills, and translates as "inhabitant of a woven basket." The songs are an important way of identifying these small, indistinct birds that often stay hidden in dense vegetation. The family also contains the apalises, camaropteras, and prinias. Eight species are threatened, of which one, Taita Apalis of the Taita Hills in southern Kenya, is Critically Endangered.

Gnatcatchers (Polioptilidae) are a small New World family of 15 species that occur in North and South America. Slender birds with a long tail and gray or brown plumage, they feed on a range of insects and other invertebrates, mainly in woodland or forests, though they are sometimes also found in dry scrubby habitats. The only threatened species, Iquitos Gnatcatcher (Critically Endangered), is known only from a tiny range in northern Peru where it has suffered from forest clearance. It was only recently described, in 2005, and is now the official bird of Iquitos—known as "La Perlita de Iquitos."

Millerbird
Acrocephalus familiaris

The Millerbird is a member of the large genus of reed-warblers, and like most other reed-warblers it has rather drab, mainly brown plumage, and a thin bill. It song is a simple phrase consisting of rapidly delivered metallic notes. This Hawaiian endemic is found only on one small rocky island, Nihoa, in the northwest of the archipelago. Its population seems to fluctuate rather wildly, and the species is therefore considered to be at considerable risk. It is found mainly in shrubby vegetation and feeds on small invertebrates such as insect larvae, beetles, and spiders. It was once also found on the island of Laysan, but disappeared from there, partly as a result of introduced rabbits, which denuded much of the suitable habitat, also diminishing the supply of insects. The small Nihoa population is at risk from extreme weather such as droughts and storms. Access to Nihoa is strictly controlled, and translocations to other suitable islands are being considered.

Family Sylviidae
Size 5 in (13 cm)
Status Critically Endangered (fluctuating)
Population 480–1,150
Range Hawaii (Nihoa) (endemic)
Habitat Scrub
Main threats Storms, fire

Usambara Hyliota
Hyliota usambara

This attractive warbler is endemic to Tanzania, where it is found only in a very small area in the foothills of the East Usambara Mountains in the northeast of the country. It is a medium-sized warbler, prettily marked with blue-black glossy upperparts, and a broad white flash on the wings. Its throat and breast are orange, shading to yellow on the belly. The Usambara Hyliota favors the canopy of the forest, but is also found along the edges of forest and in coffee plantations, though it probably depends on mature forest for breeding. The birds usually feed by picking insects from the branches and leaves high in the forest canopy, and they also chase insects in a flycatcher-like fashion. Major threats to this species are degradation of the remaining forests through pole cutting and clearing for cultivation, and efforts are being made to increase the amount of forest in protected areas, though the high local population with its demands for timber and crop plantations makes this very challenging.

Family Sylviidae
Size 5½ in (14 cm)
Status Endangered (decreasing)
Population 1,000–2,500
Range Tanzania (endemic)
Habitat Forest, plantations
Main threats Habitat degradation

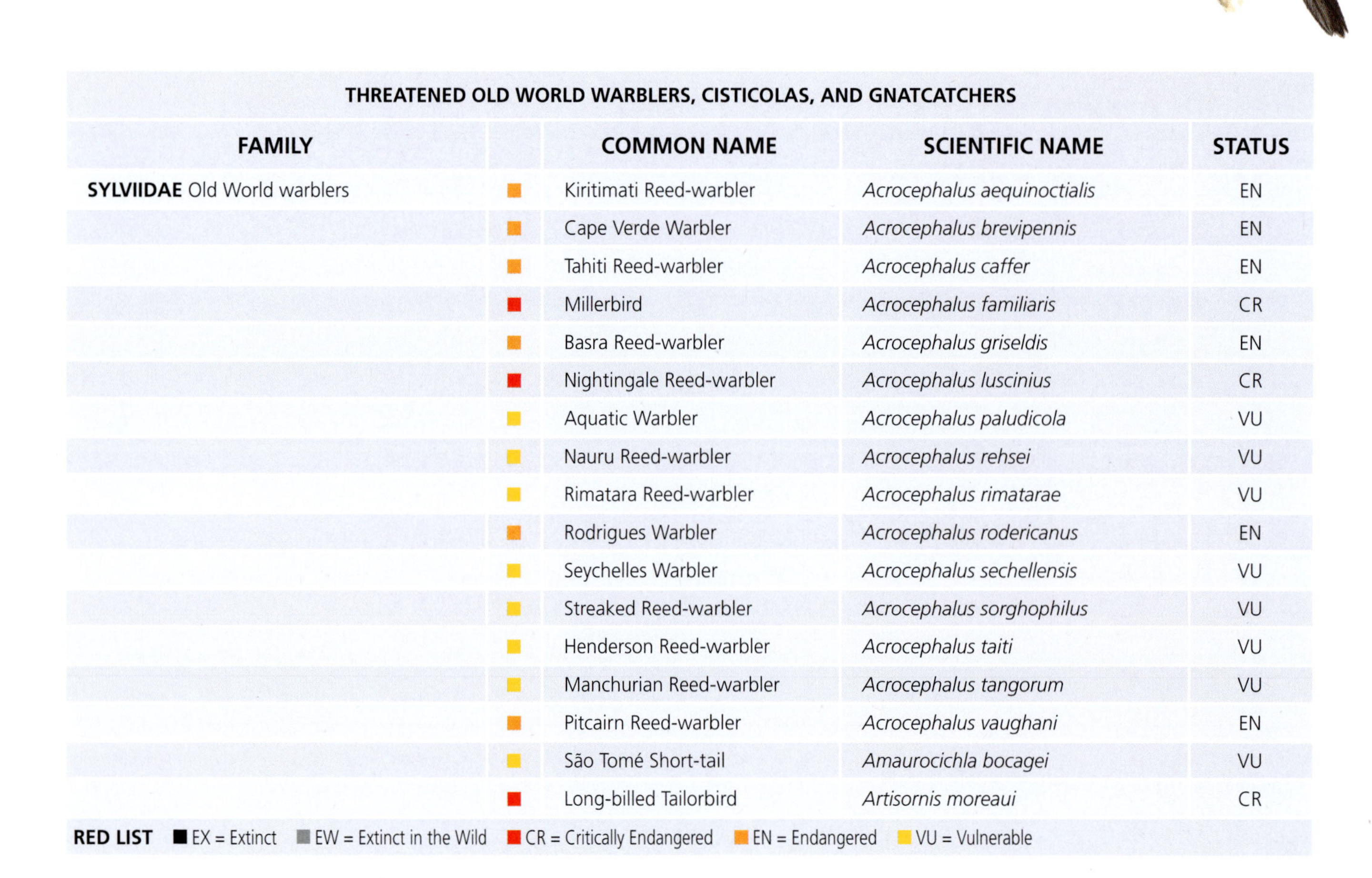

THREATENED OLD WORLD WARBLERS, CISTICOLAS, AND GNATCATCHERS

FAMILY		COMMON NAME	SCIENTIFIC NAME	STATUS
SYLVIIDAE Old World warblers		Kiritimati Reed-warbler	*Acrocephalus aequinoctialis*	EN
		Cape Verde Warbler	*Acrocephalus brevipennis*	EN
		Tahiti Reed-warbler	*Acrocephalus caffer*	EN
		Millerbird	*Acrocephalus familiaris*	CR
		Basra Reed-warbler	*Acrocephalus griseldis*	EN
		Nightingale Reed-warbler	*Acrocephalus luscinius*	CR
		Aquatic Warbler	*Acrocephalus paludicola*	VU
		Nauru Reed-warbler	*Acrocephalus rehsei*	VU
		Rimatara Reed-warbler	*Acrocephalus rimatarae*	VU
		Rodrigues Warbler	*Acrocephalus rodericanus*	EN
		Seychelles Warbler	*Acrocephalus sechellensis*	VU
		Streaked Reed-warbler	*Acrocephalus sorghophilus*	VU
		Henderson Reed-warbler	*Acrocephalus taiti*	VU
		Manchurian Reed-warbler	*Acrocephalus tangorum*	VU
		Pitcairn Reed-warbler	*Acrocephalus vaughani*	EN
		São Tomé Short-tail	*Amaurocichla bocagei*	VU
		Long-billed Tailorbird	*Artisornis moreaui*	CR

RED LIST ■ EX = Extinct ■ EW = Extinct in the Wild ■ CR = Critically Endangered ■ EN = Endangered ■ VU = Vulnerable

Long-legged Thicketbird

Trichocichla rufa

This Fiji endemic warbler is mainly brown, with rufous flanks and a whitish belly. It has a long tail and long legs. Though its population is very small, it seems to be stable and it is possible that the numbers may have been underestimated. This is mainly because it is very difficult to spot unless located by its calls—it has a loud, melodic song and distinctive, rattling alarm call. The Long-legged Thicketbird is found mainly on Viti Levu, notably in the Wabu Forest Reserve and around the Sovi Basin. It has also been recorded from Vanua Levu. It prefers old-growth forest between 600 ft (200 m) and 2,600 ft (800 m), especially close to streams, often on steep slopes with a dense understorey, including tree-ferns. The birds tend to forage on or near the ground, often in family groups. As with many ground-living birds it is at risk of predation by introduced carnivores such as Indian Mongoose (*Herpestes javanicus*) or Black Rat (*Rattus rattus*).

Family	Sylviidae
Size	6¾ in (17 cm)
Status	Endangered (stable)
Population	50–250
Range	Fiji (endemic)
Habitat	Old-growth forest
Main threats	Alien predators, logging

Yellow-throated Apalis

Apalis flavigularis

The Yellow-throated Apalis is a dainty small warbler with rather striking plumage—bright green back, yellow underparts with a black breast-band, black head, and pale eyes. Its song is not very musical, consisting of loud twittering. It builds a domed nest mainly of moss and lays a clutch of 2–3 eggs between October and December. Known only from three mountain ranges in southeastern Malawi, it is listed as Endangered as it is thought to be declining, and its range is so small and fragmented. The known sites are in high-altitude evergreen forests, on Mount Mulanje, where it is found mainly between 3,300 ft (1000 m) and 7,900 ft (2,400 m), and Mount Zomba and Mount Malosa, between 4,600 ft (1,400 m) and 6,400 ft (1,950 m). Though much of the remaining forest is protected by Forest Reserves, in practice much of the habitat continues to be lost or degraded.

Family	Cisticolidae
Size	5 in (13 cm)
Status	Endangered (decreasing)
Population	2,500–10,000
Range	Malawi (endemic)
Habitat	Montane forest
Main threats	Deforestation, fire

THREATENED OLD WORLD WARBLERS, CISTICOLAS, AND GNATCATCHERS *Continued*

FAMILY		COMMON NAME	SCIENTIFIC NAME	STATUS
SYLVIIDAE Old World warblers	■	Mrs Moreau's Warbler	*Bathmocercus winifredae*	VU
	■	Appert's Tetraka	*Bernieria apperti*	VU
	■	Dusky Tetraka	*Bernieria tenebrosa*	VU
	■	Grauer's Swamp-warbler	*Bradypterus graueri*	EN
	■	Knysna Warbler	*Bradypterus sylvaticus*	VU
	■	Bristled Grassbird	*Chaetornis striata*	VU
	■	Papyrus Yellow Warbler	*Chloropeta gracilirostris*	VU
	■	Turner's Eremomela	*Eremomela turneri*	EN
	■	Usambara Hyliota	*Hyliota usambara*	EN
	■	Pleske's Grasshopper-warbler	*Locustella pleskei*	VU
	■	Pulitzer's Longbill	*Macrosphenus pulitzeri*	EN
	■	Bismarck Thicketbird	*Megalurulus grosvenori*	VU
	■	Fly River Grassbird	*Megalurus albolimbatus*	VU
	■	Somber Leaf-warbler	*Phylloscopus amoenus*	VU
	■	Hainan Leaf-warbler	*Phylloscopus hainanus*	VU
	■	Izu Leaf-warbler	*Phylloscopus ijimae*	VU
	■	Broad-tailed Grassbird	*Schoenicola platyurus*	VU
	■	Yemen Warbler	*Sylvia buryi*	VU
	■	Long-legged Thicketbird	*Trichocichla rufa*	EN
CISTICOLIDAE Cisticolas and allies	■	Kungwe Apalis	*Apalis argentea*	EN

RED LIST ■ EX = Extinct ■ EW = Extinct in the Wild ■ CR = Critically Endangered ■ EN = Endangered ■ VU = Vulnerable

THREATENED OLD WORLD WARBLERS, CISTICOLAS, AND GNATCATCHERS *Continued*				
FAMILY		**COMMON NAME**	**SCIENTIFIC NAME**	**STATUS**
CISTICOLIDAE Cisticolas and allies	■	White-winged Apalis	*Apalis chariessa*	VU
	■	Yellow-throated Apalis	*Apalis flavigularis*	EN
	■	Taita Apalis	*Apalis fuscigularis*	CR
	■	Karamoja Apalis	*Apalis karamojae*	VU
	■	Aberdare Cisticola	*Cisticola aberdare*	EN
	■	Gray-crowned Prinia	*Prinia cinereocapilla*	VU
	■	White-eyed Prinia	*Prinia leontica*	VU
POLIOPTILIDAE Gnatcatchers	■	Iquitos Gnatcatcher	*Polioptila clementsi*	CR

RED LIST ■ EX = Extinct ■ EW = Extinct in the Wild ■ CR = Critically Endangered ■ EN = Endangered ■ VU = Vulnerable

Chats and Old World flycatchers, Australasian wrens, bristlebirds, and wattle-eyes

With about 280 species, the chats and Old World flycatchers (Muscicapidae) are one of the largest bird families. Small, mainly insectivorous songbirds, this rather varied family includes akalats, bushchats, bush-robins, jungle-flycatchers, redstarts, robins, robin-chats, magpie-robins and shamas, nightingales, niltavas, rock-thrushes, scrub-robins, stonechats, and wheatears, in addition to chats and flycatchers. The flycatchers are very agile and typically feed largely on flying insects, caught in mid-air. Thirty-two species are threatened, one of which, Rueck's Blue-flycatcher of northern Sumatra, is Critically Endangered, and may already be extinct. Akalats (*Sheppardia*) are small robinlike forest birds endemic to Africa. Of the nine species, four are threatened, mainly as a result of habitat loss.

Australasian wrens (Maluridae) are a relatively small family of 27 species, three of which are threatened—Carpentarian and White-throated Grasswrens (Vulnerable) and Mallee Emuwren (Endangered). Somewhat wrenlike in shape and behavior, they include the fairy wrens, the males of which have bright, iridescent plumage, often colored blue and black.

There are just three species of bristlebird (Dasyornithidae), found in scrub habitats in southeastern Australia. They are medium-sized ground-feeding birds with rather dull plumage, and are named for the bristlelike feathers at the base of the bill. Two are threatened: Western (Vulnerable) and Eastern (Endangered) Bristlebirds.

Wattle-eyes (Platysteiridae) consist of 31 species of African, flycatcher-like birds. The family includes the batises, and the shrike-flaycatchers. One species, the Banded Wattle-eye (Endangered) that lives in montane forests in Cameroon, is threatened, mainly as a result of habitat loss.

Seychelles Magpie-robin

Copsychus sechellarum

Bold, approachable, and very pretty, the Seychelles Magpie-robin is surely one of the world's more charismatic rarities, and is a "must-see" species for any naturalist visiting the beautiful islands that make up the granitic Seychelles. Its common name does it justice as it combines black-and-white magpielike markings with robin- or thrushlike shape and behavior. It is mostly black, with a bluish sheen and white bars on the upper half of its wings. Endemic to its namesake islands, Seychelles Magpie-robin has a small but increasing population of about 200. Its population plummeted to a handful on one island, Frégate, and by 1965 only 12–15 birds remained. By the early 1990s this had risen, but only to about 20 birds. Following better protection on Frégate and successful translocations to the islands of Aride, Cousin, and Cousine, its numbers are now increasing. Pesticides ingested when eating dead cockroaches may have killed some birds feeding near houses; introduced carnivores such as cats and rats are also a problem.

Family	Muscicapidae
Size	7–8¼ in (18–21 cm)
Status	Endangered (increasing)
Population	200
Range	Seychelles (endemic)
Habitat	Forest, plantations
Main threats	Alien predators, pesticides

Amber Mountain Rock-thrush
Monticola erythronotus

This forest bird is restricted to a single mountain range—the Amber Mountain massif in the north of Madagascar. As one of Madagascar's many endemic species it is on the wish-list of most visiting ornithologists. Most occur in just one patch of forest on the upper slopes of a single mountain—a perilously tiny distribution. It is a small thrushlike bird with, in the male, a bright blue head, nape, and throat, orange chest and belly, and brown wings and back. The tail is also bright orange with brown central feathers. The female is duller and lacks the blue hood. The Amber Mountain Rock-thrush favors mid-altitude and montane humid evergreen forest between about 2,600 ft (800 m) and 4,250 ft (1,300 m), where it forages mainly for invertebrates on the forest floor. Though it is thought to be decreasing, there seem to be few significant threats and its habitat here remains largely intact.

Family	Muscicapidae
Size	6¼ in (16 cm)
Status	Endangered (decreasing)
Population	5,000
Range	Madagascar (endemic)
Habitat	Humid forest
Main threats	Habitat degradation

Mallee Emuwren
Stipiturus mallee

With its very small body and long, filamentous tail (often cocked upward), the Mallee Emuwren is very distinctive. Its back and wings are gray-brown and streaked, but it is otherwise very colorful, with a bright blue face and bib, orange-buff belly, and a rufous cap. Its distribution, in the mallee regions of Victoria and South Australia, is very fragmented, with the main areas being Ngarkat Conservation Park, Murray Sunset National Park, and Hattah-Kulkyne National Park. As its common name suggests, it inhabits mallee scrub, but is also found in hummock grassland dominated by *Triodia* grass. Increased frequency of fires has altered some of this vegetation, though much apparently suitable habitat remains. It takes between five and ten years for mallee heath to regenerate to a state suitable for this species, and populations in fragmented patches of the habitat are at severe risk. Not enough is currently known of its precise ecology to suggest effective conservation measures.

Family	Maluridae
Size	5–5¾ in (13–14.5 cm)
Status	Endangered (decreasing)
Population	1,500–2,800
Range	Australia (endemic)
Habitat	Mallee scrub, grassland
Main threats	Habitat loss, fire

THREATENED CHATS, OLD WORLD FLYCATCHERS, AUSTRALASIAN WRENS, BRISTLEBIRDS, AND WATTLE-EYES

FAMILY		COMMON NAME	SCIENTIFIC NAME	STATUS
MUSCICAPIDAE Chats and Old World flycatchers	■	Black Shama	*Copsychus cebuensis*	EN
	■	Seychelles Magpie-robin	*Copsychus sechellarum*	EN
	■	White-headed Robin-chat	*Cossypha heinrichi*	VU
	■	Large-billed Blue-flycatcher	*Cyornis caerulatus*	VU
	■	Rueck's Blue-flycatcher	*Cyornis ruckii*	CR
	■	Matinan Flycatcher	*Cyornis sanfordi*	EN
	■	Little Slaty Flycatcher	*Ficedula basilanica*	VU
	■	Lompobatang Flycatcher	*Ficedula bonthaina*	EN
	■	Palawan Flycatcher	*Ficedula platenae*	VU
	■	Kashmir Flycatcher	*Ficedula subrubra*	VU
	■	Grand Comoro Flycatcher	*Humblotia flavirostris*	EN
	■	Black-throated Blue Robin	*Luscinia obscura*	VU
	■	Rufous-headed Robin	*Luscinia ruficeps*	VU
	■	Nimba Flycatcher	*Melaenornis annamarulae*	VU
	■	Dapple-throat	*Modulatrix orostruthus*	VU
	■	Amber Mountain Rock-thrush	*Monticola erythronotus*	EN
	■	Chapin's Flycatcher	*Muscicapa lendu*	VU
	■	Ashy-breasted Flycatcher	*Muscicapa randi*	VU
	■	White-bellied Blue Robin	*Myiomela albiventris*	EN
	■	Nilgiri Blue Robin	*Myiomela major*	EN

RED LIST ■ EX = Extinct ■ EW = Extinct in the Wild ■ CR = Critically Endangered ■ EN = Endangered ■ VU = Vulnerable

Eastern Bristlebird

Dasyornis brachypterus

The Eastern Bristlebird is a medium-sized, short-winged bird. It is mainly chestnut-brown above with rufous upper wings and gray-brown underparts. The fairly long tail is brown and rufous. It is quite vocal, making regular ringing contact calls, and it also has a loud quite melodious song. Endemic to Australia, it is found from eastern Victoria north to southeastern Queeensland. It flies reluctantly, spending most of its time on the ground, and as it is rather shy, it can be hard to spot and is usually detected by its calls or song. The favored habitats range from rain forest to tussocky clearings and heathy scrub. Such habitats are prone to burning, and the recent increased frequency of fires has had an impact on its numbers. In particular, fires destroy the understorey vegetation, which is the Eastern Bristlebird's preferred habitat.

As the ideal habitats are small and fragmented, fire damage is a considerable threat both now and in the future. The species is also threatened by alien predators and by grazing. Considerable efforts have been made to conserve the species, including controlling numbers of pigs and cats, as well as captive breeding programs aimed at reintroduction to suitable habitats. There are two other bristlebirds, the Western Bristlebird (Vulnerable) and the Rufous Bristlebird. Their name refers to the stiff feathers at the base of the bill.

Family Dasyornithidae
Size 7–8½ in (18–22 cm)
Status Endangered (decreasing)
Population 1,600
Range Australia (endemic)
Habitat Open woodland, scrub
Main threats Fire, alien predators, grazing by livestock

THREATENED CHATS, OLD WORLD FLYCATCHERS, AUSTRALASIAN WRENS, BRISTLEBIRDS, AND WATTLE-EYES *Continued*

FAMILY		COMMON NAME	SCIENTIFIC NAME	STATUS
MUSCICAPIDAE Chats and Old World flycatchers	■	White-throated Jungle-flycatcher	*Rhinomyias albigularis*	EN
	■	Brown-chested Jungle-flycatcher	*Rhinomyias brunneatus*	VU
	■	White-browed Jungle-flycatcher	*Rhinomyias insignis*	VU
	■	Luzon Water-redstart	*Rhyacornis bicolor*	VU
	■	Fuerteventura Stonechat	*Saxicola dacotiae*	EN
	■	White-throated Bushchat	*Saxicola insignis*	VU
	■	White-browed Bushchat	*Saxicola macrorhynchus*	VU
	■	Rubeho Akalat	*Sheppardia aurantiithorax*	EN
	■	Gabela Akalat	*Sheppardia gabela*	EN
	■	Iringa Akalat	*Sheppardia lowei*	VU
	■	Usambara Akalat	*Sheppardia montana*	EN
	■	Swynnerton's Robin	*Swynnertonia swynnertoni*	VU
MALURIDAE Australasian wrens	■	Carpentarian Grasswren	*Amytornis dorotheae*	VU
	■	White-throated Grasswren	*Amytornis woodwardi*	VU
	■	Mallee Emuwren	*Stipiturus mallee*	EN
DASYORNITHIDAE Bristlebirds	■	Eastern Bristlebird	*Dasyornis brachypterus*	EN
	■	Western Bristlebird	*Dasyornis longirostris*	VU
PLATYSTEIRIDAE Wattle-eyes	■	Banded Wattle-eye	*Platysteira laticincta*	EN

RED LIST ■ EX = Extinct ■ EW = Extinct in the Wild ■ CR = Critically Endangered ■ EN = Endangered ■ VU = Vulnerable

MONARCHS, THORNBILLS AND GERYGONES, FANTAILS, SHRIKE-THRUSHES, AND AUSTRALASIAN ROBINS

Monarchs (Monarchidae) are active birds with a flattened, broad bill, some species being crested and long-tailed. They are mainly insectivorous birds of forest or woodland as well as other habitats from mangroves to savanna. Several of the 96 species are brightly or strikingly colored. They are found in Africa, southern Asia, and Australasia, and also on some Pacific Ocean islands. They are often aggressive when defending their nests and will sometimes breed close to other aggressive species for additional protection. In addition to monarchs (main genus *Monarcha* with about 30 species), the family contains several species known as flycatchers, as well as paradise-flycatchers, and shrikebills. There are 12 species of paradise-flycatcher (genus *Terpsiphone*). These are dainty, attractive birds, the males of which are crested and have unusually long tail streamers. Shrikebills (five species in the genus *Clytorhynchus*) have a stouter bill, that they use to probe dead wood and leaves to find their mainly invertebrate food. They sometimes take small lizards and occasionally fruit. Twenty species are threatened, eight of which are Endangered. Five are Critically Endangered, including two paradise-flycatchers—Cerulean Paradise-flycatcher (found only on Sangihe Island in Indonesia), and Seychelles Paradise-flycatcher.

Thornbills and gerygones (Acanthizidae) are an Australasian family of 64 species. The members are small and diverse, and most have short wings and tail, and generally rather drab plumage. Thornbills (genus *Acanthiza*) live in small groups when not breeding and are acrobatic when gleaning insects from foliage. Superficially they appear similar to tits, but are not close relatives. Gerygones (genus *Gerygone*) are also insectivores, collecting food from shrubs and trees. They are known for their pleasant, simple songs and for being capable of mimicry. The family includes a number of other groups, including fieldwrens, heathwrens, mouse warblers, and scrubwrens (genus *Sericornis*). Two are threatened: the Norfolk Island Gerygone (Vulnerable) and the Yellowhead (Endangered), the latter a canary-like bird found in the forests of New Zealand's South Island.

Fantails (Rhipiduridae) are small and long-tailed, with fairly long legs. They are found from southern Asia to Australasia and the Pacific. There are 42 species, all placed in the same genus, *Rhipidura*, and the family includes the Willie Wagtail, one of Australia's best-known birds. Two species are listed as threatened.

Shrike-thrushes (Colluricinclidae) are 14 species found in Australasia and Indonesia. They are medium-sized birds of forest or scrub and the family includes the Morningbird, the pitohuis, and Crested Bellbird. One species is threatened, the Sangihe Shrike-thrush (Critically Endangered), which is endemic to Sangihe Island in Indonesia.

Australasian robins (Petroicidae) consist of 44 species of robinlike birds, some with dull plumage and some are bright and colorful. Only one is threatened—the striking Black Robin (Endangered), endemic to the Chatham Islands.

Elepaio

Chasiempis sandwichensis

A small and dainty monarch flycatcher, the Elepaio is endemic to Hawaii where it is found on the island of Hawaii itself and on two other islands—O'ahu and Kaua'i. Its plumage is quite striking: mainly gray-brown above and white below, with a prominent white rump and white wing bars. The long, dark tail is often held cocked upright. Found in a range of habitats, from high-altitude woodland to valley woods, savanna, and scrub, it feeds mainly on insects, spiders, and other invertebrates. Though its population is fairly large, its range is small and fragmented. It is still fairly common on Hawaii and Kaua'i, but it seems to be declining. The main threats are habitat loss, avian pox, malaria spread by mosquitoes, and nest predation by introduced Black Rats (*Rattus rattus*). Heavy browsing by feral ungulates has altered the natural habitats, as has the introduction of non-native grass species, and both these activities have increased the risk of fires.

Family	Monarchidae
Size	5½ in (14 cm)
Status	Endangered (decreasing)
Population	240,000
Range	Hawaii (endemic)
Habitat	Forest, scrub
Main threats	Browsing by feral ungulates, alien predators, fire, disease

Seychelles Paradise-flycatcher

Terpsiphone corvina

One of the most charismatic of rare birds, this stunning beauty is endemic to the granitic Seychelles. The male is jet black with a bluish sheen, and has very long tail streamers, extending for about 6½ in (16 cm). Females and juveniles lack the long tail feathers and are chestnut above and white below, with a black head. These beautiful birds, numbering only about 250, nest mostly in mature trees in damp forest, mainly on the island of La Digue. Water Lettuce (*Pistia stratiotes*), a floating aquatic, has been introduced to La Digue's marshes and may have reduced the supply of the flycatcher's insect prey. This alien invasive weed is therefore now routinely removed. Following protection in a dedicated reserve, these delightful birds seem to be increasing, albeit slowly. In late 2008, 23 birds were moved by helicopter to the isolated island of Denis, and two nests with eggs were recorded the following year.

Family	Monarchidae
Size	8 in (20 cm)
Status	Critically Endangered (increasing)
Population	200–300
Range	Seychelles (endemic)
Habitat	Woodland close to wetland
Main threats	Habitat loss, alien predators

Black Robin

Petroica traversi

With its pure black plumage, the Black Robin is a striking bird. A reluctant flier, it spends much of its time on the woodland floor among the leaf litter searching for invertebrates, such as crickets, worms, and cockroaches. The rescue of the Black Robin is a major success story in bird conservation. Once found throughout the Chatham Islands, east of New Zealand, it now lives only on two: Mangere and Rangatira Islands. The major cause was the introduction of rats and cats by European settlers. By 1880, the species had retreated to a small population on a single island—Little Mangere. By 1980, when there were only five birds left, and just one breeding pair, extinction seemed a distinct possibility. After strenuous efforts, the total population now stands at 250–300. It is vital to ensure that both the islands remain free of introduced predators.

Family	Petroicidae
Size	6¼ in (15 cm)
Status	Endangered (increasing)
Population	250–300
Range	Chatham Islands (endemic)
Habitat	Scrub-forest
Main threats	Habitat loss, alien predators (potential)

THREATENED MONARCHS, THORNBILLS, GERYGONES, FANTAILS, SHRIKE-THRUSHES, AND AUSTRALASIAN ROBINS

FAMILY	COMMON NAME	SCIENTIFIC NAME	STATUS
MONARCHIDAE Monarchs	Elepaio	*Chasiempis sandwichensis*	VU
	Black-throated Shrikebill	*Clytorhynchus nigrogularis*	VU
	Santa Cruz Shrikebill	*Clytorhynchus sanctaecrucis*	EN
	Cerulean Paradise-flycatcher	*Eutrichomyias rowleyi*	CR
	Celestial Monarch	*Hypothymis coelestis*	VU
	Ogea Monarch	*Mayrornis versicolor*	VU
	Chuuk Monarch	*Metabolus rugensis*	EN
	Black-chinned Monarch	*Monarcha boanensis*	CR
	Biak Monarch	*Monarcha brehmii*	EN
	White-tipped Monarch	*Monarcha everetti*	EN
	Black-backed Monarch	*Monarcha julianae*	VU
	Flores Monarch	*Monarcha sacerdotum*	EN
	Tinian Monarch	*Monarcha takatsukasae*	VU
	Samoan Flycatcher	*Myiagra albiventris*	VU
	Rarotonga Monarch	*Pomarea dimidiata*	EN
	Iphis Monarch	*Pomarea iphis*	VU
	Marquesan Monarch	*Pomarea mendozae*	EN
	Tahiti Monarch	*Pomarea nigra*	CR
	Fatuhiva Monarch	*Pomarea whitneyi*	CR

RED LIST ■ EX = Extinct ■ EW = Extinct in the Wild ■ CR = Critically Endangered ■ EN = Endangered ■ VU = Vulnerable

THREATENED MONARCHS, THORNBILLS, GERYGONES, FANTAILS, SHRIKE-THRUSHES, AND AUSTRALASIAN ROBINS *Continued*				
FAMILY		**COMMON NAME**	**SCIENTIFIC NAME**	**STATUS**
MONARCHIDAE Monarchs	■	Seychelles Paradise-flycatcher	*Terpsiphone corvina*	CR
ACANTHIZIDAE Thornbills and gerygones	■	Norfolk Island Gerygone	*Gerygone modesta*	VU
	■	Yellowhead	*Mohoua ochrocephala*	EN
RHIPIDURIDAE Fantails	■	Malaita Fantail	*Rhipidura malaitae*	VU
	■	Manus Fantail	*Rhipidura semirubra*	VU
COLLURICINCLIDAE Shrike-thrushes	■	Sangihe Shrike-thrush	*Colluricincla sanghirensis*	CR
PETROICIDAE Australasian robins	■	Black Robin	*Petroica traversi*	EN

RED LIST ■ EX = Extinct ■ EW = Extinct in the Wild ■ CR = Critically Endangered ■ EN = Endangered ■ VU = Vulnerable

WHITE-EYES

White-eyes (Zosteropidae) are a fairly large Old World family consisting of 98 species of small, mostly greenish, mainly woodland birds, found from Africa south of the Sahara through southern and eastern Asia to Australasia and Pacific islands. Their common name comes from the fact that most species have an obvious white eye-ring, though most species are otherwise rather unremarkable in plumage. Many species are superficially rather similar, making field identification something of a challenge; several have olive gray plumage with paler underparts. The white eye-ring is composed of a ring of tiny white feathers that surround the eye and is not due to skin coloration. White-eyes resemble Old World warblers in their general build and habits. They have strong legs, rounded wings, and may be up to 6 in (15 cm) in body length. Like warblers they have a mixed diet, which includes insects and berries. They also take nectar from flowers, a task that is made easier by their slender, brush-tipped tongue. These birds are gregarious out of their breeding season, often congregating in large flocks. The majority belong to the genus *Zosterops*, but the family also includes a number of other genera, including *Lophozosterops* and *Speirops*. Twenty species of white-eye are listed as threatened, no fewer than six of which are Critically Endangered. Several species are endemic to particular oceanic islands.

Rufous-throated White-eye

Madanga ruficollis

Endemic to the Indonesian island of Buru (where it is known from just two or three sites), the Rufous-throated White-eye is a small, secretive flycatcher-like bird of montane forests. It is an unusual member of the family, and is placed in its own genus. It has rather drab plumage, apart from its rufous-orange throat and a yellow-brown area under the tail, being mainly olive green above, with a gray head and underparts. Its legs and bill are black. The most distinctive feature of this bird is the orange patch on its throat. Its range and population are both small and the bird seems to be naturally rather scarce, even within apparently ideal habitat. It feeds mainly on insects, often gleaned from the bark of trees, and it clambers about mossy trunks rather in the manner of a nuthatch, sometimes joining flocks of other species. Most of the records are from between 2,800 ft (850 m) and 5,100 ft (1,550 m) and it is thought likely that it may make local altitudinal movements. Happily the montane forests of Buru remain relatively undisturbed, though logging and clearance is an ever-present potential threat. More information is needed about the precise range of this species and many of the potential sites have yet to be surveyed.

Family Zosteropidae
Size 5 in (13 cm)
Status Endangered (decreasing)
Population 2,500–10,000
Range Indonesia (Buru) (endemic)
Habitat Montane forest
Main threats Logging

Golden White-eye

Cleptornis marchei

This is a very pretty, medium-sized, warblerlike white-eye found on just two islands in the Northern Mariana Islands—Saipan and Aguijan. Its head, underside, and rump are bright golden yellow, the bill and legs are orange, and the wings and tail mainly brownish. Its eye-ring is yellowish rather than white. The majority of birds are on Saipan, where it is found in a range of wooded habitats, from native forest to stands of introduced trees to urban areas. It forages, usually in small groups, for insects and other invertebrates, and also feeds on nectar and fruit. On Saipan the major threat is the accidentally introduced Brown Tree Snake (*Boiga irregularis*), a nest-predator that has devastated local birds on the island of Guam, causing the extinction of the Guam Flycatcher. Control of the snake is therefore a conservation priority on Saipan. Aguijan holds a smaller population of Golden White-eye and remains snake-free, but has difficult access.

Family	Zosteropidae
Size	5½ in (14 cm)
Status	Critically Endangered (decreasing)
Population	58,000
Range	Mariana Islands, northwest Pacific (endemic)
Habitat	Woodland and forest
Main threats	Habitat loss, alien predator (Brown Tree Snake)

Christmas Island White-eye

Zosterops natalis

This species has the characteristic white eye-ring typical of the genus. Small and warblerlike, it is greenish above and grayish-white below, with brownish wings and tail. Originally endemic to Christmas Island, it was introduced to the Cocos-Keeling Islands toward the end of the 20th century. On Christmas Island it is still the most common bird and is found in all habitats with trees, including gardens and abandoned fields. On Cocos-Keeling it occurs mainly around the settlement. Mining operations have destroyed about one-third of the forest on Christmas Island and mining remains a threat. Another potential threat comes in the form of an introduced insect, the Yellow Crazy-ant (*Anoplolepis gracilipes*). This ant forms supercolonies and affects the habitat in complex ways. The ants upset the natural balance by killing Red Land Crabs (*Gecaroidea natalis*) that otherwise inhibit the spread of weeds. The ants also cause the forest canopy to die back through their farming of scale-insects, and they may also attack nestlings.

Family	Zosteropidae
Size	4¼–5 in (11–13 cm)
Status	Vulnerable (stable)
Population	20,000
Range	Christmas Island (introduced to Cocos Islands), Indian Ocean (endemic)
Habitat	Forest, gardens
Main threats	Habitat loss, introduced ants (may harm habitat)

THREATENED WHITE-EYES

FAMILY	COMMON NAME	SCIENTIFIC NAME	STATUS
ZOSTEROPIDAE White-eyes	Bonin White-eye	*Apalopteron familiare*	VU
	Golden White-eye	*Cleptornis marchei*	CR
	Rufous-throated White-eye	*Madanga ruficollis*	EN
	Faichuk White-eye	*Rukia ruki*	CR
	Fernando Po Speirops	*Speirops brunneus*	VU
	Mount Cameroon Speirops	*Speirops melanocephalus*	VU
	White-chested White-eye	*Zosterops albogularis*	CR
	Mauritius Olive White-eye	*Zosterops chloronothus*	CR
	Bridled White-eye	*Zosterops conspicillatus*	EN
	São Tomé White-eye	*Zosterops ficedulinus*	VU
	Annobon White-eye	*Zosterops griseovirescens*	VU
	Splendid White-eye	*Zosterops luteirostris*	EN
	Seychelles White-eye	*Zosterops modestus*	EN
	Mount Karthala White-eye	*Zosterops mouroniensis*	VU
	Christmas Island White-eye	*Zosterops natalis*	VU
	Sangihe White-eye	*Zosterops nehrkorni*	CR
	Rota Bridled White-eye	*Zosterops rotensis*	CR
	Samoan White-eye	*Zosterops samoensis*	VU
	Ranongga White-eye	*Zosterops splendidus*	VU
	Slender-billed White-eye	*Zosterops tenuirostris*	EN

RED LIST ■ EX = Extinct ■ EW = Extinct in the Wild ■ CR = Critically Endangered ■ EN = Endangered ■ VU = Vulnerable

HONEYEATERS, SUNBIRDS, FLOWERPECKERS, AND PARDALOTES

Honeyeaters (Meliphagidae) are named for the fact that they include nectar and other sweet substances in their diet. Several also eat fruit, while a few glean, probe for, or hawk insects. A fairly large family comprising about 175 species, honeyeaters are mainly restricted to Australasia and the Pacific. They inhabit forests, scrub, grassland, and mangroves. Honeyeaters are known to be important pollinators of plants in Australia and New Zealand and have been shown to make movements determined by the flowering times of their favored plant species. Ten species are threatened, three of these being Endangered, including the Regent Honeyeater, which is endemic to Australia, while one, the Crow Honeyeater of New Caledonia (another endemic), is Critically Endangered.

Sunbirds (Nectariniidae) are in many ways the Old World equivalent of hummingbirds, being specialist flower-feeders. Like hummingbirds most are brightly colored, especially the males, with beautiful iridescent feathers. Both sexes have a long, decurved bill. Though not as acrobatic, nor as prone to hover as hummingbirds, they are very agile in flight. Ten species (genus *Arachnothera*) are known as spiderhunters. These tend to be larger and less colorful and are mainly forest birds. They feed on flowers, but as their name implies, they also feed on arthropods, including spiders. Their nests are suspended from branches using plant fibers or spider web filaments. There are about 125 species in the sunbird family, of which seven are threatened.

Also specialized for feeding on nectar and fruit are the 45 species of flowerpecker (Dicaeidae), found mainly from India to the Philippines, with one species (Mistletoebird) in Australia. Mostly rather small and short-tailed, flowerpeckers have short, sharp bills and tubular tongues, the latter aiding uptake of nectar from flowers. Three species are listed as threatened, including the Critically Endangered Cebu Flowerpecker, found only on Cebu Island in the Philippines.

Pardalotes (Pardalotidae) are just four species of active sparrowlike woodland birds, found in Australia. They glean insects from the high canopy and are believed to play an important role in control of sap-sucking hemipteran bugs. These insects create protective cases of crystallized honeydew around themselves, which are readily eaten by pardalotes. One, the Forty-spotted Pardalote (Endangered), endemic to Tasmania, is threatened, mainly by habitat loss.

Stitchbird

Notiomystis cincta

This honeyeater is considered by some authorities to be more closely related to the New Zealand wattlebirds (Callaeidae), and possibly deserving of placement in its own family. The male has a black head, back, wings, and upper breast, and white ear-tufts that can be raised. It has golden-yellow patches on the shoulders and white wing-bars. The female also has white wing-bars, but is otherwise mainly gray-brown.
It lives in forests, feeding on fruit, nectar and invertebrates, and nests in holes in trees. It was once widespread on New Zealand's North Island, but is now found naturally only on Little Barrier Island, though a number of translocations have been made to other islands (notably Tiritiri Matangi and Kapiti) and to a mainland site. The translocations are supported by supply of nest boxes and also by supplementary feeding. The likely causes of extinction on the mainland were forest loss, avian disease, and possibly also predation by introduced Black Rats (*Rattus rattus*).

Family	Meliphagidae
Size	7 in (18 cm)
Status	Vulnerable (stable)
Population	500–2,000
Range	New Zealand (endemic)
Habitat	Forest
Main threats	Habitat loss, disease

Regent Honeyeater

Xanthomyza phrygia

This medium-sized mainly black bird is a more typical honeyeater. It has bold yellow and white markings, and bare skin around its dark eyes.
Its song is rather distinctive, consisting of bell-like notes. An Australian endemic, it is found mainly on the western slopes of the Great Dividing Range and in some coastal habitats, with a range from Victoria and New South Wales to southeastern Queensland. It favors damp mixed forest, including river forests of River She-oak (*Casuarina cunninghamia*), or wet coastal forests with Swamp Mahogany (*Eucalyptus robusta*) or Spotted Gum (*Corymbia maculata*). It is reliant on a mixed diet of insects and also nectar from certain species including White and Yellow Box (*Eucalyptus albens* and *E. melliodora*) and Mugga Ironbark (*E. sideroxylon*). Much of its favored habitat has been lost or fragmented, and it is also believed to be suffering from increased competition from another, more common honeyeater, the Noisy Miner (*Manorina melanocephala*).

Family	Meliphagidae
Size	8–9½ in (20–24 cm)
Status	Endangered (decreasing)
Population	1,500
Range	Australia (endemic)
Habitat	Lowland forests
Main threats	Habitat loss and fragmentation

Rufous-winged Sunbird

Nectarinia rufipennis

A medium-sized sunbird, the Rufous-winged Sunbird is very striking. The male is shiny blue on its head and upper back, with a bronze throat and a red-and-blue breast-band. Both sexes, uniquely amongst sunbirds, have the rufous patches on the wings that give the species its name, though the female is otherwise a rather drab gray-brown, with hints of color streaking on the breast. The bill is fairly long and decurved. The bird's diet consists of nectar (especially from tropical mistletoes), and also small insects. A feisty species, it energetically defends favored clumps of flowers against other sunbirds. The favored habitat is forest and forest edge, usually between about 5,000 ft (1,500 m) and 5,600 ft (1700 m). It is known only from the Udzungwa Mountains of Tanzania. Though most of the sites where it occurs are within the Udzungwa Mountains National Park, there is still a problem with logging and fires, and the remaining forests are suffering considerable fragmentation and degradation. Surveys and monitoring of the population are required, along with more effective management of the protected areas.

Family	Nectariniidae
Size	4¾ in (12 cm)
Status	Vulnerable (decreasing)
Population	10,000–20,000
Range	Tanzania (endemic)
Habitat	Highland forest
Main threats	Deforestation, logging

THREATENED HONEYEATERS, SUNBIRDS, FLOWERPECKERS, AND PARDALOTES

FAMILY		COMMON NAME	SCIENTIFIC NAME	STATUS
MELIPHAGIDAE Honeyeaters	■	Painted Honeyeater	*Grantiella picta*	VU
	■	Crow Honeyeater	*Gymnomyza aubryana*	CR
	■	Mao	*Gymnomyza samoensis*	EN
	■	Ocher-winged Honeyeater	*Macgregoria pulchra*	VU
	■	Black-eared Miner	*Manorina melanotis*	EN
	■	Long-bearded Melidectes	*Melidectes princeps*	VU
	■	Rotuma Myzomela	*Myzomela chermesina*	VU
	■	Stitchbird	*Notiomystis cincta*	VU
	■	Dusky Friarbird	*Philemon fuscicapillus*	VU
	■	Regent Honeyeater	*Xanthomyza phrygia*	EN
NECTARINIIDAE Sunbirds	■	Elegant Sunbird	*Aethopyga duyvenbodei*	EN
	■	Amani Sunbird	*Anthreptes pallidigaster*	EN
	■	Banded Sunbird	*Anthreptes rubritorques*	VU
	■	Loveridge's Sunbird	*Nectarinia loveridgei*	EN
	■	Rockefeller's Sunbird	*Nectarinia rockefelleri*	VU
	■	Rufous-winged Sunbird	*Nectarinia rufipennis*	VU
	■	Giant Sunbird	*Nectarinia thomensis*	VU
DICAEIDAE Flowerpeckers	■	Visayan Flowerpecker	*Dicaeum haematostictum*	VU
	■	Cebu Flowerpecker	*Dicaeum quadricolor*	CR
	■	Scarlet-collared Flowerpecker	*Dicaeum retrocinctum*	VU
PARDALOTIDAE Pardalotes	■	Forty-spotted Pardalote	*Pardalotus quadragintus*	EN

RED LIST ■ EX = Extinct ■ EW = Extinct in the Wild ■ CR = Critically Endangered ■ EN = Endangered ■ VU = Vulnerable

BUNTINGS, AMERICAN SPARROWS, AND ALLIES

There are some 320 species in this very large family (Emberizidae) of seed-eating, finchlike birds. Most members of this family are more heavily built than the finches (family Fringillidae), with the exception of the rather small grassquits. The main groups are the New World sparrows and the Old World buntings, but the family is quite varied and the family also contains brush-finches, grass-finches, grassquits, ground-finches, juncos, seedeaters, sierra-finches, towhees, warbling-finches, and yellow-finches. Most buntings and American sparrows have brown, streaked plumage, but many have bolder head markings, and a few are black and white. They have strong bills for feeding on seeds, though their nestlings are fed mainly on insects. The Old World buntings (main genus *Emberiza*) are familiar birds of Eurasia and Africa. They are usually found in open countryside and feed mainly on seeds, often moving around in flocks outside the breeding season. Of the 38 species, four were listed as Vulnerable, one of which, the Rufous-backed Bunting, has now been uplisted to Endangered. This pretty chestnut-headed bunting lives in grassland and scrub in the far northwest of China and North Korea, but has a very small population, threatened mainly by conversion of its habitat for agriculture and forestry. Brush-finches (main genus *Atlapetes*) comprise about 40 species of sparrowlike birds. Two brush-finches are currently listed as Critically Endangered: the Pale-headed Brush-finch of Ecuador and the Antioquia Brush-finch, a recently described species from Colombia. The American sparrows (including the genera *Aimophila*, *Ammodramus*, *Spizella*, and *Zonotrichia*) are mostly rather smaller than buntings and have brown or gray plumage with darker streaks. Seedeaters (genus *Sporophila*) number over 30 species. They are relatively small birds with short, stout, conical bills adapted for feeding on small seeds. The 15 species of famous "Darwin's finches" of the Galápagos Islands also belong in this family; they have evolved to exploit different food sources, with widely varied bill shapes and sizes, which is a fine example of adaptive radiation. Thirty-seven species in the Emberizidae are listed as threatened, of which seven are Critically Endangered, including the very rare Mangrove Finch and the rare Medium Tree-finch, both ground-finches endemic to the Galápagos Islands.

Yellow Cardinal

Gubernatrix cristata

The Yellow Cardinal is a large, handsome finch found mainly in Argentina and Uruguay. The male is mostly yellow-green with black streaking on the back, and a black tail with yellow outer feathers. Unusually for a finch, it has a prominent long black crest with feathers bordered in yellow. The female is duller and less colorful with a white-bordered crest. It favors open woodland, scrub, and savanna, and was once common over most of its range, but has suffered a drastic decline in recent decades. Habitat loss is certainly one of the factors involved, but so is trapping of this attractive species for the cagebird trade. Its song is loud and musical, which is the main reason for this trade. In both Argentina and Uruguay it is now very local, and the Uruguay population may now be as low as 300. The forests and woodland suffer timber extraction and replacement by exotic eucalypts, as well as replacement by land for cattle grazing. The integrity of the species may also be threatened through hybridizing with the Common Diuca-finch.

Family	Emberizidae
Size	8 in (20 cm)
Status	Endangered (decreasing)
Population	1,500–3,000
Range	Argentina, Uruguay, Brazil (possibly)
Habitat	Woodland, scrub
Main threats	Capture for cagebird trade, habitat loss, hybridization

St Lucia Black Finch

Melanospiza richardsoni

This is a small finch with black (male) or brown (female) plumage, and a heavy black bill. The male also has pink legs and both have a bobbing tail action. It is found only on the Caribbean island of St Lucia, mainly in the mountains, especially along ravines with dense undergrowth, though it inhabits a wide range of vegetation including rain forest, secondary forest, plantations, and scrub. It can be found in a number of forest reserves including La Sorcière and Edmond. It is mainly a ground-feeder, foraging for insects, fruit, and seeds in the leaf litter. The birds nest from April to June. Forest clearance for agriculture and introduced predators, notably rats and mongooses, are the main threats to this unusual little finch. Both probably take eggs, nestlings, and adult birds. Removal of undergrowth from plantations to ease the extraction of timber appears to make the habitat unsuitable for the St Lucia Black Finch to breed and this is considered to be the major threat.

Family	Emberizidae
Size	5–5½ in (13–14 cm)
Status	Endangered (decreasing)
Population	250–1,000
Range	St Lucia (endemic)
Habitat	Rain forest, plantations, scrub
Main threats	Habitat degradation, alien predators

Gough Bunting

Rowettia goughensis

Endemic to the South Atlantic Gough Island, this Critically Endangered bunting has a small, fluctuating, and now decreasing population on this isolated island. Large and rather chunky, it is mainly olive-green, with a striking pattern of black and yellow on its face and throat. Most common in tussock-grassland and wet heath, it is also found in peat bogs and fern-rich scrub. It feeds mostly on invertebrates, but will also eat fruit and grass seed. Nesting is on the ground, mainly on steep slopes or cliffs. Like many birds of oceanic islands, the Gough Bunting is at risk from introduced predators, in this case mainly from House Mouse (*Mus musculus*), but potentially and even more worryingly from Black Rats (*Rattus rattus*). Both mice and rats are known to predate eggs and young of ground-nesting species. Efforts are being made at this nature reserve and World Heritage Site to remove the mice and prevent colonization by rats.

Family	Emberizidae
Size	7 in (18 cm)
Status	Critically Endangered (decreasing)
Population	1,000
Range	Gough Island (endemic)
Habitat	Grassland, heath
Main threats	Alien predators

Marsh Seedeater

Sporophila palustris

Small, and with unusual attractive plumage and a jaunty, whistling song, the Marsh Seedeater is sought after as a cagebird and continues to be trapped to supply this trade. The male has a rufous body, with dark wings and tail, a white bib and collar, and a gray crown. The bill is typical for a seedeater—short and rather deep. In the wild it is known from only a few sites in northeastern Argentina (notably Entre Ríos), southern Brazil, eastern Paraguay, and southeastern Uruguay (notably Bañados del Este Biosphere Reserve), where it inhabits grassland, often in marshy areas with a mixture of dense grass and patches of open water. It feeds mainly on the seeds of grasses. Afforestation has altered many grassland sites and the species is targeted by bird-trappers. Pesticides also find their way into the marshes and may also affect the birds. No fewer than six other seedeaters are also threatened, of which one, Hooded Seedeater of Brazil, may be extinct.

Family	Emberizidae
Size	4 in (10 cm)
Status	Endangered
Population	50–250
Range	Argentina, Uruguay, Paraguay, Brazil
Habitat	Grassland
Main threats	Capture for cagebird trade, afforestation, pollution

THREATENED BUNTINGS, AMERICAN SPARROWS, AND ALLIES

FAMILY	COMMON NAME	SCIENTIFIC NAME	STATUS
EMBERIZIDAE Buntings, American sparrows and allies	Carrizal Seedeater	*Amaurospiza carrizalensis*	CR
	Saltmarsh Sharp-tailed Sparrow	*Ammodramus caudacutus*	VU
	Antioquia Brush-finch	*Atlapetes blancae*	CR
	Yellow-headed Brush-finch	*Atlapetes flaviceps*	EN
	Black-spectacled Brush-finch	*Atlapetes melanopsis*	EN
	Pale-headed Brush-finch	*Atlapetes pallidiceps*	CR
	Mangrove Finch	*Camarhynchus heliobates*	CR
	Medium Tree-finch	*Camarhynchus pauper*	CR
	Black-masked Finch	*Coryphaspiza melanotis*	VU
	Yellow-breasted Bunting	*Emberiza aureola*	VU
	Rufous-backed Bunting	*Emberiza jankowskii*	EN
	Socotra Bunting	*Emberiza socotrana*	VU
	Yellow Bunting	*Emberiza sulphurata*	VU
	Yellow Cardinal	*Gubernatrix cristata*	EN
	Gray-winged Inca-finch	*Incaspiza ortizi*	VU
	St Lucia Black Finch	*Melanospiza richardsoni*	EN
	Tristan Bunting	*Nesospiza acunhae*	VU
	Nightingale Bunting	*Nesospiza questi*	VU

RED LIST ■ EX = Extinct ■ EW = Extinct in the Wild ■ CR = Critically Endangered ■ EN = Endangered ■ VU = Vulnerable

THREATENED BUNTINGS, AMERICAN SPARROWS, AND ALLIES *Continued*				
FAMILY		COMMON NAME	SCIENTIFIC NAME	STATUS
EMBERIZIDAE Buntings, American sparrows and allies	■	Wilkin's Bunting	*Nesospiza wilkinsi*	EN
	■	Tanager Finch	*Oreothraupis arremonops*	VU
	■	Cocos Finch	*Pinaroloxias inornata*	VU
	■	Plain-tailed Warbling-finch	*Poospiza alticola*	EN
	■	Tucuman Mountain-finch	*Poospiza baeri*	VU
	■	Cinereous Warbling-finch	*Poospiza cinerea*	VU
	■	Cochabamba Mountain-finch	*Poospiza garleppi*	EN
	■	Rufous-breasted Warbling-finch	*Poospiza rubecula*	EN
	■	Yellow-green Finch	*Pselliophorus luteoviridis*	VU
	■	Gough Bunting	*Rowettia goughensis*	CR
	■	Worthen's Sparrow	*Spizella wortheni*	EN
	■	Chestnut Seedeater	*Sporophila cinnamomea*	VU
	■	Temminck's Seedeater	*Sporophila falcirostris*	VU
	■	Buffy-fronted Seedeater	*Sporophila frontalis*	VU
	■	Hooded Seedeater	*Sporophila melanops*	CR
	■	Black-and-tawny Seedeater	*Sporophila nigrorufa*	VU
	■	Marsh Seedeater	*Sporophila palustris*	EN
	■	Cuban Sparrow	*Torreornis inexpectata*	EN
	■	Sierra Madre Sparrow	*Xenospiza baileyi*	EN

RED LIST ■ EX = Extinct ■ EW = Extinct in the Wild ■ CR = Critically Endangered ■ EN = Endangered ■ VU = Vulnerable

TANAGERS AND ALLIES

Tanagers and allies (Thraupidae) are a large family of mostly very colorful, fruit-eating birds of the American tropics. Arguably the most colorful of all birds, tanagers are mainly small to medium-sized forest birds, many rather finchlike in overall shape. Most have a sturdy finch-like bill too, though in honeycreepers the bill is thin and decurved, while flowerpiercers have an upturned bill. There are about 270 species, and the family also includes the bush-tanagers, chlorophonias, conebills, dacnises, euphonias, flowerpiercers, hemispinguses, honeycreepers, and mountain-tanagers. Bush tanagers (main genus *Chlorospingus*) are small birds that are possibly closer relatives of the buntings and American sparrows. Chlorophonias (genus *Chlorophonia*), endemic to the Neotropics, are small, mostly bright green birds that live in humid forest, often at high altitude. Conebills (main genus *Conirostrum*) include the smallest tanager species, the White-eared Conebill, which has the shortest body length at only 3½ in (9 cm). Euphonias (genus *Euphonia*), are a large group of some 27 species. They are attractive birds, many with plumage that is dark metallic blue above and yellow below. Flowerpiercers (*Diglossa* and *Diglossopis*), are two genera of tanagers that feed by piercing the bottom of flowers to access nectar. Most occur in the highlands. Hemispinguses (genus *Hemispingus*) are warblerlike tanagers. The honeycreepers (main genus *Cyanerpes*) are small birds of the forest canopy, from Mexico to Brazil. They also specialize on feeding on flower nectar, using their long, curved bill. Mountain-tanagers (main genus *Anisognathus*) are mostly brightly colored birds of subtropical lowland moist forest, or moist montane forest, with a minority in tropical high-altitude grassland. Twenty-two species of the tanager family are threatened, two of which are listed as Critically Endangered: the Cone-billed Tanager and Cherry-throated Tanager, both Brazilian endemics with very small known ranges. Many tanagers are threatened by destruction or alteration of their forest habitats. The more colorful species, such as the very pretty Seven-colored Tanager, are also trapped to supply the illegal cagebird trade. The tanagers of this genus (*Tangara*) are some of the most beautiful of birds, with colors ranging from blue to green, red and yellow, often combined, and reflected in their common names—for example, Azure-rumped, Emerald, Golden-naped, Metallic-green, Red-necked, and Seven-colored. There are about 50 species in this genus, found typically in Andean forests.

Scarlet-breasted Dacnis

Dacnis berlepschi

The Scarlet-breasted Dacnis is a very colorful bird. The bright scarlet lower breast contrasts markedly with a yellow-buff belly and bright blue upper breast, throat, crown, and back. It also has silvery-blue streaks on the mantle and a silver-blue rump. It is confined to the Chocó region of southwestern Colombia and northwestern Ecuador, but its range is extremely fragmented. Here it lives mainly in humid and wet lowland and hill forest, from sea-level to about 4,000 ft (1,200 m), with the highest densities in mature or lightly logged primary forest. It forages mostly in the canopy, feeding on insects and fruits, and is often seen in mixed-species flocks. The Chocó has suffered extensive logging, with much land converted to agriculture, especially oil-palm plantations, while cattle-ranching, mining, and road building also have a significant effect. Though its population may be as high as 50,000, it is listed as Vulnerable as it seems to be slipping into rapid decline.

Family Thraupidae
Size 4¾ in (12 cm)
Status Vulnerable (decreasing)
Population 20,000–50,000
Range Colombia, Ecuador
Habitat Humid lowland forest
Main threats Logging, deforestation, mining

Azure-rumped Tanager

Tangara cabanisi

This pretty tanager is mainly pale blue, with a greenish mantle, and pale sky-blue rump and upper tail. The scapulars and inner wing feathers are black and it also has black markings on the upper breast. With a small and declining range, it is found in the Sierra Madre de Chiapas, in southwestern Mexico (notably El Triunfo Biosphere Reserve), and adjacent Guatemala (notably the Santa María and Atitlán volcanoes). It is found mainly between about 3,300 ft (1,000 m) and 5,600 ft (1,700 m), in humid, broadleaf forests, but has also been recorded in coffee plantations where these adjoin primary forest. It feeds mainly in the upper forest canopy, often in flocks, foraging for insects and fruit, such as from fig trees. Its nest is cup-shaped and built in dense foliage. Its habitat is ideal for coffee cultivation and large areas of the native forest have been converted to this crop. Ecotourism, with the emphasis on birdwatching, offers a route for protecting this and other rare species in the region.

Family Thraupidae
Size 6 in (15 cm)
Status Endangered (decreasing)
Population 2,500–10,000
Range Mexico, Guatemala
Habitat Humid forest
Main threats Habitat degradation

THREATENED TANAGERS AND ALLIES

FAMILY		COMMON NAME	SCIENTIFIC NAME	STATUS
THRAUPIDAE Tanagers	■	Gold-ringed Tanager	*Bangsia aureocincta*	EN
	■	Black-and-gold Tanager	*Bangsia melanochlamys*	VU
	■	Golden-backed Mountain-tanager	*Buthraupis aureodorsalis*	EN
	■	Masked Mountain-tanager	*Buthraupis wetmorei*	VU
	■	Chat Tanager	*Calyptophilus frugivorus*	VU
	■	Multicolored Tanager	*Chlorochrysa nitidissima*	VU
	■	Yellow-green Bush-tanager	*Chlorospingus flavovirens*	VU
	■	Tamarugo Conebill	*Conirostrum tamarugense*	VU
	■	Cone-billed Tanager	*Conothraupis mesoleuca*	CR
	■	Scarlet-breasted Dacnis	*Dacnis berlepschi*	VU
	■	Turquoise Dacnis	*Dacnis hartlaubi*	VU
	■	Chestnut-bellied Flowerpiercer	*Diglossa gloriosissima*	EN
	■	Venezuelan Flowerpiercer	*Diglossa venezuelensis*	EN
	■	Black-cheeked Ant-tanager	*Habia atrimaxillaris*	EN
	■	Slaty-backed Hemispingus	*Hemispingus goeringi*	VU
	■	Rufous-browed Hemispingus	*Hemispingus rufosuperciliaris*	VU
	■	Cherry-throated Tanager	*Nemosia rourei*	CR
	■	Azure-rumped Tanager	*Tangara cabanisi*	EN

RED LIST ■ EX = Extinct ■ EW = Extinct in the Wild ■ CR = Critically Endangered ■ EN = Endangered ■ VU = Vulnerable

THREATENED TANAGERS AND ALLIES *Continued*				
FAMILY		**COMMON NAME**	**SCIENTIFIC NAME**	**STATUS**
THRAUPIDAE Tanagers	■	Seven-colored Tanager	*Tangara fastuosa*	VU
	■	Green-capped Tanager	*Tangara meyerdeschauenseei*	VU
	■	Black-backed Tanager	*Tangara peruviana*	VU
	■	Orange-throated Tanager	*Wetmorethraupis sterrhopteron*	VU

RED LIST ■ EX = Extinct ■ EW = Extinct in the Wild ■ CR = Critically Endangered ■ EN = Endangered ■ VU = Vulnerable

Seven-colored Tanager

Tangara fastuosa

One of the most colorful of the tanagers, this species is endemic to Brazil where it is found in about 50 sites in the northeast: in Alagoas, Paraíba, and Pernambuco. It lives in the Atlantic forest, a habitat that is now severely fragmented and disturbed. It requires dense habitat, and forages both in the forest canopy and along forest edges, often within mixed-species flocks. Breeding tends to be in areas where the forest trees support a good growth of epiphytic bromeliads, the cuplike clumps of which provide relatively safe nesting sites. With its turquoise-green head and mantle, black bill and throat, blue breast, bright orange-red rump, and pale blue wing coverts, it is not surprising that this beautiful species is threatened in part by trapping to supply the cagebird trade. It has already become locally extinct in a number of its former sites, and this trade, combined with forest clearance, is a major problem. Only about 2 percent of the original cover of Brazil's Atlantic forest remains, so it is vital that as much as possible of the remainder is protected. A number of reserves provide nominal protection, but encroachment and forest loss are prevalent in some of these. In some areas, as in the Pedra Talhada Biological Reserve, protection is better, with some areas being reforested.

Family	Thraupidae
Size	5¼ in (13.5 cm)
Status	Vulnerable (decreasing)
Population	2,500–10,000
Range	Brazil (endemic)
Habitat	Atlantic forest, humid forest, scrub (cerrado)
Main threats	Capture for cagebird trade, deforestation, logging, fire

New World Warblers

The New World warblers (Parulidae) consist of some 118 species of small, mostly brightly colored or boldly marked insectivorous birds found in the Americas, reaching their greatest diversity in Central America. Sometimes referred to as wood warblers, most species do indeed spend much of their time in the branches and treetops. The waterthrushes and Wrenthrush, however, feed mainly on the ground. Though they are similar to Old World warblers in general habits and behavior, they differ in that most species have colorful plumage, and in general they have less musical and less varied songs than their Old World counterparts. Many of the North American species have yellow and gray plumage and these can be quite a challenge to separate in the field. Some members of this family are confusingly known as chats, others as redstarts, and the family also contains the yellowthroats (*Geothlypis*), three species in the genus *Seiurus*—two species of waterthrush and the Ovenbird (not to be confused with the true ovenbirds, family Furnariidae), and the Wrenthrush. In all, 15 species are threatened, three of which are Critically Endangered: Belding's Yellowthroat (endemic to Mexico's Baja California Peninsula), Semper's Warbler (endemic to St Lucia), and Bachman's Warbler, the latter two possibly already extinct. Habitat loss is a major problem for the threatened species.

About half the New World warblers are migratory, breeding in North America, and moving south to the warmer regions of Central and South America for the winter. Spring sees large movements of such migrants northward again, an event awaited with great anticipation by birdwatchers. Certain sites are reliable stopovers for these migrants, and one of the most famous is Central Park in New York. This large expanse of green is an oasis of insect-rich vegetation and hosts a constant stream of warblers traveling through each year.

Gray-headed Warbler

Basileuterus griseiceps

This New World warbler is endemic to Venezuela, where it is found only in the northeast, in the Cordillera de Caripe. Its underparts are bright yellow, contrasting with its dark olive-green upperparts, and gray head. It also has a white spot just above the eye at the base of its bill. Primarily a bird of undisturbed subtropical forest with a dense understorey, it is normally seen between about 3,300 ft (1,000 m) and 8,000 ft (2,400 m). Conversion of the forest to plantations of coffee, citrus, banana, and mango is extensive, and the pressures on the remaining forest are considerable. One of the most important protected areas here is El Guácharo National Park, which contains much suitable habitat, but even in this reserve there is forest clearance, involving burning and removal of the understorey, often to facilitate coffee cultivation.

Family Parulidae
Size 5½ in (14 cm)
Status Endangered (decreasing)
Population 2,500–10,000
Range Venezuela (endemic)
Habitat Subtropical forest
Main threats Forest degradation, understorey plantations

Bachman's Warbler

Vermivora bachmanii

Bachman's Warbler is listed as Critically Endangered, but sadly may already be extinct. Last sighted for certain in 1988, it may still cling on in certain seasonally flooded swamp forests of the southern states of the U.S.A. The favored habitats were swampy sites with 'canebreaks' of Giant Cane (*Arundinaria gigantea*), most of which have been cleared or drained. The wintering range is or was mainly Cuba. It may perhaps survive in suitable sites such as Congaree Swamp in South Carolina, and tantalizingly there have been a number of recent unconfirmed reports. To many keen birders Bachman's Warbler has, like the Ivory-billed Woodpecker, become something of a Holy Grail. Possible sightings in Cuba in the 1980s raised hopes that it may still survive, but this dainty warbler may indeed now have vanished for good.

Family Parulidae
Size 4¾ in (12 cm)
Status Critically Endangered (possibly extinct)
Population Fewer than 50
Range U.S.A. (endemic breeder); Cuba (winter)
Habitat Swamp-forest
Main threats Habitat loss

THREATENED NEW WORLD WARBLERS

FAMILY		COMMON NAME	SCIENTIFIC NAME	STATUS
PARULIDAE New World warblers	■	Santa Marta Warbler	*Basileuterus basilicus*	VU
	■	Gray-headed Warbler	*Basileuterus griseiceps*	EN
	■	Pirre Warbler	*Basileuterus ignotus*	VU
	■	Whistling Warbler	*Catharopeza bishopi*	EN
	■	Elfin-woods Warbler	*Dendroica angelae*	VU
	■	Cerulean Warbler	*Dendroica cerulea*	VU
	■	Golden-cheeked Warbler	*Dendroica chrysoparia*	EN
	■	Pink-headed Warbler	*Ergaticus versicolor*	VU
	■	Belding's Yellowthroat	*Geothlypis beldingi*	CR
	■	Altamira Yellowthroat	*Geothlypis flavovelata*	VU
	■	Black-polled Yellowthroat	*Geothlypis speciosa*	EN
	■	Semper's Warbler	*Leucopeza semperi*	CR
	■	Paria Redstart	*Myioborus pariae*	EN
	■	Bachman's Warbler	*Vermivora bachmanii*	CR
	■	White-winged Warbler	*Xenoligea montana*	VU

RED LIST ■ EX = Extinct ■ EW = Extinct in the Wild ■ CR = Critically Endangered ■ EN = Endangered ■ VU = Vulnerable

Golden-cheeked Warbler
Dendroica chrysoparia

This pretty New World warbler is small and mainly black and white, with distinctive golden-yellow cheeks, a black eye-stripe, and yellow supercilium. The black wings have two white bars, while the underparts are mainly white with black streaks on the flanks. As a breeding species it is endemic to Texas, but winters to southern Mexico (Chiapas), Guatemala, El Salvador, Nicaragua, and Honduras, and occasionally to Costa Rica and Panama. It breeds in oak and juniper woods in certain sites in Texas, notably in Balcones Canyonlands National Wildlife Refuge at the east of Edwards Plateau. It relies heavily on a local species of juniper, using its bark for nest building. In winter it forages in mixed-species flocks and prefers habitats with an abundance of oak. A number of small reserves are also managed for this species. Logging and extraction of firewood are threats to its specialized habitat and it is also threatened in some of its wintering sites by conversion of pine-oak woodland to land for cattle-ranching. The recovery plan for this species includes a community education initiative and plans for habitat restoration.

Family	Parulidae
Size	5 in (12.5 cm)
Status	Endangered (decreasing)
Population	21,000
Range	U.S.A. (Texas) (endemic breeder), Mexico to Panama (winter)
Habitat	Woodland
Main threats	Logging, habitat fragmentation

NEW WORLD BLACKBIRDS AND VIREOS

The New World blackbirds (Icteridae) consist of about 100 species of medium-sized birds, restricted to the Americas. The majority are tropical, and they reach high species diversity in southern Mexico and Colombia. A very varied family, it includes the Bobolink, caciques, cowbirds, blackbirds, grackles, meadowlarks, orioles and troupials, and oropendolas. Many species are social and move about in flocks, especially outside the breeding season. Blackbirds are often found in open, sometimes marshy habitats. The cowbirds, like many cuckoos, are brood parasites, leaving their eggs in the nests of other species. Grackles are rather crowlike with harsh calls and an omnivorous diet. Oropendolas are large, gregarious birds of the Central and South American forests. They build woven nests that hang down from the branches of trees, giving them some protection from monkeys and snakes. The orioles are medium-sized with bright plumage, often orange or yellow and black. Thirteen species are listed as threatened, with one, Montserrat Oriole, endemic to the Caribbean island of Montserrat, Critically Endangered.

Vireos (Vireonidae) are also restricted to the Americas. They are mainly rather small, warblerlike birds of forest and scrub, with rather dull plumage and relatively heavy bills, and they feed mainly on insects. There are 52 species, of which three are threatened.

Montserrat Oriole
Icterus oberi

The New World Orioles (genus *Icterus*) should not be confused with those of the Old World (genus *Oriolus*). Endemic to the small Caribbean island of Montserrat, the male Montserrat Oriole has striking black-and-yellow plumage, being mostly black with a yellow breast, belly, undertail, and lower back. The female is a much duller yellow-green. With a range that has always been very small, it has suffered marked declines in recent decades. The main cause has been volcanic activity, especially between 1995 and 1997 when some two-thirds of this bird's habitat was destroyed, and again in 2001, 2003, and 2006 when there were heavy ash falls that affected the breeding sites. Predation by introduced rats and by native Pearly-eyed Thrashers (*Margarops fuscatus*) is also a problem. It lives in a range of forest habitats, preferably those in the higher altitude wetter forests. The woven nests are suspended from the leaves of bananas and especially those of *Heliconia* plants. The birds feed mainly on insects, but also on fruits and nectar.

Family	Icteridae
Size	8¼ in (21 cm)
Status	Critically Endangered (decreasing)
Population	500–5,200
Range	Montserrat (U.K.) (endemic)
Habitat	Forest
Main threats	Volcanic activity, possibly climate change, predators

Pampas Meadowlark

Sturnella defilippii

This meadowlark has very distinctive plumage. The male is mainly black, with brown feather-edges, but the throat and breast are an impressive bright red, and there is also a red flash between the eye and upper mandible, and a cream eyebrow stripe. The female is more streaked and browner, and pinkish-buff on the belly and throat. It is found mostly in natural grassland with high growth, tending to avoid planted grassland and areas that are intensively grazed. It feeds on insects, seeds, and shoots. Its range stretches from eastern central Argentina to Uruguay and south Brazil. It has suffered a massive decrease since about 1900, with a 90 percent reduction in range. The bulk of the population is in Argentina, mainly in southwestern Buenos Aires and La Pampa. Huge areas that once held natural grassland have been given over to grazing—mainly cattle-ranching—making the habitat unsuitable for this species. There is also some evidence of capture for the cagebird market.

Family	Icteridae
Size	8½ in (21 cm)
Status	Vulnerable (decreasing)
Population	28,000
Range	Argentina, Uruguay, Brazil
Habitat	Grassland
Main threats	Habitat loss, trampling

Saffron-cowled Blackbird

Xanthopsar flavus

This brightly colored bird has very striking plumage—the male has a yellow head, face, breast, belly, and rump, but is otherwise black. The female is mainly olive-brown and streaked. It is a vocal bird, producing harsh calls and a short, high-pitched, trilling song. A social bird, it breeds in colonies, usually in grass-dominated habitats, often in dense, marshy vegetation, but sometimes in drier agricultural sites. It is found in certain sites in Paraguay (notably Itapúa abd Misiones), Argentina (notably northeastern Corrientes), Uruguay (mainly in the wetlands of the south and west), and southern Brazil. This attractive bird seems to be declining rather rapidly in the face of habitat degradation and loss. Many wetland sites have been drained, or in some cases dammed and flooded, while burning and grazing have altered the grassland. It is also trapped for the cagebird trade. It occurs in a number of protected areas, and there has been some success with reintroduction programs.

Family	Icteridae
Size	7 in (18 cm)
Status	Vulnerable (decreasing)
Population	2,500–10,000
Range	Brazil, Paraguay, Uruguay, Argentina
Habitat	Grassland, marshland, agricultural land
Main threats	Habitat loss, fire, trapping

Black-capped Vireo

Vireo atricapilla

This small, rather compact bird is found in the U.S.A. and Mexico. The male Black-capped Vireo has a black head with a white flash between the red eyes and black bill, giving the appearance of white-framed spectacles. It has mostly olive upperparts, blackish wings with two yellowish bars, and mainly whitish to olive underparts. The female is generally duller and has a gray head. Once much more widespread, breeding from Kansas and Oklahoma to Texas and south into central Mexico, it is now known only from a few much more fragmented sites. Nesting is in dense, shrubby habitats, where grassland and mixed forests meet and intermingle. Both male and female help to build the nest and incubate the eggs. The male then cares for nestlings while the female produces another clutch, sometimes partnered by another male. In Mexico it breeds mainly between about 3,300 ft (1,000 m) and 6,600 ft (2,000 m). Black-capped Vireos winter mostly to western and southwestern Mexico. Along with the Golden-cheeked Warbler, it is one of the star birds of the Balcones Canyonlands National Wildlife Refuge in Texas, and these two species provide a local focus for habitat restoration projects in the central region of the state. Threats include conversion to pasture, intensive grazing, and nest predation by snakes and mammals. Increases in the population of Brown-headed Cowbird (*Molothrus ater*), a brood parasite, also impact on this species, and fire is also a significant threat.

Family	Vireonidae
Size	4¾ in (12 cm)
Status	Vulnerable (decreasing)
Population	8,000
Range	U.S.A., Mexico
Habitat	Scrub
Main threats	Fire, habitat loss, predation

THREATENED NEW WORLD BLACKBIRDS AND VIREOS				
FAMILY		**COMMON NAME**	**SCIENTIFIC NAME**	**STATUS**
ICTERIDAE New World blackbirds	■	Tricolored Blackbird	*Agelaius tricolor*	EN
	■	Yellow-shouldered Blackbird	*Agelaius xanthomus*	EN
	■	Selva Cacique	*Cacicus koepckeae*	VU
	■	Forbes's Blackbird	*Curaeus forbesi*	EN
	■	Rusty Blackbird	*Euphagus carolinus*	VU
	■	Red-bellied Grackle	*Hypopyrrhus pyrohypogaster*	EN
	■	Martinique Oriole	*Icterus bonana*	VU
	■	Montserrat Oriole	*Icterus oberi*	CR
	■	Mountain Grackle	*Macroagelaius subalaris*	EN
	■	Jamaican Blackbird	*Nesopsar nigerrimus*	EN
	■	Baudo Oropendola	*Psarocolius cassini*	EN
	■	Pampas Meadowlark	*Sturnella defilippii*	VU
	■	Saffron-cowled Blackbird	*Xanthopsar flavus*	VU
VIREONIDAE Vireos	■	Black-capped Vireo	*Vireo atricapilla*	VU
	■	San Andres Vireo	*Vireo caribaeus*	VU
	■	Choco Vireo	*Vireo masteri*	EN

RED LIST ■ EX = Extinct ■ EW = Extinct in the Wild ■ CR = Critically Endangered ■ EN = Endangered ■ VU = Vulnerable

FINCHES

There are about 165 species of finches (Fringillidae). Compact, mostly small birds, with a sturdy, conical bill, finches show a wide range of plumage, some being drab gray or brown, and others, such as goldfinches and grosbeaks, very colorful. Many finches are regular visitors to gardens and have adapted well to feeding at bird tables. The members of this large family go by a variety of common names, including canaries, crossbills, goldfinches, grosbeaks, redpolls, rosefinches, bullfinches, seedeaters, serins, and siskins. In several species, the basic bill shape has evolved into different forms, relating mainly to different methods of feeding. Crossbills have an unusual bill in which the mandibles cross at the tip, an adaptation for prizing open pine cones.

Serins (genus *Serinus*) are some of the smallest of all finches, and some species are known as canaries. Indeed, this genus, one of the largest of the family with over 40 species, contains the Island Canary as well as a number of species called seedeaters (not to be confused with the seedeaters of the Emberizidae). The Island Canary, found wild on the Canary Islands, is the ancestor of domestic canaries, the latter now found in a wide range of artificially selected breeds and known for their impressive songs.

Another large genus is *Carduelis*, which is made up of over 30 species, including goldfinches, greenfinches, linnets, redpolls, and siskins. These are mainly delicate finches with a short, fine bill used for extracting small seeds from a range of plants, including thistles and cones (depending on the species). The rosefinches (genus *Carpodacus*) are bulkier finches with a deeper bill and typically with rose-pink plumage. The grosbeaks and the hawfinch have massive bills that are capable of cracking open even the very tough stones of cherries.

The locally named finches of Hawaii, some of which are also known as honeycreepers, are also remarkable mainly for the shapes of their bill. The Akialoa has a long, narrow decurved bill, while in the Akiapolaau it is the longer upper mandible that curves down. The bright red Iiwi also has a strongly, decurved bill, used to probe into flowers for nectar. The threatened species number 30, most notably those of Hawaii, ten of which are Critically Endangered: Akekee, Akikiki, Akohekohe, Maui Parrotbill, Nihoa Finch, Nukupuu, Oahu Alauahio, Ou, Palila, and Po'o-uli, the latter not thought to be extinct. Several further Hawaiian endemic finches have already become extinct. Also Critically Endangered is a finch from another island, namely the São Tomé Grosbeak, and the Azores Bullfinch is Endangered.

Red Siskin

Carduelis cucullata

This small finch is one of many small songbirds suffering at the hands of cagebird trappers. It is undergoing a rapid decline, largely as a result of this trade, and its population is now very fragmented. The male is unmistakable, with his gaudy black-and-red plumage. His song is a pleasant series of trills and twittering sounds. The main known sites are scattered in northern Venezuela, northern Colombia, and southwestern Guyana. Numbers are very difficult to estimate, hence the large range for the population figures. It inhabits a variety of habitats, moving between moist evergreen forest, dry deciduous woodland, and grassland and scrub. The nest is typically built in clumps of epiphytic bromeliads toward the crown of a tree. The diet includes fruits, notably mistletoes and figs, as well as flower buds and seeds. One reason for its popularity as a cagebird is its ability to hybridize with canaries, allowing the introduction of red into the canaries' plumage. Though it has legal protection in Venezuela and Guyana, it is still threatened in the wild, partly through habitat loss.

Family	Fringillidae
Size	4 in (10 cm)
Status	Endangered (decreasing)
Population	2,500–10,000
Range	Venezuela, Colombia, Guyana
Habitat	Forest, woodland and grassland
Main threats	Capture for cagebird trade, habitat loss

Akohekohe

Palmeria dolei

The Akohekohe is a large honeycreeper with a slightly decurved sharp bill. It is mainly black, with a spotted and streaked chest and an unusual tuft of stiff white feathers on its forehead. The eye has an orange ring and there is an orange-red patch at the back of its neck. Its call is a clear whistle and its song is a series of rather low-pitched notes recalling its local Hawaiian name. Endemic to Hawaii, it is found only on Maui, where it is relatively common on the northeastern slopes of Haleakala. Formerly it also occurred on Moloka'i, but is now extinct there. It favors damp and wet forests between about 5,000 ft (1,500 m) and 6,600 ft (2000 m). In addition to nectar, it also feeds on invertebrates. Feral pigs have degraded its habitat and also encouraged the spread of mosquitoes, which carry diseases. Predation by introduced cats, rats, Barn Owls and possibly Indian Mongoose (*Herpestes javanicus*) also affect this species.

Family	Fringillidae
Size	7 in (18 cm)
Status	Critically Endangered (decreasing)
Population	3,800
Range	Hawaii (Maui) (endemic)
Habitat	Moist and wet forest
Main threats	Habitat loss, alien predators, disease

THREATENED FINCHES

FAMILY		COMMON NAME	SCIENTIFIC NAME	STATUS
FRINGILLIDAE Finches	■	Red Siskin	*Carduelis cucullata*	EN
	■	Warsangli Linnet	*Carduelis johannis*	EN
	■	Saffron Siskin	*Carduelis siemiradzkii*	VU
	■	Yellow-faced Siskin	*Carduelis yarrellii*	VU
	■	Oahu Amakihi	*Hemignathus flavus*	VU
	■	Kauai Amakihi	*Hemignathus kauaiensis*	VU
	■	Nukupuu	*Hemignathus lucidus*	CR
	■	Akiapolaau	*Hemignathus munroi*	EN
	■	Anianiau	*Hemignathus parvus*	VU
	■	Hispaniolan Crossbill	*Loxia megaplaga*	EN
	■	Palila	*Loxioides bailleui*	CR
	■	Akekee	*Loxops caeruleirostris*	CR

RED LIST ■ EX = Extinct ■ EW = Extinct in the Wild ■ CR = Critically Endangered ■ EN = Endangered ■ VU = Vulnerable

Iiwi

Vestiaria coccinea

With its bright scarlet plumage and long, pink, strongly decurved bill, the Iiwi is among the most extraordinary of Hawaii's honeycreepers. Its wings and tail are black and the legs bright pink, the vivid colors of the adult making it easy to spot against a habitat of green foliage. The young birds have a very different plumage—they are golden and spotted, with an ivory-colored bill, and were assumed by some early explorers to belong to a different species. Its common name is a representation of its normal song, though it often mimics the calls of other species as well. Its bill, aided by the bird's tubular, brush-tipped tongue, is well adapted for gathering nectar from native plants, some of which have similarly shaped flowers. Sometimes it will access the nectar by piercing a hole at the base of the flower, but like hummingbirds it is also able to hover while feeding. It also feeds to some extent on invertebrates. Once found throughout the Archipelago, the only remaining stable populations are now on Mau'i and Hawaii (Hakalau). It is found in varied forests, including disturbed habitats from about 1,000 ft (300 m) to 9,500 ft (2,900 m). Like the Akohekohe, it too has suffered from introduced mammals, including cattle, pigs, cats, and rats. It is also known to be very susceptible to avian malaria, carried by introduced mosquitoes, with populations in lower altitude forests being apparently most affected. Though still relatively abundant, the Iiwi is undergoing a marked population decline and is found in fewer places than previously.

Family Fringillidae
Size 6 in (15 cm)
Status Vulnerable (decreasing)
Population 350,000
Range Hawaii (endemic)
Habitat Forest
Main threats Habitat loss, alien predators, disease

THREATENED FINCHES *Continued*

FAMILY		COMMON NAME	SCIENTIFIC NAME	STATUS
FRINGILLIDAE Finches	■	Akepa	*Loxops coccineus*	EN
	■	Po'o-uli	*Melamprosops phaeosoma*	CR/EX
	■	São Tomé Grosbeak	*Neospiza concolor*	CR
	■	Akikiki	*Oreomystis bairdi*	CR
	■	Hawaii Creeper	*Oreomystis mana*	EN
	■	Akohekohe	*Palmeria dolei*	CR
	■	Oahu Alauahio	*Paroreomyza maculata*	CR
	■	Maui Alauahio	*Paroreomyza montana*	EN
	■	Maui Parrotbill	*Pseudonestor xanthophrys*	CR
	■	Ou	*Psittirostra psittacea*	CR
	■	Azores Bullfinch	*Pyrrhula murina*	EN
	■	Ankober Serin	*Serinus ankoberensis*	VU
	■	Yellow-throated Seedeater	*Serinus flavigula*	EN
	■	Syrian Serin	*Serinus syriacus*	VU
	■	Salvadori's Serin	*Serinus xantholaemus*	VU
	■	Laysan Finch	*Telespiza cantans*	VU
	■	Nihoa Finch	*Telespiza ultima*	CR
	■	Iiwi	*Vestiaria coccinea*	VU

RED LIST ■ EX = Extinct ■ EW = Extinct in the Wild ■ CR = Critically Endangered ■ EN = Endangered ■ VU = Vulnerable

WAXBILLS, MUNIAS, AND ALLIES, AND WEAVERS, FODIES, AND ALLIES

Waxbills, munias, and allies (Estrildidae) comprise about 140 species of small, active, finchlike birds, many of which have bright, colorful plumage. It is no surprise, therefore, that many have become popular cagebirds, especially as most species breed readily in captivity, adapting well to life in an aviary. Many species also have pleasant, rather soft songs. Some of the other groups included in this family are cordon-bleus, crimson-wings, firefinches, firetails, olivebacks, and parrotfinches—the names indicating their features, especially their bright plumage. Waxbills are native to Africa and Asia, munias are mainly found in southern Asia and the southwestern Pacific, while others such as Zebra Finch, Star Finch, Gouldian Finch, and relatives are native to Australia. They tend to form very stable pair-bonds and many species may mate for life. They are small birds with bodies 4¼–5 in (11–13 cm) in length and weighing just ¼–½ oz (7–10 g). The name Waxbill comes from the bill color, said to resemble red sealing wax. Munias are found in Africa, southern Asia, India, and east as far as the Philippines. Mostly seed-eaters, they prefer an open habitat with reeds and grasses. Similar in size to waxbills they are less colorful (often brown), but are also popular cagebirds. A total of 9 species are threatened, with the Gouldian Finch Endangered.

Weavers, fodies, and allies (Ploceidae) represent a total of 106 species, of which 13 are threatened. Weavers (main genus *Ploceus*) are sparrowlike, but unlike sparrows male weavers are mostly quite brightly colored. As their name suggests, they weave complex nests from grasses or strips of other plant material, and some breed in large colonies. Fodies (genus *Foudia*) are members of the weaver family found mainly on islands in the southern Indian Ocean, notably Madagascar, the Seychelles, Mascarene Islands, and Comoros Islands. These are small birds, 4¾–6 in (12–15 cm) long with a short, conical bill suited to feeding on seeds and some insects. They nest in woodland or scrub and construct a domed nest of grass and other plant material, with a side opening; the nest is suspended from a palm leaf or branch. The family also includes bishops, malimbes, queleas, and widowbirds. Many species move about in flocks and the Red-billed Quelea has the distinction of being the world's most abundant bird, with an estimated population of 1.5 billion. These birds sometimes form huge flocks and move about in a similar way to locusts, feeding on seeds and grain, when they can be a serious pest to farmers.

The Old World sparrows (*Passeridae*) are related to weavers and include about 40 species, one of which, Abd'Al Kuri Sparrow is Vunerable.

Mauritius Fody
Foudia rubra

The Mauritius Fody is a small, mainly brown weaver found only on the Indian Ocean island of Mauritius. In male breeding plumage, the dark brown back, wings, and tail contrast sharply with the bright red head, neck, and breast. The rump is also reddish. Though its population is very small, it does appear to be stable, and the species is not in immediate danger of extinction. A bird of various types of forest, it feeds on insects, fruit, and nectar. Once found in suitable habitats across the island, it declined rapidly, partly through habitat loss, but more importantly because of direct predation by introduced predators, particularly the Black Rat (*Rattus rattus*) and Crab-eating Macaque (*Macaca fascicularis*). On the mainland of Mauritius it is now mainly found within the Black River National Park in the southwest. In 2005, birds were translocated to a small offshore island (Ile aux Aigrettes), which now holds a healthy population as an insurance.

Family	Ploceidae
Size	5½ in (14 cm)
Status	Endangered (stable)
Population	200–350
Range	Mauritius (endemic)
Habitat	Forest, plantations
Main threats	Habitat loss, alien predators

Yellow Weaver
Ploceus megarhynchus

The Yellow Weaver has a rather fragmented distribution, mainly found in northern India, and also in eastern India (Assam) and in eastern Nepal. Large, and rather chunky in build, with a strong, sturdy bill, it is mainly yellow on the head and underparts, with a greenish streaked back and wings. It is found in "terai" marshy grassland, especially those subject to seasonal flooding. Like most weavers, it is a social species, feeding in flocks and breeding in colonies. It nests either in tall grasses and reeds or in trees. These grasslands have been much reduced in area through deliberate draining to convert the land to crop cultivation—mainly rice, sugarcane, mustard, and tea plantations—and for grazing. The reeds and other tall grasses are also cut for thatching. An additional pressure is capture for sale into the cagebird trade. In India, this trade has been outlawed since 1991 and the species protected.

Family	Ploceidae
Size	6¾ in (17 cm)
Status	Vulnerable (decreasing)
Population	2,500–10,000
Range	India, Nepal
Habitat	Marshland, wet woodland
Main threats	Habitat loss, livestock grazing, capture for cagebird trade

Gouldian Finch

Erythrura gouldiae

Well known to cagebird enthusiasts, the dainty Gouldian Finch is a brightly colored Australian endemic finch with a long, pointed tail. The adult is multicolored, with a shiny green back and nape, yellow belly, blue chest, and pale blue collar. There are two variants—one with a black head and face, and one with a bright red face, bordered black. In the early 20th century this finch was much more common and has suffered partly through collection for the cagebird trade, but probably mainly because its habitat has been altered by grazing pressures as a result of increased densities of cattle and other livestock. Change in the frequency of fires is another possible cause.

The Gouldian Finch survives in small populations in northern Australia, mainly in the north of Northern Territory, with a handful of records from north Queensland. The total population lies in the range 2,000–10,000, mainly in the Kimberley region. The favored habitat is woodland with a grass-dominated understorey, and plenty of hollow trees for nesting. These delightful birds feed mainly on the seeds of a range of grass species, switching from dry- to wet-season species at different times of the year.

Family	Estrildidae
Size	4¼–5 in (11–12.5 cm)
Status	Endangered (decreasing)
Population	2,000–10,000
Range	Australia (endemic)
Habitat	Open grassy woodland
Main threats	Grazing, fire

THREATENED WAXBILLS, MUNIAS, AND ALLIES, AND WEAVERS, FODIES, AND ALLIES

FAMILY		COMMON NAME	SCIENTIFIC NAME	STATUS
ESTRILDIDAE Waxbills, munias, and allies	■	Green Avadavat	*Amandava formosa*	VU
	■	Shelley's Crimson-wing	*Cryptospiza shelleyi*	VU
	■	Gouldian Finch	*Erythrura gouldiae*	EN
	■	Pink-billed Parrotfinch	*Erythrura kleinschmidti*	VU
	■	Royal Parrotfinch	*Erythrura regia*	VU
	■	Green-faced Parrotfinch	*Erythrura viridifacies*	VU
	■	Anambra Waxbill	*Estrilda poliopareia*	VU
	■	Gray-banded Munia	*Lonchura vana*	VU
	■	Java Sparrow	*Padda oryzivora*	VU
PLOCEIDAE Weavers, fodies, and allies	■	Rodrigues Fody	*Foudia flavicans*	VU
	■	Mauritius Fody	*Foudia rubra*	EN
	■	Gola Malimbe	*Malimbus ballmanni*	EN
	■	Ibadan Malimbe	*Malimbus ibadanensis*	EN
	■	Golden-naped Weaver	*Ploceus aureonucha*	EN
	■	Bannerman's Weaver	*Ploceus bannermani*	VU
	■	Bates's Weaver	*Ploceus batesi*	EN
	■	Kilombero Weaver	*Ploceus burnieri*	VU
	■	Yellow-legged Weaver	*Ploceus flavipes*	VU
	■	Clarke's Weaver	*Ploceus golandi*	EN
	■	Yellow Weaver	*Ploceus megarhynchus*	VU
	■	Usambara Weaver	*Ploceus nicolli*	EN
	■	Loango Weaver	*Ploceus subpersonatus*	VU
PASSERIDAE Old World Sparrows	■	Abd'Al Kuri Sparrow	*Passer hemileucus*	VU

RED LIST ■ EX = Extinct ■ EW = Extinct in the Wild ■ CR = Critically Endangered ■ EN = Endangered ■ VU = Vulnerable

STARLINGS, ORIOLES, DRONGOS, AND BIRDS OF PARADISE

Essentially an Old World family, the 110 species of starlings (Sturnidae) are opportunistic birds that thrive in a range of habitats. Small to medium-sized and rather dumpy, they have a long, straight bill and the species vary in plumage, some being shiny and brightly colored. Many are social birds, feeding and roosting in flocks. Some starlings, notably the Common Starling, often form flocks that can reach vast numbers, especially when gathering to roost. Mainly ground-feeding birds, they have a preference for open habitats. The family also includes mynahs and two species of oxpecker, both mainly tropical groups. Some mynahs and starlings are known for their ability to mimic sounds, including the calls of other birds. Oxpeckers feed mainly by plucking parasitic insects from the bodies of large herbivores, including cattle, to the benefit of both bird and mammal. Eight species of the starling family are threatened.

Orioles (Oriolidae) are 30 species of graceful, slim-bodied, usually colorful birds known for their musical, fluting calls and songs. Species are found in Africa, Europe, Asia, and Australia. The males are more brightly colored than the females. Orioles have a slightly curved hooked bill and tend to feed high in the forest canopy, where they are opportunistic omnivores. The family also contains three species of figbird, found in Australasia. Three orioles are threatened, of which one, Isabela Oriole, endemic to Luzon in the Philippines, is Critically Endangered.

Drongos (Dicruridae) are found from Africa to Southeast Asia and Australasia. There are 23 species, of which two are threatened, Grand Comoro Drongo and Mayotte Drongo, both endemic to the Comoro Islands. Rather crowlike, most drongos have dark or black plumage and their tails are forked (some elaborately decorated), square-ended, or elongated, and some species have crests. Some drongos are also good mimics and can imitate birds as well as other sounds. Drongos are found in open forest or bush and feed on insects. Some feed in the manner of flycatchers. They are particularly aggressive to any interference when on their nests, often taking on birds much larger than themselves.

Birds of paradise (Paradisaeidae) are some of the world's most fascinating and beautiful birds. Medium-sized and rather crow-shaped, they range in size from 6 to 17½ in (15 to 44 cm). Most species are found only in New Guinea, with a few in northeast Australia and some in Indonesia. Several species have elaborate displays and the males are adorned with magnificent plumed feathers. In their complex mating rituals, they use special meeting areas in lekking displays. These may include dances in which the males use their wonderful plumage to full effect. The territorial and mating displays are accompanied by strange and unusual sounds, including loud cracking noises, some of which are produced by modifications of their wings. They are truly extraordinary birds and this may be why, historically, they were believed by some to be the mythical phoenix. Of the 40 species, three are threatened, mainly by loss of their tropical forest habitats, and also through being hunted for their beautiful feathers.

Bali Starling

Leucopsar rothschildi

This beautiful almost pure white starling has declined almost to the point of extinction, mainly through illegal trapping for the cagebird trade. A medium-sized bird, its head is adorned with a drooping crest, and it has blue bare skin around its eye and a rather stout yellow bill. The tips of its wings and tail are black. Like many starlings, its calls are rather sharp and chattering. Confined to the Indonesian island of Bali, it has long been the target of collectors, and is now on the very brink of extinction as a wild species, even though the entire small remnant population lies within the Bali Barat National Park. It breeds in monsoon forest and also open scrub and savanna, dispersing into open forest and flooded woodland outside the breeding season. The Bali starling was also the victim of illegal poaching until the 1990s, when numbers fell as low as six individuals. Efforts are made to boost the wild population by releasing captive-bred birds to suitable habitats, and a special bird sanctuary for this species has been set up on the small offshore island of Nusa Penida.

Family	Sturnidae
Size	10 in (25 cm)
Status	Critically Endangered (decreasing, possibly extinct in the wild)
Population	50-100
Range	Bali (endemic)
Habitat	Scrub, savanna
Main threats	Capture for cagebird trade, habitat loss

Blue Bird-of-paradise

Paradisaea rudolphi

This magnificent endemic forest bird is one of 38 species in the bird-of-paradise family found in New Guinea. Birds of paradise are threatened both by habitat loss and by being hunted traditionally for their remarkable feathers. The Blue Bird-of-paradise is one of the most beautiful, but sadly also one of the rarest. It is still found in a few scattered patches of hill forest between about 3,300 ft (1,000 m) and 6,600 ft (2,000 m) in the east of the central mountains of Papua New Guinea, notably on Mt Sisa and in the Owen Stanley range. Estimates of its population put the numbers left at not more than 10,000, but there may be as few as 2,500, and it is classed as Vulnerable. The numbers seem to be declining as the human population increases and more and more of its habitat is converted to agriculture. It has generally dark plumage, with a rather thick, ivory-colored bill, bright blue back, wings, and tail and a white partial eye-ring. The male has an unusual courtship dance in which he hangs inverted on a branch, spreading out a fan of blue feathers overtopped by a pair of long, arching tail streamers. This display is accompanied by bizarre, mechanical sounding, buzzing and nasal chattering calls.

Family Paradisaeidae
Size 11¾ in (30 cm)
Status Vulnerable (decreasing)
Population 2,500–10,000
Range Papua New Guinea (endemic)
Habitat Montane forest
Main threats Habitat loss, hunting

THREATENED STARLINGS, ORIOLES, DRONGOS, AND BIRDS OF PARADISE

FAMILY		COMMON NAME	SCIENTIFIC NAME	STATUS
STURNIDAE Starlings	■	White-eyed Starling	*Aplonis brunneicapillus*	EN
	■	Rarotonga Starling	*Aplonis cinerascens*	VU
	■	Pohnpei Starling	*Aplonis pelzelni*	CR
	■	Santo Starling	*Aplonis santovestris*	VU
	■	Abbott's Starling	*Cinnyricinclus femoralis*	VU
	■	Bali Starling	*Leucopsar rothschildi*	CR
	■	White-faced Starling	*Sturnus albofrontatus*	VU
	■	Black-winged Starling	*Sturnus melanopterus*	CR
ORIOLIDAE Orioles	■	São Tomé Oriole	*Oriolus crassirostris*	VU
	■	Isabela Oriole	*Oriolus isabellae*	CR
	■	Silver Oriole	*Oriolus mellianus*	VU
DICRURIDAE Drongos	■	Grand Comoro Drongo	*Dicrurus fuscipennis*	EN
	■	Mayotte Drongo	*Dicrurus waldenii*	VU
PARADISAEIDAE Birds of paradise	■	Black Sicklebill	*Epimachus fastuosus*	VU
	■	Blue Bird-of-paradise	*Paradisaea rudolphi*	VU
	■	Wahnes's Parotia	*Parotia wahnesi*	VU

RED LIST ■ EX = Extinct ■ EW = Extinct in the Wild ■ CR = Critically Endangered ■ EN = Endangered ■ VU = Vulnerable

Crows and jays, and wattled crows

Crows and jays (Corvidae) number some 115 species. They are medium-sized to large birds, many with black or dark plumage, but several, notably the jays, are colorful. Intelligent and adaptable, they are found in a wide variety of habitats and take a range of food. Crows are known to be some of the cleverest and most adaptable of birds, and many have adapted to live alongside people, often exploiting the sources of food available in and around houses and fields. Many crow species are highly social, breeding and roosting in colonies. There are 13 threatened species, of which two, the Mariana Crow and the Banggai Crow, are Critically Endangered, and one, the Hawaiian Crow, is Extinct in the Wild. In addition to crows and jays, this family also includes choughs, nutcrackers, ground-jays, jackdaws, magpies, ravens, scrub-jays, and treepies. The Rook, with a wide distribution across Europe and Asia, is a common and highly social member of the crow family. They roost and breed colonially, nesting in rookeries in robust twiggy nests created high in the forks of tree branches. Rooks and other crows are omnivores, consuming a range of plant and insect food and may also be seen at rubbish dumps. Rooks often join with Eurasian Jackdaws, in mixed flocks. Choughs are agile colonial birds of cliff and mountain sites. They have glossy black plumage and distinctive bright red legs. There are two species: Red-billed Chough and Yellow-billed Chough. Neither is currently under global threat, but both have suffered local declines, partly due to habitat fragmentation. There are also two species of nutcracker—Clark's Nutcracker of the western U.S.A., and the Spotted Nutcracker of Europe and Asia. Their name derives from their main food source of nuts and seeds, which their heavy bills are adapted to exploit. The four species of ground-jay inhabit high altitude semi-desert areas from central Asia to Mongolia. They have long strong legs, and are capable of agile leaps and bounds between rocks and boulders. The Xinjiang Ground-jay lives in the Taklimakan Desert of western China and is listed as Near Threatened as its population is rather small and fragmented. Magpies and treepies are active and long-tailed, many with bright plumage. The largest members of the family, indeed the largest of all passerines (perching birds), are the ravens, of which there are nine species. These powerful birds are mostly found in mountainous habitats.

Wattled crows (Callaeatidae) are a tiny family from New Zealand with just two species, one of which, the Kokako is threatened (Endangered). The other species is the Saddleback. They are crowlike, but have characteristic fleshy wattles at the sides of the base of their bill.

Florida Scrub-jay

Aphelocoma coerulescens

Like most jays, this member of the crow family has quite colorful plumage—mainly blue and pale gray, with a dark facial mask. Noisy and gregarious, the birds call frequently as they work the bushes for food. As its common name suggests, the species' favored habitat is scrub. It is found mainly in dry, oak-dominated shrubby woodland and scrub with a sparse ground-layer, on sandy soils. This is a community maintained partly by fire, and is optimal for the species between five and 15 years after burning. Florida Scrub-jays feed on acorns, berries, invertebrates, and small vertebrates such as frogs, lizards, and young birds. Much of the dry scrub has been lost to housing, cultivation, and grazing, while fire frequency has also been reduced, allowing pine forest to develop, a habitat unsuitable for this species. Carefully controlled burning of selected sites on a rotational basis may be a key to its survival.

Family	Corvidae
Size	10¾–12¼ in (27–31 cm)
Status	Vulnerable (decreasing)
Population	6,500
Range	U.S.A. (Florida) (endemic)
Habitat	Scrub, woodland
Main threats	Habitat loss, disturbance

Kokako

Callaeas cinereus

Endemic to New Zealand, the Kokako shares its family with just one other species, the Saddleback (*Philesturnus carunculatus*). The Kokako is a large bird with long legs and tail, broad, rounded wings and a short, thick bill with blue wattles hanging down from the base. Its plumage is mainly blue-gray. It has a remarkable slow song with loud, clear tones that have been likened to notes produced by an organ. The song is preceded by mewing and buzzing calls. Birds often duet and nearby pairs may sing together. It is now found only on the North Island, mainly in scattered sites in the north and northwest. It inhabits tall lowland forest and feeds on fruit, leaves, buds, flowers, nectar, and invertebrates. Habitat loss was the main cause of early declines, while introduced predators, including Black Rat (*Rattus rattus*), Brush-tailed Possum (*Trichosurus vulpecula*), and Stoat (*Mustela erminea*), are now thought to pose the major threat. Most of the remaining sites are protected and many are managed to control the alien predators, and some birds have also been translocated to nearby offshore islands.

Family	Callaeatidae
Size	15 in (38 cm)
Status	Endangered (decreasing)
Population	1,500
Range	New Zealand (endemic)
Habitat	Lowland forest
Main threats	Habitat loss, alien predators, grazing

THREATENED CROWS, JAYS, AND WATTLED CROWS				
FAMILY		**COMMON NAME**	**SCIENTIFIC NAME**	**STATUS**
CORVIDAE Crows and jays	■	Florida Scrub-jay	*Aphelocoma coerulescens*	VU
	■	Flores Crow	*Corvus florensis*	EN
	■	Hawaiian Crow	*Corvus hawaiiensis*	EW
	■	Mariana Crow	*Corvus kubaryi*	CR
	■	White-necked Crow	*Corvus leucognaphalus*	VU
	■	Banggai Crow	*Corvus unicolor*	CR
	■	White-throated Jay	*Cyanolyca mirabilis*	VU
	■	Dwarf Jay	*Cyanolyca nana*	VU
	■	Amami Jay	*Garrulus lidthi*	VU
	■	Pinyon Jay	*Gymnorhinus cyanocephalus*	VU
	■	Sichuan Jay	*Perisoreus internigrans*	VU
	■	Sri Lanka Magpie	*Urocissa ornata*	VU
	■	Ethiopian Bush-crow	*Zavattariornis stresemanni*	EN
CALLAEATIDAE Wattled crows	■	Kokako	*Callaeas cinereus*	EN

RED LIST ■ EX = Extinct ■ EW = Extinct in the Wild ■ CR = Critically Endangered ■ EN = Endangered ■ VU = Vulnerable

CONSERVATION

CONSERVATION IS VITALLY IMPORTANT, AND THERE ARE MANY REASONS WHY WE SHOULD STRIVE TO PROTECT WILD BIRDS AND THEIR HABITATS. BIRDS ARE AMONG THE MOST VISIBLE AND ATTRACTIVE OF ALL OUR WILDLIFE—THEY INSPIRE AND DELIGHT US, WHETHER SEEN AND WATCHED IN GARDENS AND PARKS, OR SPOTTED ON OUR TRAVELS AND ON VACATIONS ABROAD.

Sensitive and alert, birds respond remarkably quickly to changes in their environment and are accurate indicators of the health of ecosystems. In this way, they can offer early warning if things are going wrong. In short, if conditions become unsuitable for our birdlife then the chances are that they are going to be bad for other wildlife and humans too, either immediately or in the future.

Studies of birds and their population changes have frequently alerted us to dangers to the environment. In Costa Rica for example, lowland forest birds are extending their ranges up mountain slopes because, it seems, the high-altitude cloud forests are drying out through global warming. In another example, Common Whitethroat numbers in the U.K. fell sharply in the late 1960s as a result of desertification of their wintering grounds in sub-Saharan Africa.

Birds are beautiful and fascinating in their own right and give huge pleasure to all those interested in nature and wildlife, but there are more tangible and pressing reasons to justify bird conservation.

Ensuring the protection of habitats around the world is essential if we are to preserve an environment with a high biodiversity. Such richness provides us with a range of products, from timber and paper to food crops and pharmaceuticals. Conservation has many aspects, and it is often at its most effective when it involves the support and efforts of local people—especially those that depend on a close relationship with wild habitats—and where these can be exploited in a sustainable way.

The number of people belonging to bird and conservation societies continues to increase worldwide. In the U.S.A., for example, the National Audubon Society has about 60,000 members and in the U.K., the membership of the RSPB (Royal Society for the Protection of Birds) stands at more than one million. The World Wide Fund for Nature (WWF), with offices in over 40 countries, now has a membership of 1.2 million in the U.S.A. and nearly 5 million supporters worldwide.

Ecotourism, especially where it features birds, is growing in popularity, and funds received from visitors can be channeled to support bird conservation projects in the field. One example is the Indian Ocean island of Cousin in the Seychelles. This island is now visited by up to 8,000 people a year, with tourist revenue as well as the local communities sustaining the reserve.

◁ *A Gentoo Penguin chick hatching. Although there are several populations, mainly on sub-Antarctic islands, this species is suffering a general decline and is listed as Near Threatened.*

CONSERVATION WORKS

Though the threats to rare birds and to other animals and plants are very real and on the increase—occasionally even catastrophic—the picture is not all gloomy and it should be emphasized that there have been many success stories. Careful research and recording, followed up by targeted local and political initiatives, such as habitat protection and education, have saved many species from dangerous declines and the threat of extinction. In short, conservation works.

Thanks to targeted conservation efforts, 18 bird species have increased significantly in the last 20 years to the extent that they have qualified for downlisting to a lower category of threat. Many more species have also benefited and may cross the threshold for downlisting in the future. In 1994, 16 of the species then listed as Critically Endangered would have been lost forever in the following ten years in the absence of conservation initiatives.

One of the key approaches to successful conservation is to establish priority action for all those species that are listed as Critically Endangered—a total currently standing at 190. Such actions include the protection of sites where the species occur in the wild (including the establishment of practical habitat management), control of invasive species, restoration of habitat where this has been degraded or removed, raising awareness both locally and internationally, and the reintroduction of the species to suitable sites. In fact, most species usually require a combination of some or all of these activities. In terms of the percentage of species affected, the most important action is the protection and management of sites, notably the Important Bird Areas (IBAs). The next most effective measures are management of the habitats and sites concerned, raising awareness, and the control of invasive species.

Since 2004, some 88 percent of all Critically Endangered bird species have received targeted conservation action, and over half of these have involved the BirdLife International Partnership, working with other organizations, agencies, and governments. Encouragingly, more than 70 percent of the species that have received conservation action have already benefited, through reduction of threat, slowing of population declines, or population increases. These are heartening

△ *A Black Rat Snake (*Elaphe obsoleta*) raiding a bird's nest in eastern North America. Introduced snakes are a threat to several birds, especially those found on oceanic islands.*

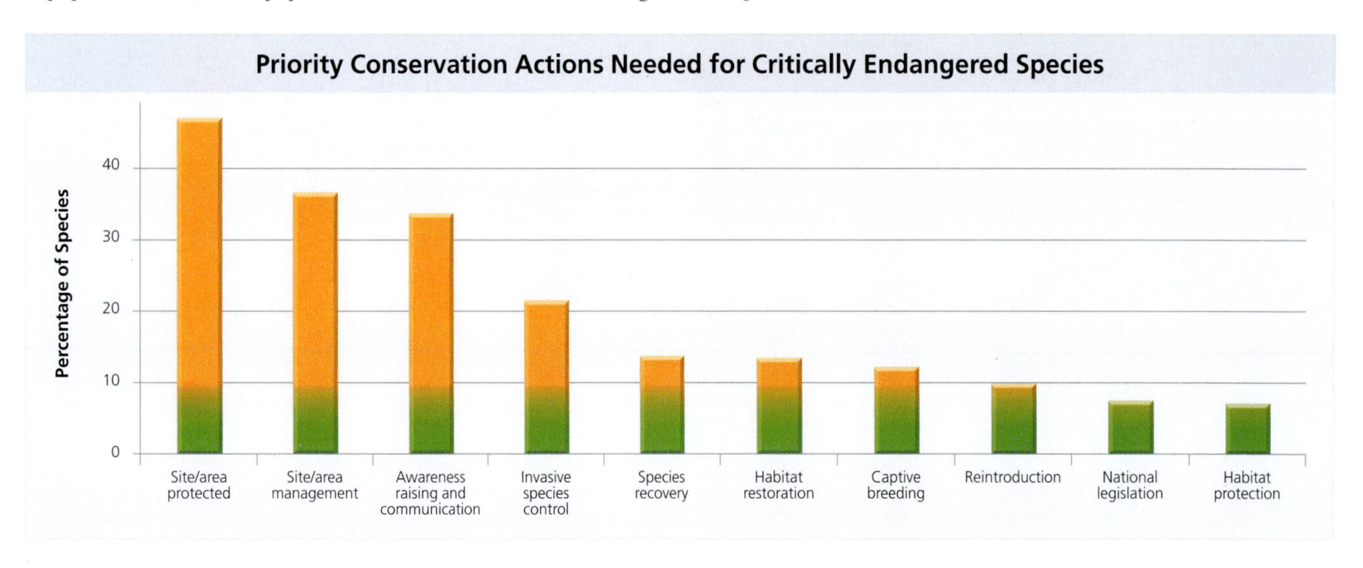

statistics that offer hope and encouragement to all who care about preserving nature and the marvellous variety of birds and other wildlife around the world.

The following four species are classic examples of birds that were once reduced to tiny numbers of individuals, but that were saved from extinction by timely conservation action. In each species, the declines were reversed, and all now have substantially larger populations that are increasing or stable. All four of the species are listed as Endangered.

BLACK ROBIN

Endemic to the Chatham Islands, New Zealand, this jet-black robin was down to the last five birds in 1980. It staged a spectacular recovery after targeted conservation efforts and its numbers have continued to increase, and now stand at some 200.

MAURITIUS PARAKEET

Though its numbers are still small—estimated at fewer than 250—this attractive, long-tailed parakeet has steadily increased following conservation management. It is found only in the southwest of the Indian Ocean island of Mauritius.

RAROTONGA MONARCH

Once rated as one of the world's rarest birds, this dainty flycatcher is endemic to the Cook Islands, New Zealand. From about 20 birds in 1983, it now numbers 250–300 birds. Predation by introduced Black Rats (*Rattus rattus*) and domestic cats is a serious threat.

SEYCHELLES MAGPIE-ROBIN

This thrushlike, black-and-white bird has a small but increasing population of about 200. Its population plummeted to only a handful on one island, Frégate in the Seychelles. Following the introduction of protection on Frégate and translocations to some other islands its numbers are now increasing.

△ *Future targets for the conservation of the Endangered Seychelles Magpie-robin are to introduce it to more islands.*

▽ *The Endangered Black Robin has staged a spectacular recovery from near extinction to some 200 individuals, and rising.*

Threats to Critically Endangered Birds by Alien Species

Source of Alien Species threat
Fish
Reptile
Invertebrate
Bird
Disease
Plant
Mammal
0
20
40
60
Number of Species Threatened

Saving Seabirds

Of all groups of birds, seabirds are the most threatened, and albatrosses are becoming so at a faster rate globally than all other groups. In their extensive forays over the oceans, many seabirds spend much time over international waters, so the conservation issues need addressing globally.

Through its Global Seabirds Program, BirdLife International seeks to identify and conserve marine IBAs, taking into account such criteria as foraging grounds, migration bottlenecks (where migrating birds are channelled, for example, through narrow straits or around headlands), and favored feeding grounds in the high seas (often associated with upwellings of currents and the consequent raised productivity and food supply). Certain areas are especially productive, notably where cold ocean currents rich in nutrients well upward. Examples of this are the Antarctic Polar Frontal Zone, the Benguela Current in the eastern South Atlantic, the Humboldt Current along the west coast of South America, and the Patagonian Shelf. The nutrients support rapid growth of the phytoplankton fed on by zooplankton and larger animals including squid and fish. Seabirds congregate at such sites to take advantage of the high concentrations of prey, but then often come into contact with fishing fleets.

Key to effective conservation efforts is the identification of those areas where marine fisheries are impacting most seriously on species such as albatrosses and petrels. The overlap between albatross distribution and areas of longline fishing is a major concern and more than 300,000 seabirds, including 100,000 albatrosses, die every year as a bycatch of the longline fishing fleets. The birds perish when they try to take fish bait from the hooks on the lines, becoming ensnared and often drowning. Some also die when they collide with trawling gear as they try to feed on discarded fish. Of the 22 species of albatross, 17 are threatened and five are Near Threatened. Of the threatened species of albatross, six are Endangered and three are Critically Endangered.

BirdLife has set up the Albatross Task Force to work alongside fishermen in seven countries—Argentina, Brazil, Chile, Ecuador, South Africa, Namibia, and Uruguay—to find workable solutions to this problem. For example in southern Chile, modifications to the fishing gear reduced the annual capture of seabirds from 1,500 to zero, whilein Brazil, the use of bird-scaring lines has reduced the bycatch by as much as 56 percent.

Another major threat to seabirds that affects their breeding success is invasive alien species, introduced plants and animals. More than half of the world's threatened birds face pressures from alien invasive species—especially on islands, where seabirds often nest. Many seabirds nest in colonies and are easy prey to introduced

◁ *A Laysan Albatross sits on top of fishing nets and other marine debris on Midway Atoll in the northwestern Hawaiian Islands. This species is Near Threatened.*

△ *A Red-footed Booby and chick on the Galápagos. While still listed as Least Concern, this species is threatened by the fishing industry and coastal development.*

carnivores such as mice, rats, and cats. Having evolved in the absence of native predators, the island birds lack the strategies for countering this threat and many species suffer great losses.

Vatu-i-Ra, a small island in Fiji, is a good example of a success story in countering the alien threat. This island supports more than 10,000 breeding pairs of six species of seabird, including the Black Noddy, Brown Noddy, Red-footed Booby, Lesser Frigatebird, and White-tailed Tropicbird. These seabirds were threatened by rats, which predate the eggs and chicks of seabirds. Following a sustained program of rat eradication, the island was declared rat free. Similar projects have removed alien rats from 16 islands with major seabird populations, protecting colonies of 17 species.

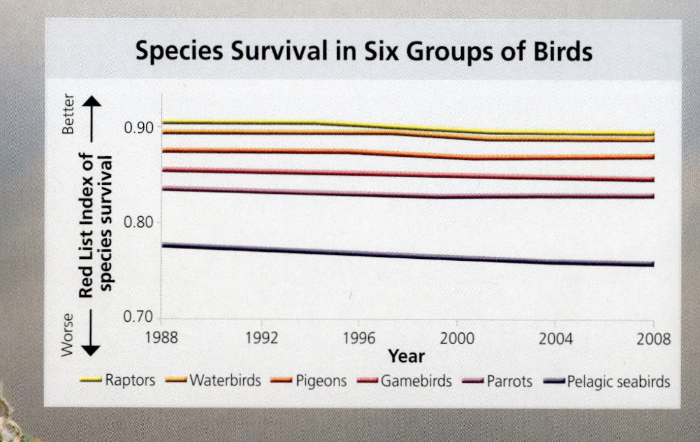

Invasive Species on Islands

In recent times the extinction rate for birds has accelerated greatly—estimates suggest this rate has risen to between 1,000 and 10,000 times what is considered to be the natural background rate. Most of these extinctions have been to species that were restricted to small islands. Since 1500, more than 150 bird species have been lost, 18 of these during the last quarter of the 20th century.

As the graph shows, island species are especially vulnerable. The major threat to these species comes from introduced alien species. As mentioned, many seabirds suffer in this way, but competition and predation from introduced species affect many other island birds, especially several that are flightless or near-flightless.

Once the key threats have been identified, it is possible to reverse the declines of island birds and there have been a number of notable successes, including Black Robin, Mauritius Kestrel, Mauritius Parakeet, Rarotonga Monarch, Seychelles Magpie-robin, Seychelles Warbler, and Kakapo.

A recent success is the Azores Bullfinch, a species now downlisted from Critically Endangered to Endangered. In this case, the main threat was loss of habitat to alien invasive plants. The Azores Bullfinch is found only on the tiny island of São Miguel in the Azores (Portugal) in native laurel forest (*laurisilva*) in a small area in the east of the island.

Clearance for forestry and agriculture reduced its population, but this was compounded by the spread of alien invasive plants, especially *Hedychium garnerianum*, *Clethra arborea*, and *Pittosporum undulatum*. These have invaded the natural woodland, reducing the bird's main food supply (fruits, seeds, and buds of the native species). Rats and mustelids (weasels) may also affect nesting success. As part of the BirdLife International Preventing Extinctions program, Species Guardian SPEA (Sociedade Portuguesa Para o Estudo Das Aves) implemented the clearance of the invasive plants, combined with planting of native species, which has reversed this bullfinch's decline.

▷ *Once in serious decline, the population of the Endangered Azores Bullfinch had reached at least 1,300 individuals by 2008.*

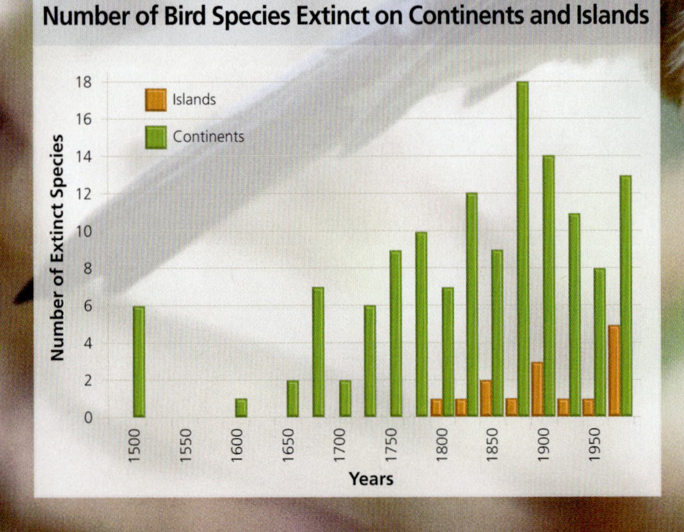

Conserving Forest Birds

The rate of bird extinctions is also increasing on the continents of the world, the main threat being habitat destruction. The richest of all terrestrial habitats are the forests, especially those in the tropics. These are home to 70 percent of the world's plants and animals—more than 13 million species—and contain 70 percent of the world's vascular plants and 90 percent of invertebrates. They also support 30 percent of all bird species. The protection of tropical forests has high priority and BirdLife International has created the Forests of Hope Program to coordinate conservation projects throughout the tropics. The aim is to prevent deforestation and restore natural forests at up to 20 sites covering over 12 million acres (5 million hectares) of tropical forest by 2015. The Forests of Hope Program links forest conservation to advocacy work at national and international levels.

Conserving natural forest is also an essential method of reducing greenhouse gas emissions and this project also makes a contribution to the maintenance and restoration of carbon stocks and helps counter climate change.

An important aspect of this initiative is to work in partnership with local people who live in and around tropical rain forests and depend upon them. Some 4.6 billion people, mostly in the tropics, depend on water from forest systems, and the forests are also a source of food, timber, and fuel. The promotion of conservation and ecologically sustainable use of natural resources is essential.

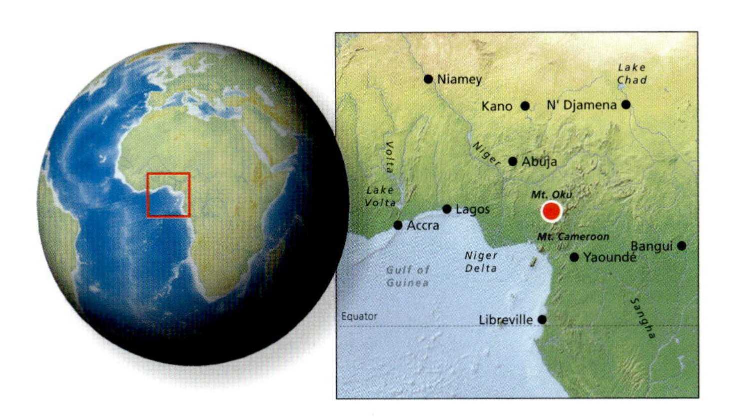

One such initiative concerns the vegetation of the Mount Oku Important Bird Area (IBA) in the Bamenda Highlands of northwestern Cameroon (right), the last stronghold of the beautiful Bannerman's Turaco and Banded Wattle-eye, both listed as Endangered. These and other species have suffered through loss to agriculture of the montane forest in which they live. A community-managed forest project started in 1987 has resulted in considerable forest regeneration. Studies of changes to the forest cover—using satellite imagery and aerial photographs—reveal that regeneration has been strong, and from 1995 it exceeded the rate of deforestation, reversing forest loss and protecting these two "flagship" species. This community-led project demonstrates that conservation can be compatible with protection of the natural resources that are vital to the livelihoods of local people.

△ *Bannerman's Turaco is a large fruit-eating bird with striking plumage. It is mainly restricted to Cameroon's Bamenda Highlands, and is listed as Endangered.*

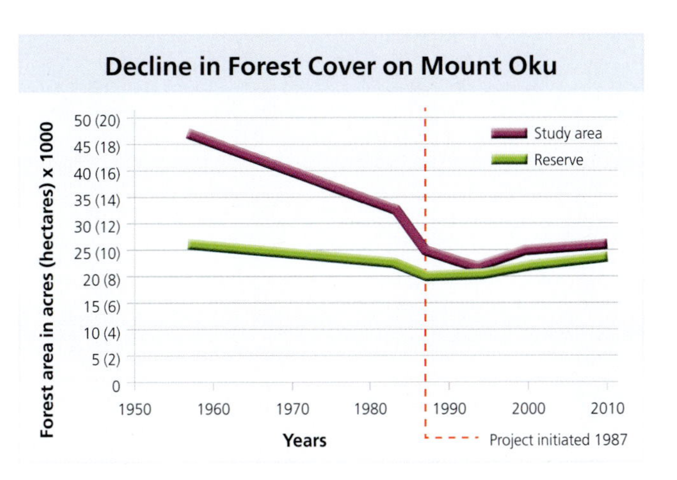

What is BirdLife?

BirdLife International is a unique global partnership of national conservation organizations that share the same mission: to conserve wild birds, their habitats, and global biodiversity, working locally to deliver long-term conservation for the benefit of nature and people. As a worldwide community, we work together across a wide range of conservation actions—cooperating with governments on global issues such as climate change, or local actions such as protecting specific species and their habitats. Together, we are the leading authority on the status of birds, their habitats, and the issues and threats facing them. With one in eight of the world's birds threatened with global extinction, many species need a helping hand to prevent them from becoming extinct.

It is easy to feel discouraged by the destruction and degradation of habitats that continues apace across the globe, but we can all do our bit to highlight such losses and make a positive contribution to conservation, and we can do this in many different ways.

Governments, non-governmental organizations, academic institutions, and the corporate (business) sector all have a role to play, but so does each individual. Everyone can help in tackling the currently unsustainable impacts of human activities on the planet.

Individually, we need to minimize the negative effects of our lifestyles on biodiversity, particularly through consumption and carbon emissions. Governments have a particular responsibility to implement policies that will lead to environmentally sustainable development. As well as advocating general actions, such as mitigating climate change and protecting the global network of IBAs, BirdLife has highlighted specific actions needed for all threatened species (see chart below).

▷ *Oiled African Penguins rescued from islands along the southern Namibian coastline in rehabilitation in Cape Town, South Africa.*

Key Actions for Threatened Species

BirdLife has identified the following priorities for safeguarding the future of the world's Critically Endangered bird species and preventing further extinctions.

Priority 1	Remove the veterinary drug diclofenac from the supply chain in the Indian subcontinent and Southeast Asia, and prevent its veterinary use in Africa, to halt the catastrophic declines of several vulture species.
Priority 2	Implement appropriate mitigation measures to reduce seabird bycatch by commercial longline fishery fleets in the world's oceans. This will benefit many albatross and petrel species—for example, the Tristan Albatross—that are declining significantly due to accidental drowning.
Priority 3	Implement adequate measures to restrict the further spread of alien invasive species, and eradicate or control these on a priority suite of oceanic islands— for example, the Brown Tree Snake in the Northern Mariana Islands; rats and cats on Niau and rats on Fatuhiva (French Polynesia); rats and cats in the Balearic Islands (Spain); cats, rats, and plants in Juan Fernández Islands (Chile); and cats, pigs, sheep, and rabbits on Socorro (Mexico).
Priority 4	Strengthen the control and management of hunting and the cagebird trade (including through national laws and CITES), for the Yellow-crested Cockatoo and Bali Starling (Indonesia), the Philippine Cockatoo and Rufous-headed Hornbill (Philippines), the Blue-billed Curassow (Colombia), the Blue-throated Macaw (Bolivia), and the Gray-breasted Parakeet (Brazil).
Priority 5	Substantially scale up efforts to tackle the interlinked threats of habitat degradation, invasive species, and climate change for the eight Critically Endangered species found only on Hawaii (U.S.A.), and for endemic species facing similar threats elsewhere, such as those on French and U.K. overseas territories.
Priority 6	Adequately safeguard and manage the remaining forests on two island groups in Africa and one in Asia, each of which supports three endemic Critically Endangered birds: São Tomé, the Comoro Islands, and Sangihe in Indonesia.
Priority 7	In the Atlantic Forest of Brazil, adequately safeguard and manage the remaining fragments, in particular those IBAs supporting Critically Endangered species, such as Chapada do Araripe (for Araripe Manakin); ESEC Murici (White-collared Kite, Alagoas Foliage-gleaner, Alagoas Antwren); Complexo Pedra Azul/Forno Grande (Cherry-throated Tanager); and Restinga de Maçambaba (Restinga Antwren).
Priority 8	Protect and appropriately manage IBAs conserving tropical forest, which is increasingly threatened by inappropriate expansion of biofuel cultivation in addition to the well-established threats of clearance for agriculture and logging, for example in Indonesia, Philippines, Colombia, Ecuador, Peru, and Mexico, each of which support high numbers of forest-dependent Critically Endangered bird species.
Priority 9	In Asia, strengthen wetland conservation efforts—including the protection of key tidal wetlands—under the Asia–Pacific Flyway Partnership for the benefit of species such as the Critically Endangered Spoon-billed Sandpiper and Chinese Crested Tern.
Priority 10	Mount appropriately targeted surveys and searches for the suite of "lost" and possibly Extinct species, such as the Hooded Seedeater (Brazil), the Himalayan Quail (India), the Slender-billed Curlew (Russia), and the Samoan Moorhen (Samoa).

Data from Birdlife's Critically Endangered Birds: A Global Audit *(www.birdlife.org/crbirds/)*

△ *A preserved egg of the Elephant Bird* (Aepyornis maximus), *a giamt flightless bird that was once native to Madagascar, but which became extinct, possibly as late as the 17th century.*

PREVENTING EXTINCTIONS PROGRAM

We have a responsibility to act now to prevent the ongoing extinction crisis, and BirdLife has launched a special program—Preventing Extinctions—to help focus such aims. This program involves identifying actions required for each species and appointing Species Champions to provide the resources and Species Guardians to implement priority actions.

BirdLife Species Champions are a global community of businesses, institutions, and individuals who provide funding needed for the vital conservation measures identified by BirdLife International to help prevent bird extinctions.

BirdLife Species Guardians are individuals or organizations that take on a responsibility to implement and/or stimulate conservation action for a particular threatened species in a defined geographical area, usually a particular country. They also monitor the status of the species and identify the key actions needed. Species Guardians' activities typically include some of the following:

- Implementing priority actions for the species.
- Developing a Species Action Plan, if one does not yet exist.
- Facilitating the implementation of priority actions by other individuals or organizations.
- Liaising and communicating with other individuals and organizations involved in carrying out research and taking action for the species.
- Advocating for appropriate conservation measures to relevant authorities and institutions.
- Monitoring the status of the species and the implementation and effect of actions by all parties.

Ways to Protect Birds and the Environment

Protection can take place both locally and internationally. This means that each and every individual and nation can take responsibility, even in a small way. This table gives some suggestions.

Locally	Shop locally, and try to purchase sustainably produced products Buy organic cotton clothes—most others use pesticides that impact birds Do not encourage the trade in wild birds by keeping pet birds Encourage birds to your garden by regular feeding of suitable food Put up nest boxes Plant bird-friendly plants, especially native species Do not use pesticides or other chemicals Support local wildlife and bird conservation organizations
Globally	Support conservation organizations such as BirdLife International and WWF Support efforts to combat human-induced global warming
Education	Teachers and parents can educate children to appreciate the value of the natural world and a healthy environment, encouraging an interest in birds and other wildlife
Ecotourism	Support and use ecologically friendly tour operators when planning a holiday, especially where support is given to local people and local conservation projects
Database	BirdLife, with the RSPB and Audubon, is creating a network of internet systems for the collection, storage, and retrieval of bird observations worldwide. This enables individuals to submit bird observations by selecting a country using the map portal at www.worldbirds.org. As more and more countries come online the data collected will increase, leading to better coverage and inputs from amateur birdwatchers as well as professionals

BirdLife Global Partners

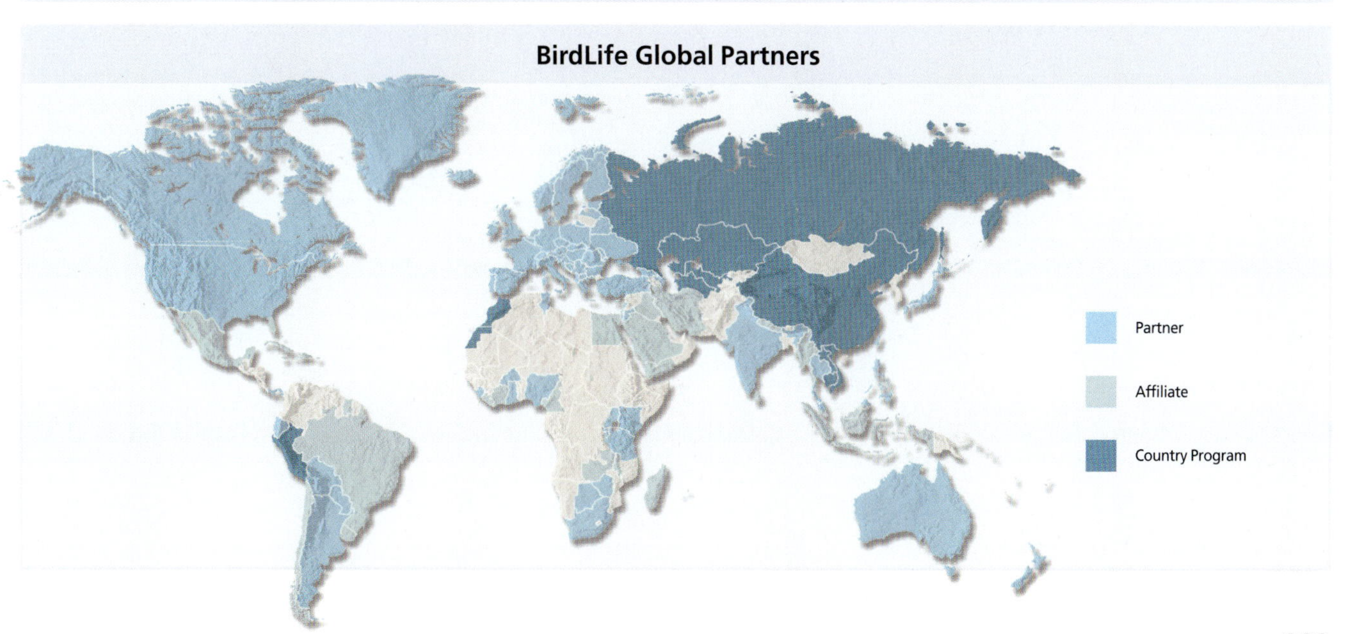

Birdwatching Hot Spots of the World

This section highlights more than 40 of the world's best birding sites, offering a selection that represents a wide range of habitats and regions. Although this is very much a subjective choice, it is aimed at including those birding sites that are known for the great diversity of species that can be found there, or for the fact that they are home to some unusual or rare birds.

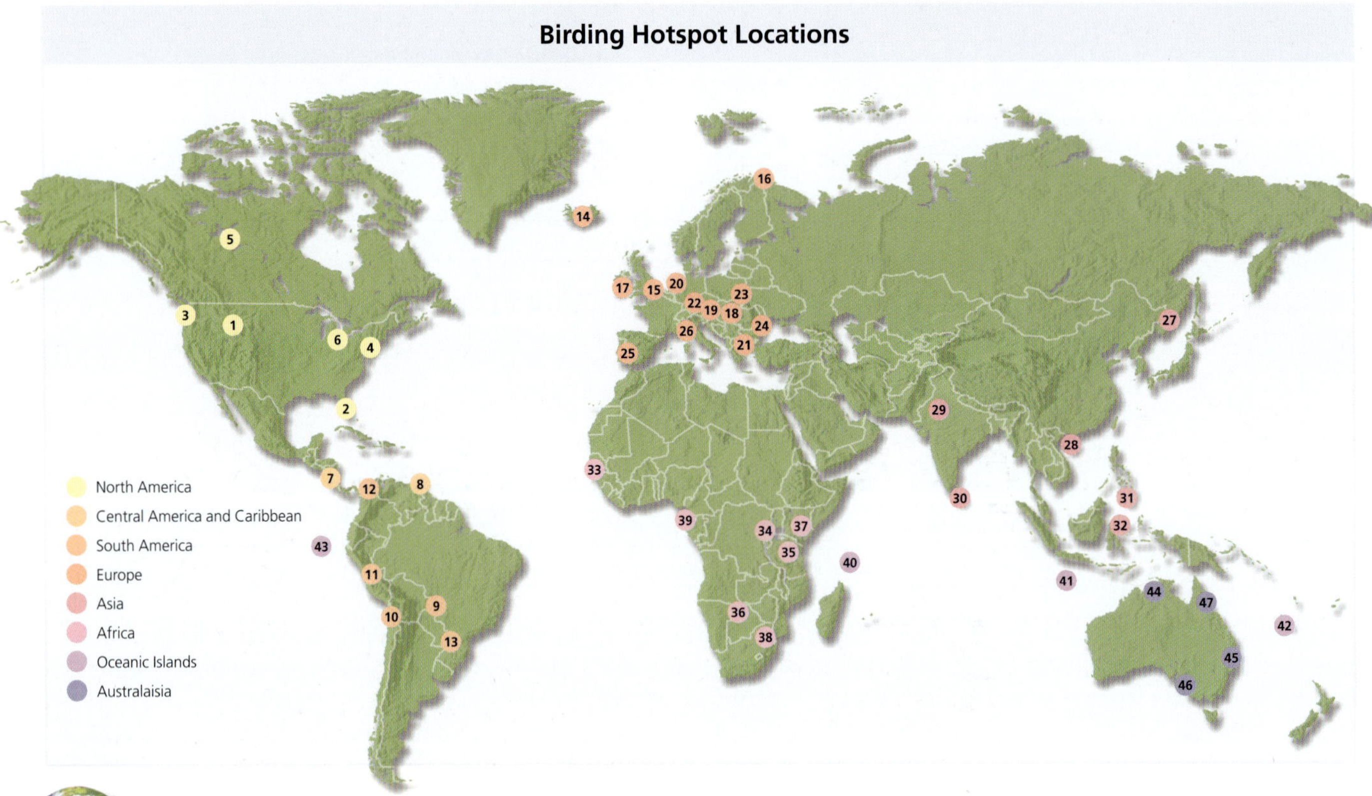

North America

❶ Yellowstone National Park (U.S.A.)
Main Habitats: Meadows, rivers, lakes, forest
Notable Birds: Bald Eagle, Great Gray Owl, American White Pelican, Trumpeter Swan, Harlequin Duck, Black Rosy-finch, American Dipper
Best Season: Summer

❷ Everglades National Park (U.S.A.)
Main Habitats: Freshwater marsh, estuary, mangroves
Notable Birds: Wood Stork, Roseate Spoonbill, Reddish Egret, Limpkin, Snail Kite, Short-tailed Hawk, Mottled Duck, Anhinga, Black Skimmer, Red-cockaded Woodpecker
Best Season: Year-round

❸ Olympic Peninsula (U.S.A.)
Main Habitats: Shore, temperate rain forest, coniferous forest
Notable Birds: Spotted Owl, Dusky Grouse, Marbled Murrelet, Tufted Puffin, Rhinoceros Auklet, Black Oystercatcher, Varied Thrush
Best Season: Spring, summer

❹ Cape May (U.S.A.)
Main Habitats: Shore, lagoons, marshes, meadows, woodland, scrub
Notable Birds: Waders: Red Knot, Short-billed Dowitcher. Migrant raptors, seabirds and passerines, Orchard Oriole, Summer Tanager
Best Season: Spring and fall migration times

❺ Wood Buffalo National Park (Canada)
Main Habitats: Meadow, boreal forest, wetland
Notable Birds: Whooping Crane, Sandhill Crane, Bald Eagle, Peregrine Falcon, Snow Goose
Best Season: Spring, summer

❻ Point Pelee (Canada)
Main Habitats: Lakeshore, woodland, marsh
Notable Birds: Migrant warblers: Blackburnian, Cerulean, Golden-winged, Hooded, Prairie, Prothonotary, Warblers, wildfowl
Best Season: Spring and fall migration times

Central America and Caribbean

❼ Monteverde Cloud Forest (Costa Rica)

Main Habitats: Montane tropical forest

Notable Birds: Resplendent Quetzal, Emerald Toucanet, Black Guan, Three-wattled Bellbird, tanagers, hummingbirds

Best Season: Year-round

❽ Arima Valley (Asa Wright Nature Center) Trinidad (Trinidad & Tobago)

Main Habitats: Tropical forest, caves

Notable Birds: Oilbird (breeding in caves), Golden-headed Manakin, White-bearded Manakin, Bearded Bellbird, Channel-billed Toucan, Bananaquit, tanagers, hummingbirds, Blue-headed Parrot, Orange-winged Amazon, Chestnut Woodpecker

Best Season: Year-round

South America

❾ Pantanal (Brazil)

Main Habitats: Freshwater wetland, forest, scrub

Notable Birds: Herons, storks, ibises including Plumbeous Ibis; raptors such as the Snail Kite, Black-collared Hawk; Sunbittern, Sungrebe; macaws (including Hyacinth and Yellow-collared macaws)

Best Season: Summer, fall

❿ Lauca National Park (Chile)

Main Habitats: High-altitude bogs, lakes, puna grassland

Notable Birds: Andean, Chilean, and Puna flamingos, Giant Coot, Silvery Grebe, Lesser Rhea, Diademed Sandpiper-plover, seedsnipes

Best Season: October–November (southern spring)

⓫ Manu (Peru)

Main Habitats: Tropical forest including *varzea* (seasonally flooded forest), river, lakes

Notable Birds: Huge diversity, notably antbirds, hummingbirds, tanagers, parrots (including macaws), Andean Cock-of-the-rock

Best Season: Year-round

⓬ Sierra Nevada de Santa Marta (Colombia)

Main Habitats: Tropical forest (wide range of types), grassland

Notable Birds: Local or endemic species including Santa Marta Parakeet, Santa Marta Warbler, Santa Marta Brush Finch, Santa Marta Bush-tyrant, Santa Marta Sabrewing, Santa Marta Woodstar, Santa Marta Tapaculo, Santa Marta Antpitta, Santa Marta Mountain-tanager, White-tipped Quetzal

Best Season: Year-round

⓭ Iguazú National Park (Argentina/Brazil)

Main Habitats: Rivers, waterfalls, tropical humid forest

Notable Birds: Solitary Tinamou, Black-fronted Piping-guan, Saffron Toucanet, Toco Toucan, Great Dusky Swift, Chestnut-headed Tanager

Best Season: Year-round

Europe

⓮ Myvatn (Iceland)

Main Habitats: Lakes, rivers, volcanic landscape, lava cliffs, bogs

Notable Birds: Harlequin Duck, Barrow's Goldeneye, Long-tailed Duck, Horned Grebe, Red-necked Phalarope, Gyr Falcon, Merlin, Rock Ptarmigan

Best Season: Spring, summer

⓯ North Norfolk Coast (UK)

Main Habitats: Coastal dunes, shingle and mudflats, saltmarsh, marsh, woodland, scrub

Notable Birds: Wintering geese (notably Pink-footed, Brent), waders (e.g. Red Knot, godwits, Pied Avocet), Western Marsh-harrier, Great Bittern, Eurasian Spoonbill, Bearded Parrotbill, Cetti's Warbler, also migrants in spring and fall

Best Season: Year-round

⓰ Varanger (Norway)

Main Habitats: Tundra, cliffs, sea

Notable Birds: Steller's Eider, King Eider, Yellow-billed Loon, Thick-billed Guillemot, Eurasian Dotterel, Gyr Falcon, Leach's Storm-petrel

Best Season: Spring, summer

⓱ Wexford Slobs (Republic of Ireland)

Main Habitats: Mudflats, wet meadows

Notable Birds: Barnacle Goose, Bean Goose, Brent Goose, Greater White-fronted Goose, Snow Goose, Whooper Swan, Tundra (Bewick's) Swan

Best Season: Fall, winter

⓲ Hortobagy (Hungary)

Main Habitats: Grassland, river, lakes, some open woodland

Notable Birds: Great Bustard, Saker and Red-footed falcons, Eastern Imperial Eagle, herons, egrets, Common Crane, European Roller, Lesser Gray Shrike, warblers (including Savi's Warbler and Great Reed-warbler)

Best Season: Year-round

⓳ Neusiedlersee (Austria)

Main Habitats: Lakes, reedbeds

Notable Birds: Red-crested Pochard, Ferruginous Duck, Eurasian Spoonbill, Great Bittern, Little Bittern, Purple Heron, Great Egret, Moustached Warbler, Savi's Warbler

Best Season: Spring

⓴ Waddensee (Netherlands, Germany, Denmark)

Main Habitats: Coastal mudflats, islands, saltmarsh, dunes, meadows, scrub

Notable Birds: Waders (notably godwits, plovers, stints, Red Knot, Pied Avocet), wildfowl (notably Pink-footed, Barnacle, and Brent geese, Common Shelduck, Northern Pintail, Common Eider, scoters), raptors (notably harriers, buzzards)

Best Season: Year-round

㉑ Dadia-Lefkimi-Soufli Forest (Greece)

Main Habitats: Forest, meadows, rivers, cultivated land

Notable Birds: Egyptian, Griffon, and Cinereous vultures, Short-toed Snake-eagle, Golden, Greater Spotted, Lesser Spotted, and Booted eagles, Levant Sparrowhawk, Eurasian Eagle-owl, Masked Shrike

Best Season: Spring through fall

Europe *continued*

㉒ Berchtesgaden National Park (Germany)

Main Habitats: Alpine foothills and peaks: mixed and subalpine spruce forest, alpine meadows, rocky habitats, scree

Notable Birds: Griffon Vulture, Golden Eagle, European Honey-Buzzard, Peregrine Falcon, Rock Ptarmigan, Western Capercaillie, Black Grouse, Hazel Grouse, Boreal Owl, Eurasian Pygmy-Owl, Black, Eurasian Green, Gray-faced, White-backed- and Eurasian Three-toed woodpeckers, Eurasian Crag Martin, Water Pipit, Alpine Accentor, Ring Ouzel, Bonelli's Warbler, Red-breasted Flycatcher, White-throated Dipper, Wallcreeper (rare), Spotted Nutcracker, Common Raven, Yellow-billed Chough, White-winged Snow Finch

Best Season: Late spring, summer

㉓ Białowieza Forest (Poland)

Main Habitats: Forest, clearings, bog, marsh, meadows

Notable Birds: Woodpeckers (including Black, Gray-faced, Middle Spotted, White-backed and Eurasian Three-toed), flycatchers (including European Pied, Spotted, Collared and Red-breasted), Hawfinch, Eurasian Golden Oriole, Black Stork, raptors (including European Honey-buzzard, Booted and Lesser Spotted eagles, Short-toed Snake-eagle,

Best Season: Year-round

㉔ Danube Delta (Romania)

Main Habitats: River delta, ponds, marshes, reedbeds, coast

Notable Birds: Dalmatian and Great White pelicans, herons, Glossy Ibis, Great and Pygmy cormorants, crakes, Red-breasted Goose, warblers, Eurasian Penduline-tit, raptors (including White-tailed, Eastern Imperial and Greater Spotted eagles, Lesser Kestrel, Saker and Red-footed falcons)

Best Season: Year-round

㉕ Monfragüe National Park (Spain)

Main Habitats: River, cliffs, forest, scrub

Notable Birds: Raptors (Cinereous, Griffon and Egyptian Vultures, Spanish Imperial, Golden, Bonelli's and Booted eagles, Short-toed Snake-eagle, Red and Black kites), White and Black storks, Alpine Swift, Red-billed Chough, Blue Rock-thrush

Best Season: Year-round

㉖ Camargue (France)

Main Habitats: Coastal lagoons, reedbeds

Notable Birds: Greater Flamingo, Black-crowned Night-heron, Squacco Heron, Purple Heron, Black-winged Stilt, Collared Pratincole, Slender-billed Gull, Mediterranean Gull, Gull-billed Tern, Caspian Tern, European Roller

Best Season: Spring and summer

Asia

㉗ Ussuriland (Russia)

Main Habitats: Forest, lakes, rivers, coastal mudflats, islands

Notable Birds: Red-crowned, White-naped and Hooded cranes, Oriental Stork, Blakiston's Fish-owl, Scaly-sided Merganser, Mandarin Duck, Azure Tit

Best Season: Spring, summer

㉘ Ba Wang Ling Nature Reserve (Hainan, China)

Main Habitats: Tropical forest, scrub

Notable Birds: Hainan Partridge, Hainan Peacock-pheasant, Silver Pheasant, Black Eagle, Blue-bearded Bee-eater, Silver-breasted Broadbill, Hainan Leaf-warbler, Rufous-cheeked Laughingthrush, Pale Blue-flycatcher, Yellow-billed Nuthatch, White-winged Magpie

Best Season: Year-round

㉙ Bharatpur (India)

Main Habitats: Woodland, grassland, scrub, marshes

Notable Birds: Herons, egrets, storks (including Painted and Black-necked storks), ducks (including Spot-billed Duck and Lesser Whistling-duck), Bar-headed Goose, raptors (including Greater and Indian Spotted Eagles), Dusky Eagle-owl, migrant passerines

Best Season: Winter

㉚ Sinharaja Forest (Sri Lanka)

Main Habitats: Lowland rain forest

Notable Birds: Sri Lanka Magpie, Sri Lanka Spurfowl, Sri Lanka Wood-pigeon, Sri Lanka Whistling-thrush, White-faced Starling, Sri Lanka Junglefowl, Red-faced Malkoha, Serendib Scops-owl

Best Season: Year-round

㉛ Mount Apo Natural Park (Mindanao, Philippines)

Main Habitats: Montane forest

Notable Birds: Philippine Eagle, Giant Scops-owl, Celestial Monarch, Philippine Cockatoo, Philippine Eagle-owl, Blue-naped Parrot, Philippine Hawk-eagle, Philippine Dwarf-kingfisher, Rufous-lored Kingfisher, Silvery Kingfisher, Spotted Imperial-pigeon, Black-bibbed Cicadabird

Best Season: Year-round

㉜ Bogani Nani Wartabone National Park (Sulawesi, Indonesia)

Main Habitats: Lowland rain forest, rivers, ponds

Notable Birds: Maleo, Sulawesi Owl, Sulawesi and Knobbed hornbills, Purple-bearded Bee-eater, Purple-winged Roller, kingfishers (including Black-billed, Lilac-cheeked and Sulawesi kingfishers), Ivory-backed Woodswallow

Best Season: Year-round

Africa

㉝ Gambia River (The Gambia)

Main Habitats: River, forest, scrub, marsh, mangroves

Notable Birds: African Finfoot, Red-throated and Northern Carmine bee-eaters, Egyptian Plover, Caspian and Lesser Crested terns, Goliath Heron, Blue-breasted Kingfisher, Spur-winged Goose, Yellow-billed and Woolly-necked storks, Pel's Fishing-owl

Best Season: Year-round

㉞ Bwindi Impenetrable Forest (Uganda)

Main Habitats: Tropical forest, scrub, swamp

Notable Birds: African Green Broadbill, Grauer's Swamp-warbler, Neumann's Warbler, Black Bee-eater, White-headed Woodhoopoe, Purple-breasted and Regal sunbirds, Red-throated Alethe, Equatorial Akalat, White-starred Robin

Best Season: Year-round

Africa *continued*

35 Serengeti (Tanzania)

Main Habitats: Grassland, savanna, woodland, lakes, hills

Notable Birds: Ostrich, Secretarybird, bustards, oxpeckers, Rufous-tailed Weaver, Gray-breasted Spurfowl, Fischer's Lovebird, Karamoja Apalis, Gray-crested Helmet-Shrike, Black-winged Pratincole, Southern Ground-hornbill, raptors (notably Hooded, Lappet-faced, Rüppell's and White-backed vultures, Martial Eagle)

Best Season: Year-round

36 Okavango Delta (Botswana)

Main Habitats: Rivers, swamp, pools, papyrus and reed beds, scrub, semi-desert

Notable Birds: Herons (18 species, including Slaty Egret), Wattled Crane, storks, crowned-cranes, African Sacred Ibis, raptors (including African Fish-eagle, snake-eagles, Martial and Wahlberg's Eagles), bee-eaters, cuckoos, shrikes, kingfishers

Best Season: Year-round

37 Rift Valley Lakes (Kenya)

Main Habitats: Freshwater and soda lakes

Notable Birds: Lesser and Greater flamingos, African Spoonbill, Goliath Heron, Yellow-billed Stork, Great White Pelican, African Fish-eagle, hornbills, waders, ducks (including Cape Teal and White-backed Duck)

Best Season: Year-round

38 Kruger National Park (South Africa)

Main Habitats: Savanna, scrub, forest, grassland, rivers

Notable Birds: Eagles (including Martial, Tawny, Wahlberg's and Lesser Spotted), Bateleur, Black-chested and Brown snake-eagles, African Fish-eagle, vultures (including Cape, White-backed, White-headed and Lappet-faced), Ostrich, Secretarybird, Kori Bustard, Goliath and Black-headed Herons, storks (including White, Black, Abdim's, Saddle-billed and Woolly-necked), nightjars (including Pennant-winged), kingfishers, bee-eaters, rollers, hornbills

Best Season: Year-round

Oceanic Islands

39 São Tomé (Democratic Republic of São Tomé and Príncipe) (Atlantic Ocean)

Main Habitats: Rain forest, shore, cultivation

Notable Birds: São Tomé Scops-owl, São Tomé Fiscal, São Tomé Oriole, São Tomé Paradise-flycatcher, São Tomé Short-tail, São Tomé White-eye, São Tomé Grosbeak

Best Season: June though September (dry season)

40 Granitic Seychelles (Seychelles) (Indian Ocean)

Main Habitats: Woodland, scrub, mangroves, cultivation

Notable Birds: Seychelles Warbler, Seychelles Magpie-robin, Seychelles Paradise-flycatcher, Seychelles White-eye, Seychelles Swiftlet, Seychelles Kestrel, Black Parrot, Common White Tern

Best Season: Year-round

41 Christmas Island National Park (Christmas Island) (Indian Ocean)

Main Habitats: Rain forest, mangrove forest, beaches, cliffs, caves

Notable Birds: Abbott's, Brown, and Red-footed boobies, Christmas Island and Greater frigatebirds, Red-tailed and White-tailed tropicbirds, Christmas Island Hawk-owl

Best Season: Year-round

42 Rivière Bleue (New Caledonia, France) (Pacific Ocean)

Main Habitats: Rain forest

Notable Birds: Kagu, New Caledonian Imperial-pigeon, Crow Honeyeater, New Caledonian Lorikeet, Horned and New Caledonian Parakeet, New Caledonian Owlet-nightjar, New Caledonian Rail, New Caledonian Friarbird, New Caledonian Cuckooshrike, New Caledonian Crow, New Caledonian Grassbird, New Caledonian Myzomela, New Caledonian Whistler

Best Season: Year-round

43 Galápagos Islands (Ecuador) (Pacific Ocean)

Main Habitats: Islands, cliffs, scrub

Notable Birds: "Darwin's" finches—e.g. ground finches, tree finches, warbler finches—Woodpecker Finch, mockingbirds (including Floreana Mockingbird), Galápagos Penguin, Flightless Cormorant, Waved Albatross, Lava and Swallow-tailed Gulls, Galápagos Hawk, Galápagos Martin, Galápagos Rail, Galápagos Petrel

Best Season: Year-round

Australasia

44 Kakadu (Australia)

Main Habitats: Lakes, rivers, coastal wetlands, ponds, savanna, woodland

Notable Birds: White-bellied Sea-eagle, Whistling Kite, egrets, Black-necked Stork, Brolga, Royal Spoonbill, White-necked Heron, Rufous Night-heron, Comb-crested Jacana, Chestnut Rail, Magpie Goose, Radjah Shelduck, Australian Pelican, Green Pygmy-goose, Plumed Whistling-duck, Red Goshawk, Gouldian Finch, Azure and Little kingfishers

Best Season: Year-round

45 Lamington (Australia)

Main Habitats: Temperate and subtropical rain forest, scrub

Notable Birds: Albert's Lyrebird, Noisy Pitta, Paradise Riflebird, Regent and Satin bowerbirds, Rufous Scrub-bird, Wonga Pigeon, Australian King-parrot, Crimson Rosella, Yellow-throated Scrubwren, Australian Logrunner

Best Season: Year-round

46 Coorong National Park (Australia)

Main Habitats: River, lake, coastal lagoons, sand dunes, sea

Notable Birds: Waders (including Masked Lapwing, Hooded Plover, Far Eastern Curlew, Sooty Oystercatcher), wildfowl (including Black Swan, Cape Barren Goose, Gray Teal, Pacific Black Duck, Australian Shelduck, Pink-eared Duck, Musk Duck), Australian Sacred, Glossy, and Straw-necked ibises, Royal and Yellow-billed spoonbills

Best Season: Year-round

47 Daintree Valley, Queensland Wet Tropics (Australia)

Main Habitats: Rain forest, wetland, mangroves

Notable Birds: Southern Cassowary, Australian Brush-turkey, Orange-footed Megapode, Chowchilla, Fernwren, Golden and Tooth-billed bowerbirds, Victoria's Riflebird, Atherton and Large-billed scrubwrens, Noisy Pitta

Best Season: Year-round

GLOSSARY

Words in **bold** refer to terms defined elsewhere in the glossary.

Afrotropical The ecological zone (realm/ecozone) that includes Africa south of the Sahara, the south and east of the Arabian Peninsula, southern Iran and southwestern Pakistan, Madagascar, and the islands of the western Indian Ocean.

alien An organism not native to a particular place, usually introduced by human action, either deliberately or by accident. Alien animals and plants are sometimes a threat to local birds, especially on oceanic islands.

altiplano The high altitude plateau of the central Andes Mountains of South America in parts of Peru, Bolivia, Chile, and Argentina.

Australasia A region that includes Australia, New Zealand, the island of New Guinea, and neighboring Pacific islands.

billabong A type of small lake found in Australia that is normally an oxbow lake formed in a river valley.

biodiversity The range of life forms found in a particular **habitat** or region. High biodiversity is typical of relatively undisturbed habitats and is highest in tropical forests.

biofuel Fuel produced directly from biological sources, usually from plants grown especially for the purpose such as sugar cane and oil palm.

biome A region, defined mainly with reference to the dominant vegetation, as well as climatic and geographic criteria, that has a characteristic assemblage of organisms. Tundra, subtropical moist broadleaf forests, and deserts are examples of biomes.

bird of prey Birds that catch kill and feed on (mainly) other vertebrates. They typically have keen eyesight, and sharp claws and bills. Normally refers to birds in the order Falconiformes (eagles, hawks, falcons, and relatives), but often used to include also the order Strigiformes (owls). Birds of prey are also known as raptors.

Boreal Northern, often used to describe the mainly coniferous forests of the north.

brackish Water with salinity intermediate between fresh water and sea water, as found in estuaries, saltmarshes, and other places where fresh and salt water intermingle.

breeding plumage The plumage of birds in the breeding season, often more elaborate and colorful than the plumage in the rest of the year. Birds usually molt into their breeding plumage in spring and summer, although many ducks attain their breeding plumage in the winter.

brood parasites Birds that lay their eggs in the nest of another **species**, leaving the latter to hatch and rear their young. This breeding strategy has evolved in a number of bird groups, notably cuckoos and cowbirds.

bushveldt Part of the southern African **savanna** that is composed of a mixture of **habitats**—grassland, open woodland, and **scrub**.

bycatch The accidental capture of other organisms by fishing fleets. Bycatch is a serious threat to seabirds, especially to albatrosses and petrels that often drown in fishing nets or get caught on the baited hooks used in longline fishing.

cage bird Any **species** of bird that is commonly kept in captivity. Many tropical birds, notably parrots and some songbirds, are illegally trapped to supply the cage-bird trade, and this threatens many species.

campo cerrado Often known simply as *cerrado*, this is a type of South American **savanna**, found mainly in Brazil and Paraguay, grading with intermediate **habitats** into open grassland.

campo limpo A type of South American grassland, found mainly in Brazil and Paraguay, grading with intermediate **habitats** into open **savanna**.

canopy The upper level of woodland or forest, formed from the branches and foliage of the tallest stratum of trees.

casque A natural growth on the upper mandible of a bird, notably found on hornbills. Also used for the natural growth on the skull of cassowaries.

cere A fleshy swelling at the base of the bill in certain birds, notably parrots, pigeons, and hawks.

chaparral Scrub vegetation developed under a Mediterranean climate in southern California, dominated mainly by broadleaved evergreen **species**.

climate change Alteration of the global climate. Climate change has occurred many times in the Earth's history, through natural causes, notably variations in the Earth's orbit around the Sun. However, recent change, involving **global warming**, is regarded by most scientists to be caused by human influence, including the release of large quantities of gases such as carbon dioxide that increase the **greenhouse effect**.

cloud forest Almost permanently damp forest developed under conditions of high humidity, and often enveloped in mist or clouds. This rather special **habitat** is found in certain highland regions in the tropics. It is particularly well developed in Central and South America and has a very high bird diversity.

coverts Short feathers in a bird's plumage that help to cover the bases of larger feathers and also act to smooth airflow over the body.

Critically Endangered (CR) A category of threat describing a **species** that is considered to be facing an extremely high risk of extinction in the wild (see also **IUCN**; **Red List**).

crown The top of a bird's head.

Data Deficient (DD) A category describing a **species** about which there is inadequate information to make an assessment of its category of threat (see also **IUCN**; **Red List**).

decurved Describes the bill of a bird if it curves downward toward the tip (see also **recurved**).

deforestation Removal of forest by various direct or indirect human actions, notably felling and fire. Deforestation is a major cause of threat to birds and other wildlife, especially in the tropics, and also historically over much of the temperate regions.

dehesa An open, sustainable managed forest **habitat** found mainly in Spain and Portugal that has scattered trees of evergreen oaks. The ground beneath the trees is grazed and sometimes cultivated. Known as *montado* in Portugal, they are rich in wildlife.

ecosystem Rather an imprecise term used to describe a living assemblage of animals and plants, together with the non-living elements including the local soil and climate.

ecotourism Tourism that highlights the wildlife of a particular region. Arranged properly, it can channel much-needed funds to benefit the wildlife and thus be a positive force for conservation.

Endangered (EN) A category of threat describing a **species** that is considered to be facing a very high risk of extinction in the wild (see also **IUCN**; **Red List**).

endemic An organism (animal or plant) that is found only within a particular restricted range. For example, it may be endemic to a country or a particular island; many of the world's rare bird **species** are endemic.

Endemic Bird Area (EBA) A "hot-spot" of bird **species** diversity, also called a "center of bird endemism." An EBA is defined on the basis of the areas of overlap between the ranges of the local **endemic** birds.

eutrophication A process causing over-enrichment of chemical nutrients in a body of water. The causes include pollution, such as run-off from fertilizers and sewage. This process may result in overgrowth of algae (an algal bloom), major changes to the organisms inhabiting the water, and major loss of **biodiversity**. In extreme cases, it can cause severe changes in water quality, notably oxygen reduction, leading to the death of some aquatic organisms including fish.

Extinct (EX) A category describing a **species** for which there is no reasonable doubt that the last individual has died (see also **IUCN**; **Red List**).

Extinct in the wild (EW) A category describing a **species** known only to survive in captivity (see also **IUCN**; **Red List**).

family A category in the classification hierarchy, containing one or more genera (see also **genus**; **species**).

fynbos A **scrub habitat** of South Africa's southern Cape region dominated by low shrubs with hard, evergreen foliage, including **species** of heather (*Erica*), proteas (*Protea*), and the curious rushlike restios (*Restio*). *Fynbos* is Afrikaans for "fine bush" and refers to the needlelike leaves of many of the shrubs. Nearly 70 percent of *fynbos* plant species are **endemic**.

garrigue A Mediterranean **scrub** community made up of low-growing plants, many of which are aromatic, often the result of further degradation of ***maquis***.

genus (pl. genera) A category in the classification hierarchy, containing one or more **species** (see also **family**).

global warming An increase in the average temperature of the Earth, which is usually based on increases since the mid-20th century. It is caused by human activities, especially the increasing concentrations of **greenhouse gases**. Such changes are already having effects on the **ecosystems** of the world and the **species** they support.

greenhouse effect The natural process by which radiation is trapped in the atmosphere and which keeps the surface of the Earth at a temperature suitable for life. In recent decades, this effect has been strengthened by human-induced production of **greenhouse gases** (see also **climate change**).

greenhouse gas A gas responsible for the **greenhouse effect**. The major gases are carbon dioxide, methane, nitrous oxide, and ozone, as well as water vapor. In recent decades, the concentration of carbon dioxide in particular has increased considerably, in turn increasing the greenhouse effect and leading to **global warming**.

habitat The typical natural environment for a particular bird, or other animal or plant.

heath A **scrub habitat** usually dominated by members of the heather family, typical of sandy, well-drained sites in northwestern Europe, where it generally replaces woodland and is maintained by a combination of grazing and fire.

Important Bird Area (IBA) An area identified by BirdLife International as a key site for bird conservation. To qualify, a site must do one or more of the following: hold significant numbers of one or more globally **threatened species**; be one of a set of sites that together hold a suite of restricted-range species; have exceptionally large numbers of migratory or congregatory species. Almost 11,000 sites in some 200 countries have now been recognized as IBAs.

Indomalayan The ecological zone (realm/ecozone) that stretches from Afghanistan in the west, through southern China and south to India, Sri Lanka, Indonesia, Philippines, and Borneo. This region holds nearly 60 percent of the world's population and has suffered substantial forest loss, affecting many bird **species**.

invasive A **species** that aggressively colonizes a **habitat**, typically an introduced **alien**. Invasive species may impact through direct predation or competition.

IUCN The World Conservation Union (formerly known as the International Union for the Conservation of Nature), an organization that gathers information about the status of **threatened species** (see also **Red List**).

kwongan Local term for a type of floristically rich scrub found on the coastal plains of Western Australia, developed under a Mediterranean type of climate.

Least Concern (LC) A category describing a **species** known to be widespread and abundant (see also **IUCN**; **Red List**).

mandible The beak of a bird, divided into the upper mandible and lower mandible.

mantle The region between the neck and back of a bird.

maquis A French term widely used for Mediterranean **scrub** vegetation dominated by evergreen shrubs and maintained mainly through grazing and fire, usually replacing original evergreen oak woodland. It is known as *macchia* in Italy.

matorral Spanish term for Mediterranean **scrub habitat**, also used for similar vegetation in Mexico and Chile.

Melanesia Part of **Oceania** in the western Pacific Ocean, stretching from the New Guinea region south and east to New Caledonia and Fiji.

Micronesia Part of **Oceania** in the western Pacific Ocean to the north of **Melanesia**, containing scattered archipelagos of small islands, stretching east to the Marshall Islands and Kiribati.

migration Regular, usually annual movement, of a **species** from one region to another, following a seasonal rhythm. In regions with cold winters, many birds migrate to spend the winter in a region with a warmer climate where food is still abundant.

miombo A type of **savanna** woodland found over much of south–central Africa, from Angola in the west across to Tanzania and Mozambique in the east. It is dominated mainly by trees of the genus *Brachystegia*, for which the Swahili word is *miombo*.

montado See ***dehesa***

nape The back of the neck of a bird.

Near Threatened (NT) A category describing a **species** that is close to qualifying for, or is likely to qualify for, a **threatened** category in the near future (see also **IUCN**; **Red List**).

Nearctic The ecological zone (realm/ecozone) that covers North America (except southern Florida), and the adjacent highlands of Mexico and Greenland.

Neotropical The ecological zone that covers South and Central America, the lowlands of Mexico, the Caribbean islands, and the southern tip of Florida.

Oceania A loose term often taken to mean the region including the islands of the Pacific Ocean, but sometimes also used to include **Australasia** and New Zealand (see also **Melanesia**; **Micronesia**; **Polynesia**).

Palearctic The ecological zone (realm/ecozone) that includes Europe, and north and central Asia. In the north, it stretches from Iceland to Japan, and in the south from North Africa across the Middle East, and along the southern foothills of the Himalayas to central China.

pampas A type of temperate grassland found mainly in Argentina, but also in parts of Uruguay and southern Brazil. Dry pampas is found in central Argentina, while the more fertile humid pampas dominates farther east.

passerine A member of the largest order of birds, the *Passeriformes*, also known as perching birds, as their feet are especially well adapted to gripping twigs and the like. The passerines contain the **songbirds**.

pelagic Found in the open oceans. Pelagic birds, notably albatrosses, shearwaters, and petrels, spend much of their time at sea, coming ashore only to breed.

phrygana Greek term (equivalent to ***garrigue***) for a Mediterranean scrub vegetation, usually found on rocky soils, and dominated by low, often cushionlike spiny shrubs. It is maintained mainly through grazing and fire.

phytoplankton These microscopic, photosynthetic plants and plantlike organisms float in water, and include algae, bacteria, and diatoms. They form the basis of oceanic food chains (see also **zooplankton**).

Polynesia Part of **Oceania** in the central and southern Pacific Ocean to the east of **Melanesia**, containing scattered archipelagos of small islands, stretching from Tonga and Samoa in the west, to Hawaii in the north, and French Polynesia and the Pitcairn Islands in the east.

Ramsar List A list of wetlands of international importance, compiled under the "Convention on Wetlands," signed in Ramsar, Iran in 1971, with the aim of maintaining their ecological integrity. So far, 1,896 sites have been designated in 160 countries.

raptor See **bird of prey**

recurved Of a bird's bill that curves upward toward the tip (see also **decurved**).

Red List Abbreviated form of the "IUCN Red List of Threatened Species," which documents those species of animals and plants most in need of conservation. The **Red List** categories are: **Least Concern (LC)**, **Near Threatened (NT)**, **Vulnerable (VU)**, **Endangered (EN)**, **Critically Endangered (CR)**, **Extinct in the wild (EW)**, and **Extinct (EX)** (see also **IUCN**; **threatened species**).

rump The area between the back of a bird and the base of its tail.

Sahel A transition belt of varying width immediately to the south of the Sahara Desert. It is mainly a mosaic of semi-desert, grassland, and open **savanna**.

savanna A habitat of open grassland with scattered trees in tropical areas, typically between tropical forest and hot deserts.

scrub A general term for vegetation dominated by low woody plants (shrubs) and stunted trees, which typically forms an intermediate community between grassland and forest; the reduced growth may be caused by lack of water and extreme hot or cold temperatures.

songbird A suborder (*Passeri* or *Oscines*) of **passerine** birds known for their often elaborate and sometimes musical songs.

species A category in the classification hierarchy that groups similar organisms that are capable of interbreeding in the wild and producing fertile offspring. Each species is given an internationally recognized double name known as a binomial, the latter conventionally written in italics. The binomial is Latin in form and consists of the generic (**genus**) name, followed by the specific epithet. For example, the Barn Owl, one of 13 species in the genus *Tyto*, has the name *Tyto alba*.

steppe A type of temperate grassland stretching in a band of highly variable width from the plains of eastern Europe right across Asia to Mongolia in the east.

threatened species A **species** that faces a risk of extinction in the wild—those listed as **vulnerable**, **endangered**, or **critically endangered**.

tomillares Spanish term (equivalent to ***garrigue***) for a Mediterranean scrub community made up of low-growing plants, many of which are aromatic, often the result of further degradation of ***maquis***.

tundra Mainly treeless vegetation found in polar regions, especially around the Arctic, dominated by dwarf shrubs, sedges, grasses, mosses, and lichens. Tundralike vegetation is also found at high altitudes on many mountain ranges.

vertical migration The movement of **species** upward or downward on mountain ranges, normally in response to changes in the availability of food.

Vulnerable (V) A category of threat describing a **species** that is considered to be facing a high risk of extinction in the wild (see also **IUCN**; **Red List**).

wader A member of the order Charadriiformes, including plovers and sandpipers. Also known as shorebirds.

World Bird Database (WBDB) A body of data through which the BirdLife Partnership manages, analyzes, and reports on the state of the world's birds. In addition to detailed information about **species**, and notably the **threatened species**, this database also includes two important categories of defined regions—**Endemic Bird Areas (EBAs)** and **Important Bird Areas (IBAs)**.

World Heritage List A listing by UNESCO of important sites, either for their cultural or natural value. In 2010, the list included 911 sites, of which 180 are natural.

zooplankton Tiny animals that live floating in water. Marine zooplankton are either permanent members, such as amphipods, copepods, and krill, or the larvae of other marine animals such as crabs and starfish. They are a fundamental component of marine food chains (see also **phytoplankton**).

INDEX

Page numbers in **bold** type indicate references to illustrated profiles; those in *italics* refer to illustrations. Page numbers underlined relate to tables.

ACKNOWLEDGMENTS

Marshall Editions would like to thank the following for their kind permission to reproduce their images.

Key: t = **top** b = **bottom** c = **center** r = **right** l = **left**

Cover credits: Front cover design by Linda Cole
Jacket photos: Shutterstock

Pages: 1 Corbis/Visuals Unlimited/John Cornell; **2-3** Corbis/Epa/Nic Botha; **4-5** Shutterstock/Vishnevskiy Vasily; **6** Corbis/Frans Lanting; **7** BirdLife International; **8-9** Corbis/Arthur Morris; **10** Corbis/All Canada Photos/Rolf Hicker; **11t** FLPA/Minden Pictures/Michael & Patricia Fogden; **11b** FLPA/Shem Compion; **13t** FLPA/Minden Pictures/Claus Meyer; **13b** Shutterstock/Sue Robinson; **14-15b** Corbis/Momatiuk-Eastcott; **14tr** FLPA/Minden Pictures/Konrad Wothe; **15t** Corbis/Science Faction/Rainbow/Dan McCoy; **16** Corbis/Philip Gould; **17t** FLPA/Frans Lanting; **17b** FLPA/Malcolm Schuyl; **18t** FLPA/Steve Trewhella; **18b** Corbis/Bob Sacha; **19** Corbis/Paul Souders; **20** FLPA/Neil Bowman; **21** Shutterstock/Francois Loubser; **22-23b** Corbis/Reuters/John Kolesidis; **23tr** FLPA/John Eveson; **26-27** FLPA/Minden Pictures/Tui De Roy; **29tr** Nature/Karl Ammann; **29b** Nature/Aflo; **30** Ardea/Andrey Zvoznikov; **31t** Nature/Morley Road; **31b** Corbis/Aurora Photos/John Lee; **32** FLPA/Frans Lanting; **33** Corbis/Paulo Whitaker; **34t** Ardea/John S. Dunning; **34b** Nature/Luiz Claudio Marigo; **35t** Shutterstock/Neale Cousland; **36** Getty Images/National Geographic/Klaus Nigge; **37** Photolibrary/age fotostock/Patricio Robles Gil; **38** Photoshot/NHPA/Jany Sauvanet; **39** FLPA/Neil Bowman; **40** FLPA/Minden Pictures/Konrad Wothe; **41t** Corbis/Frans Lanting; **41b** FLPA/Minden Pictures/Mark Moffett; **42-43** FLPA/Minden Pictures/Carr Clifton; **45** Corbis/Robert Y. Kaufman; **46t** FLPA/S & D & K Maslowski; **46b** Ardea/Wardene Weisser; **47** Corbis/David Muench; **48** Photolibrary/John Warburton-Lee Photography/Paul Harris; **49** Franz Immoos, Amsterdam; **50** FLPA/Minden Pictures/Kevin Schafer; **51** Nature/Roger Powell; **52-53** Shutterstock/AP Design; **54** Shutterstock/Yoann Combronde; **55** Science Photo Library/Bernhard Edmaier; **56** FLPA/Minden Pictures/Winfried Wisniewski; **57** FLPA/Imagebroker; **58** FLPA/Richard Brooks; **59** Photolibrary/Imagebroker/Fabian von Poser; **60-61** Photolibrary/Image Source; **63** Corbis/Hans Reinhard; **64** FLPA/Minden Pictures/Patricio Robles Gil; **65** FLPA/Jurgen & Christine Sohns; **66t** Photolibrary/Russian Look/Konstantin Mikhailov; **66-67** Corbis/Latitude/Galen Rowell; **68-69** FLPA/Minden Pictures/Yva Momatiuk & John Eastcott; **69br** Corbis; **70** Photolibrary/Flirt Collection/Joe McDonald; **71** Science Photo Library/Bruce F Molina/National Snow & Ice Data Center/World Data Center for Glaciology; **72-73** Corbis/All Canada Photos/Russ Heinl; **74** Photolibrary/age fotostock/Andoni Canela; **75** Photolibrary/Picture Press/Klaus Nigge; **76** Peter van Zoest; **77** Photolibrary/Animals Animals/ABL/Roger de la Harpe; **79t** Corbis/Eric and David Hosking; **79b** FLPA/Winfried Wisniewski; **80-81** Photolibrary/Cuboimages/Paroli Galperti; **82** Corbis/Peter Johnson; **83** Photolibrary/Saga Photo/Patrick Forget; **84** Photolibrary/age fotostock/Nigel Dennis; **85** Getty Images/Robert Harding Picture Library/Steve & Ann Toon; **86-87** Photolibrary/Robert Harding Picture Library/Jochen Schlenker; **87tr** Jon Thornton; **88-89** Corbis/Kevin Schafer; **90-91** Corbis/Martin Harvey; **91t** Corbis/JAI/Nigel Pavitt; **92-93** FLPA/Minden Pictures/Jim Brandenburg; **93tr** Corbis/Wolfgang Kaehler; **94tl** FLPA/Winfried Wisniewski; **94r** Corbis/EPA/Pawel Supernak; **95** FLPA/Minden Pictures/Martin Woike; **96-97** Ardea/Jean Paul Ferrero; **97tr** FLPA/Winfried Wisniewski; **98** Corbis/JAI/Nigel Pavitt; **99t** Nature/Gary K. Smith; **99b** Corbis/Reuters/Diario Panorama; **100** Getty Images/National Geographic/Steve Winter; **101** Miguel A. Torres; **102-103** Corbis/Macduff Everton; **105** Alamy/Martin Harvey; **106** Getty Images/Buena Vista Images; **107t** Bill Clark/U.S Geological Survey; **107b** Photolibrary/age fotostock/Kevin O'Hara; **109** Photolibrary/Juniors Bildarchiv; **110t** Photoshot/World Pictures/Jean Du Boisberranger; **110b** FLPA/Frans Lanting; **111t** FLPA/Terry Whittaker; **111b** Alamy/Andre Jenny; **112** Photoshot/NHPA/ANT; **113** Photolibrary/age fotostock/Andoni Canela; **114bl** Ardea/Jean Paul Ferrero; **114-115** Corbis/Matthieu Colin/Sygma; **115tr** Ardea/Jean Paul Ferrero **116** Shutterstock/Dean Bertoncij **117** Shutterstock/H.Damke **118** Shutterstock/George Lamson **119t** Shutterstock/beltsazar **119b** FLPA/Tui de Roy Minden **228-229** Tim Laman **230** Corbis/Jim Merli/Visuals Unlimited, Inc. **231t** FLPA/David Hosking **231b** FLPA/Geogg Moon **232-233** FLPA/Frans Lanting **233t** FLPA/Tuit de Roy/Minden Pictures **234** Gonçalo M. Rosa **235t** Nik Borrow **237** Corbis/epa/Nic Bothma/**238** Corbis/Frans Lanting

New bird illustrations by Gill Tomblin